International Workshop on Statistical-Mechanical Informatics 2010

(IW-SMI 2010)

Journal of Physics: Conference Series Volume 233

Kyoto, Japan
7-10 March 2010

Editors:

Jun-Ichi Inoue
Yoshiyuki Kabashima
Kazuyuki Tanaka
Toshiyuki Tanaka

ISBN: 978-1-61738-775-3
ISSN: 1742-6588

Printed from e-media with permission by:

Curran Associates, Inc.
57 Morehouse Lane
Red Hook, NY 12571

Some format issues inherent in the e-media version may also appear in this print version.

Printed by Curran Associates, Inc. (2010)

Institute of Physics
Dirac House, Temple Back
Bristol BS1 6BE UK

Phone: 44 1 17 929 7481
Fax: 44 1 17 920 0979

techtracking@iop.org

Additional copies of this publication are available from:

Curran Associates, Inc.
57 Morehouse Lane
Red Hook, NY 12571 USA
Phone: 845-758-0400
Fax: 845-758-2634
Email: curran@proceedings.com
Web: www.proceedings.com

TABLE OF CONTENTS

METHODOLOGIES

CLASSICAL AND QUANTUM SPIN-GLASSES

INFORMATION AND COMMUNICATION

BIOINFORMATICS

Preface

Everything that exists in the natural world is made up of several types of elementary particles. However, we cannot understand nature simply by identifying the properties of these particles. This is because collections of the particles sometimes exhibit completely unexpected collective phenomena, quite independently of the individual particles' properties. In the physical sciences, the importance of focusing on the properties of objects composed of a large number of constituents is reflected in the phrase, "More is different."

The main concept of the research project, the Grant-in-Aid for Scientific Research on Priority Areas `Deepening and Expansion of Statistical Mechanical Informatics (DEX-SMI)' (Head Investigator: Yoshiyuki Kabashima, Tokyo Institute of Technology) (Project Webpage DEX-SMI), launched in 2006, was to introduce this perspective into information science under the common slogan, "More is different in informatics as well." As milestones in the research activity, the International Workshop on Statistical-Mechanical Informatics (IW-SMI) was held annually, featuring studies of information and communication (2007), quantum information (2008), and bioinformatics (2009). The workshop series provided fruitful opportunities for leading researchers of various disciplines to interact with one another, which led to several collaborative studies.

The final workshop, IW-SMI2010, was held at Shiran Kaikan, Kyoto University, Kyoto, Japan, on 7-10 March 2010 to wrap up the achievements of the four years of activity in the DEX-SMI research project. This workshop also aimed to bring together leading researchers in the physical and information sciences to discuss possible future directions for further exploring the successes of DEX-SMI.

We would like to thank the contributors of the workshop as well as all the participants. We hope that the successes of IW-SMI2010 and DEX-SMI will lead to further development of this highly vigorous interdisciplinary field between statistical mechanics and information science.

Editors
Jun-ichi Inoue
Yoshiyuki Kabashima
Kazuyuki Tanaka
Toshiyuki Tanaka

The IW-SMI 2010 Organizing Committee:
Masato Okada (University of Tokyo)
Yoshiyuki Kabashima, General Chair (Tokyo Institute of Technology)
Shin Ishii (Kyoto University)
Jun-ichi Inoue, Publications Chair (Hokkaido University)
Kazuyuki Tanaka (Tohoku University)
Toshiyuki Tanaka, Vice-General Chair (Kyoto University)

All papers published in this volume of *Journal of Physics: Conference Series* have been peer reviewed through processes administered by the proceedings Editors. Reviews were conducted by expert referees to the professional and scientific standards expected of a proceedings journal published by IOP Publishing.

International Workshop on Statistical-Mechanical Informatics 2010 IOP Publishing
Journal of Physics: Conference Series **233** (2010) 012001 doi:10.1088/1742-6596/233/1/012001

Cavity approach to the first eigenvalue problem in a family of symmetric random sparse matrices

Yoshiyuki Kabashima[†1], **Hisanao Takahashi**[†2] **and Osamu Watanabe**[‡3]

[†]Department of Computational Intelligence and Systems Science, Tokyo Institute of Technology, Yokohama 226-8502, Japan
[‡]Department of Mathematical and Computing Science, Tokyo Institute of Technology, Tokyo 152-8552, Japan

E-mail: [1]`kaba@dis.titech.ac.jp`, [2]`takahashi@sp.dis.titech.ac.jp`, [3]`watanabe@is.titech.ac.jp`

Abstract. A methodology to analyze the properties of the first (largest) eigenvalue and its eigenvector is developed for large symmetric random sparse matrices utilizing the cavity method of statistical mechanics. Under a tree approximation, which is plausible for infinitely large systems, in conjunction with the introduction of a Lagrange multiplier for constraining the length of the eigenvector, the eigenvalue problem is reduced to a bunch of optimization problems of a quadratic function of a single variable, and the coefficients of the first and the second order terms of the functions act as cavity fields that are handled in cavity analysis. We show that the first eigenvalue is determined in such a way that the distribution of the cavity fields has a finite value for the second order moment with respect to the cavity fields of the first order coefficient. The validity and utility of the developed methodology are examined by applying it to two analytically solvable and one simple but non-trivial examples in conjunction with numerical justification.

1. Introduction

The first (largest) eigenvalue and its eigenvector (first eigenvector) play key roles in many problems in information science. In multivariate data analysis, the first eigenvector of the variance-covariance matrix represents the most significant component that underlies a set of data under the assumption that the data are generated from a multivariate Gaussian distribution, and the first eigenvalue indicates its relevance [1]. The well-known Google PageRank™ ranks World Wide Web pages based on the first eigenvector of a transition matrix over a huge network of pages [2]. The first eigenvectors of certain matrices expressing a given graph can also be utilized as a practical solution to combinatorial problems such as graph three coloring- [3] and graph bisection problems [4].

The first eigenvalue problem is of significance in physics as well. The assessment of the ground state is generally formulated as a first eigenvalue problem in quantum mechanics [5]. It is generally difficult to accurately assess the correlations in spin systems in statistical mechanics. However, when the temperature is sufficiently high, spin correlations can be evaluated by replacing spin variables with so-called spherical spins in a class of systems [6], which is practically reduced to the eigenvalue analysis of the interaction matrix. In particular, that for the first eigenvalue is of great relevance because it directly leads to the assessment of the critical temperature/mode for the emergence of spontaneous magnetization [7].

Many properties have thus far been clarified for the first eigenvalue problem for the ensembles of large dense matrices. For $N \times N$ symmetric matrices $\boldsymbol{S}$ whose entries are independently and identically distributed (i.i.d.) random variables of a zero mean and a variance of N^{-1}, the first eigenvalue asymptotically converges to 2 and the first eigenvector has no preferential directions in the N-dimensional space as N tends toward infinity [8]. However, when a projection operator of a certain direction, $\boldsymbol{d}$, is added to $\boldsymbol{S}$ with an amplitude, B, the asymptotic eigenvalue is switched to $B + 1/B$ for $B > 1$ having a first eigenvector of a non-zero overlap with $\boldsymbol{d}$. Similar results are also obtained for the correlation matrix, $P^{-1}\boldsymbol{X}^{\mathrm{T}}\boldsymbol{X}$, of $P \times N$ matrices $\boldsymbol{X}$ whose entries i.i.d. random variables of a zero mean and variance N^{-1}, where T denotes the operation of the matrix transpose; the first eigenvalue is offered as $(1+\alpha^{-1/2})^2$ in the limit of $N, P \to \infty$ keeping $\alpha = P/N$ finite providing no preferential directions of the first eigenvector [9], and the analysis can also be generalized to cases in which the projection operators of several directions with arbitrary strengths are added [10]. However, as far as we know, relatively little knowledge has been gained on ensembles of sparse matrices in which the density of non-zero entries vanishes as the size of the matrices tends to infinity although details on eigenvalue distributions have recently been unraveled for several cases [11, 12, 13, 14, 15, 16].

Based on such an understanding, we will herein examine the first eigenvalue and the distribution of components of the first eigenvector for a family of large symmetric random sparse matrices. To achieve this purpose, we employ the cavity method of statistical mechanics [17]. A major advantage of this scheme is its capability for deriving equations for macroscopically characterizing objective systems without having to resort to complicated computation that is generally demanded in an alternative approach termed the replica method [18], on the basis of a tree approximation, which is intuitively plausible for large sparse matrices

This paper is organized as follows. The next section introduces the model that will be explored. In section 3, we develop a scheme for examining the eigenvalue problem in a large system limit based on the cavity method. The scheme is applied to two analytically solvable and one simple but nontrivial examples in section 4. The final section is devoted to a summary.

2. Model definition

We consider ensembles of $N \times N$ real symmetric sparse matrices $\boldsymbol{J} = (J_{ij})$ that are characterized by a distribution, $p(k)$, of *degree* $k(= 0, 1, 2, \ldots)$, which denotes the number of non-zero entries for a column/row in the limit of $N \to \infty$. For simplicity, we assume that the diagonal elements of the matrices are always constrained to zero. For aspects other than degrees, we assume that the matrices are randomly constructed. Let us denote d_i as the degree of index $i(= 1, 2, \ldots, N)$. For $k = 0, 1, 2, \ldots$, we set $d_i = k$ for $Np(k)$ indices of $i = 1, 2, \ldots, N$. A practical scheme for generating a random configuration of non-zero entries characterized in the above is basically as follows [19]:

(S) Make a set of indices U to which each index i attends d_i times. Accordingly, we iterate the following (A)–(C).

(A) Choose a pair of two different elements from U randomly.

(B) Let us denote the values of the two elements as i and j. If $i \neq j$ and the pair of i and j has not been chosen up to the moment, make a link between i and j, and remove the two elements from U. Otherwise, we return them back to U.

(C) If U becomes empty, finish the iteration. Otherwise, if there is no possibility that any more links can be made by (A) and (B), return to (S).

Links generated by the above procedure stand for pairs of indices i and j to which the non-zero entries of J_{ij} are assigned. There may be other algorithms for randomly generating a configuration of the non-zero entries. However, we expect that the properties, which we will investigate after this, do not depend on the details of the generation schemes.

When the support of $p(k)$ is not bounded from the above and values of the entries are kept finite, the first eigenvalue generally diverges as $N \to \infty$ [11, 12, 13, 14, 15, 16]. To avoid this possibility, we assume that $p(k) = 0$ for k, which is larger than a certain value, $k_{\max}$, unless infinitesimal entries are assumed.

We also assume that for a given configuration of the non-zero entries, a value of the entry $J_{ij} = J_{ji}$ is determined as a statistically independent sample following an identical distribution, $p_J(J_{ij})$, for each pair of indices i and $j(\neq i)$ of the non-zero entries. In cases where $p_J(J_{ij})$ is provided by a simple binary distribution,

$$p_J(J_{ij}|\Delta, J) = \frac{1+\Delta}{2}\delta(J_{ij} - J) + \frac{1-\Delta}{2}\delta(J_{ij} + J) \tag{1}$$

$(1 \geq \Delta \geq 0,\ J > 0)$, where $\delta(x)$ denotes Dirac's delta function, these guarantee that the first eigenvalue Λ is bounded from the above by $k_{\max}J$.

In matrix ensembles of Erdös-Rényi type [20], which have been widely studied in network science, a non-zero value of $J_{ij} = J_{ji}$ is assigned to each pair of indices i and $j(\neq i)$ with a certain probability, c/N, where $c > 0$ is $O(1)$. This eventually leads to a Poissonian degree distribution as $p(k) = e^{-c}c^k/k!$, whose support is not bounded from the above. Therefore, the following analysis does not cover such ensembles as long as the matrix entries are $O(1)$.

3. Cavity approach to first eigenvalue problem

3.1. Message passing in fixed graph

Formulating the first eigenvalue problem as a constrained quadratic optimization problem

$$\min_{\boldsymbol{v}} \left\{ -\boldsymbol{v}^{\mathrm{T}} \boldsymbol{J} \boldsymbol{v} \right\} \text{ subject to } |\boldsymbol{v}|^2 = N \tag{2}$$

is the basis of our analysis, where $\min_X\{\cdots\}$ denotes minimization with respect to X. The solution to this problem $\boldsymbol{v}^* = (v_i^*)$ represents the first eigenvector that is normalized to $\sqrt{N}$ and the first eigenvalue Λ is assessed as $\Lambda = (\boldsymbol{v}^*)^{\mathrm{T}}\boldsymbol{J}\boldsymbol{v}^*/N$. The introduction of a Lagrange multiplier λ converts (2) to a saddle point problem with respect to an objective function,

$$\begin{aligned} \mathcal{L}(\boldsymbol{v}, \lambda) &= -\boldsymbol{v}^{\mathrm{T}}\boldsymbol{J}\boldsymbol{v} + \lambda(|\boldsymbol{v}|^2 - N) \\ &= \lambda \sum_{i=1} v_i^2 - 2\sum_{i>j} J_{ij} v_i v_j - N\lambda. \end{aligned} \tag{3}$$

Minimizing this function with respect to $\boldsymbol{v}$ and determining λ so as to satisfy the stationary condition, $|\boldsymbol{v}|^2 = N$, indicate that the solution is provided by the first eigenvalue, $\lambda = \Lambda$, and the first eigenvector, $\boldsymbol{v} = \boldsymbol{v}^*$.

The key idea underlying our approach is to approximate the minimization problem of (3) with respect to $\boldsymbol{v}$ for fixed λ by a bunch of those for effective quadratic functions of a single variable

$$\mathcal{L}_i(v_i|A_i, H_i) = A_i v_i^2 - 2H_i v_i \tag{4}$$

$(i = 1, 2, \ldots, N)$, where the coefficients of the second and first order terms, A_i and H_i, are determined in a certain self-consistent manner. In the cavity method, these coefficients are evaluated by a message-passing algorithm that yields the exact solution when the connectivity of J_{ij} is pictorially represented by a cycle free graph (tree).

To explain this, let us focus on an arbitrary index, i, assuming that matrix connectivity is specified by a tree. We use a notation, ∂i, to represent the set of indices j that are connected directly to i with non-zero entries of J_{ij}. We also introduce auxiliary variables $A_{j\to i}$ and $H_{j\to i}$ to represent the second and first order coefficients of $j \in \partial i$ in the *i-cavity system*, which is defined by removing index i from the original system. In physics, such variables are occasionally termed *cavity fields*.

A general and distinctive feature of trees is that all indices $j \in \partial i$ are completely disjointed from one another by removing the index, i. Let us assume that we add i into the i-cavity system retaining one link J_{li} for an index, $l \in \partial i$, removed. This creates a problem with minimizing a function with respect to the v_i and v_j of $j \in \partial i \backslash l$, where $S\backslash l$ indicates the removal of l from a set, S, as

$$\mathcal{L}_{i,\partial i\backslash l}(v_i, \{v_{j\in\partial i\backslash l}\}) = \lambda v_i^2 - 2v_i \sum_{j\in\partial i\backslash l} J_{ij}v_j + \sum_{j\in\partial i\backslash l} \left(A_{j\to i}v_j^2 - 2H_{j\to i}v_j\right). \tag{5}$$

Since i is connected to l only with J_{li} in the original tree, minimizing this function with respect to the v_j of $\forall j \in \partial i\backslash l$ fixing v_i yields cavity fields concerning i in the l-cavity system as

$$A_{i\to l} = \lambda - \sum_{j\in\partial i\backslash l} \frac{J_{ij}^2}{A_{j\to i}}, \tag{6}$$

$$H_{i\to l} = \sum_{j\in\partial i\backslash l} \frac{J_{ij}H_{j\to i}}{A_{j\to i}}. \tag{7}$$

Given an initial condition, these equations are capable of assessing sets of the cavity fields, $A_{j\to i}$ and $H_{j\to i}$, which are defined for all directed pairs of the connected indices, moving entirely over the graph. $J_{ij}^2/A_{j\to i}$ and $J_{ij}H_{j\to i}/A_{j\to i}$ in (6) and (7) represent the influences of j to i conveyed through connection J_{ij} and are sometimes referred to as *cavity biases*. After the cavity fields are determined, A_i and H_i in (4) are assessed by summing up the cavity biases from $\forall j \in \partial i$ as $A_i = \lambda - \sum_{j\in\partial i} J_{ij}^2/A_{j\to i}$ and $H_i = \sum_{j\in\partial i} J_{ij}H_{j\to i}/A_{j\to i}$. This yields $v_i^* = H_i/A_i$. Finally, adjusting λ so that the normalization constraint, $|\boldsymbol{v}^*|^2 = \sum_{i=1}^N (v_i^*)^2 = \sum_{i=1}^N (H_i/A_i)^2 = N$, is satisfied provides Λ.

The above procedure can be regarded as a variant of the belief propagation developed in research on probabilistic inference [21] and is guaranteed to offer the exact result when the graph is free from cycles. When the graph contains cycles, the algorithm in (6) and (7) can still be performed; unfortunately, the result obtained is just an approximation. However, for our current purpose, we can generally utilize a much simpler and exact method that just repeats matrix multiplication as $\boldsymbol{v}^{t+1} = (a\boldsymbol{I}_N + \boldsymbol{J})\boldsymbol{v}^t$, which provides $\Lambda = \lim_{t\to\infty} |\boldsymbol{v}^{t+1}|/|\boldsymbol{v}^t| - a$ and $\boldsymbol{v}^* = \sqrt{N}\lim_{t\to\infty} \boldsymbol{v}^t/|\boldsymbol{v}^t|$, where a is a sufficiently large positive number and $\boldsymbol{I}_N$ denotes the $N \times N$ identity matrix. Therefore, our approach may seem less competitive in practice. However, the cavity approach is still useful for examining the properties of the eigenvalue problem introducing a macroscopic description as is explained below.

3.2. Description by distribution of cavity fields

Let us examine the behavior of the algorithm in (6) and (7). To do this, we need to pay attention to the property of randomly constructed sparse matrices, which implies that the lengths of cycles in the connectivity graph of a matrix typically grow as $O(\ln N)$ as $N \to \infty$ [22]. This naturally motivates us to ignore the effects of self-interactions in the updates of (6) and (7). We also introduce a macroscopic characterization of the cavity fields utilizing the distribution, $q(A, H) = (\sum_{i=1}^N d_i)^{-1} \sum_{i=1}^N \sum_{j\in\partial i} \delta(A - A_{j\to i})\delta(H - H_{j\to i})$. When a link in the connectivity graph is chosen randomly, the probability that the index of one terminal has a degree, k, is provided as

$$r(k) = \frac{kp(k)}{\sum_{k=0}^{k_{\max}} kp(k)}. \tag{8}$$

This and dealing with (6) and (7) as the elemental process for updating the cavity field distribution lead to an equation that determines $q(A, H)$ in a self-consistent manner as

$$q(A,H) = \sum_{k=1}^{k_{\max}} r(k) \int \prod_{j=1}^{k-1} dA_j dH_j q(A_j, H_j) \left\langle \delta\left(A - \lambda + \sum_{j=1}^{k-1} \frac{\mathcal{J}_j^2}{A_j}\right) \delta\left(H - \sum_{j=1}^{k-1} \frac{\mathcal{J}_j H_j}{A_j}\right) \right\rangle_{\mathcal{J}}, \tag{9}$$

where $\langle \cdots \rangle_{\mathcal{J}}$ represents the operation of taking averages with respect to all relevant $\mathcal{J}_j$'s following $p_J(\mathcal{J}_j)$. After $q(A, H)$ is determined with this equation, the distribution of the auxiliary variables in the original system, $Q(A, H) = N^{-1} \sum_{i=1}^{N} \delta(A - A_i)\delta(H - H_i)$, is assessed as

$$Q(A,H) = \sum_{k=0}^{k_{\max}} p(k) \int \prod_{j=1}^{k} dA_j dH_j q(A_j, H_j) \left\langle \delta\left(A - \lambda + \sum_{j=1}^{k} \frac{\mathcal{J}_j^2}{A_j}\right) \delta\left(H - \sum_{j=1}^{k} \frac{\mathcal{J}_j H_j}{A_j}\right) \right\rangle_{\mathcal{J}}. \tag{10}$$

This makes it possible to evaluate the quadratic norm per element of $\boldsymbol{v}^*$ as $T = N^{-1}|\boldsymbol{v}^*|^2 = \int dAdHQ(A, H)(H/A)^2$. Moreover, the distribution of elements of the first eigenvector, $\boldsymbol{v}^*$, can be assessed as $P(v) = N^{-1} \sum_{i=1}^{N} \delta(v - v^*) = \int dAdHQ(A, H)\delta(v - H/A)$.

The argument provided in the preceding subsection implies that λ should be adjusted so that T accords with unity. However, here we adopt an alternative approach to determining Λ. Let us assume a situation where (9) is solved by a method of successive iteration, which leads to an exact result when the graph is free from cycles. Equation (9) guarantees that for large λ, A is distributed over large values in $q(A, H)$. This means that for sufficiently large λ, the absolute values of H are mostly reduced by each iteration and, therefore, the marginal distribution, $q(H) = \int dAq(A, H)$, converges to $\delta(H)$ implying $T \to 0$ as the number of iterations tends to infinity. However, as λ is decreased from sufficiently large values, A can take smaller values in the distribution and larger values of $|H|$ could appear with considerable frequency in each iteration. This indicates that the frequencies with which $|H|$ is reduced and enlarged are balanced at a certain value of λ, for which non-trivial distribution $q(H) \neq \delta(H)$ becomes invariant under the cavity iteration of (9) and T is kept as a finite constant. If λ is lowered further, the frequency of enlargements will overcome that of reductions, which will eventually make T tend to infinity. These imply that one can characterize the first eigenvalue, Λ, as the value of λ for which non-trivial distribution $q(H) \neq \delta(H)$ emerges by solving (9). The resulting cavity field distribution, $q(A, H)$, offers the distribution of the first eigenvector elements, $P(v)$, of any finite value of T by applying appropriate rescaling with respect to H. This is also reasonable because eigenvalues are generally irrelevant to the values of the normalization constraint with respect to eigenvectors.

The validity and utility of this characterization of the eigenvalue problem are examined by applying it to three examples in the next section.

4. Application to three examples

4.1. Single degree model

In the first example, we consider cases where all indices have a certain identical degree, K, which is characterized by $p(k) = \delta_{k,K}$, where $\delta_{i,j}$ denotes Kroeneker's delta, and the non-zero entries are i.i.d. following the binary distribution of (1). In such cases, the marginal distribution, $q(A) = \int dHq(A, H)$, is provided in the form of $q(A) = \delta(A - A^*)$ since $\mathcal{J}^2 = J^2$ always holds for

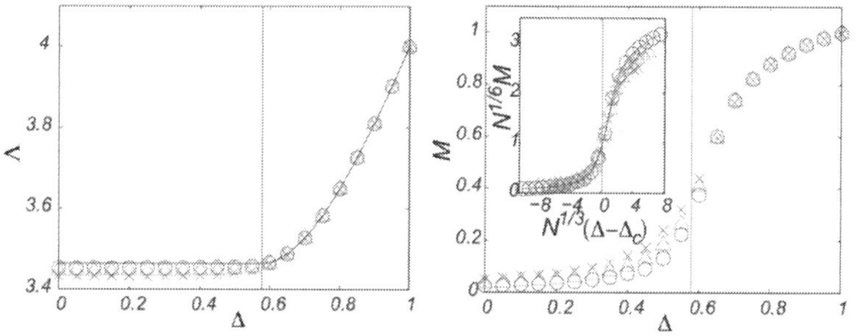

Figure 1. (a): First eigenvalue Λ versus Δ for single degree model of $K = 4$ and $J = 1$. The full curve represents the theoretical prediction of (17) while the markers denote the averages of 2000 experimental results for $N = 256$ (cross), 512 (triangle) and 1024 (circle) systems. (b): $M = |N^{-1}\sum_{i=1}^{N} v_i^*|$ versus Δ for the same experiments as (a). Markers of cross, triangle and circle correspond to $N = 256$, 512, and 1024 as well. The inset has scaling plots obtained by conversion of $M \to N^{1/6}M$ and $\Delta \to N^{1/3}(\Delta - \Delta_c)/\Delta_c$. In both (a) and (b), the vertical straight lines stand for $\Delta_c = 1/\sqrt{3} = 0.577\ldots$.

$\forall \mathcal{J}$ that are sampled from (1). Equation (9) indicates that A^* satisfies $A^* = \lambda - (K-1)J^2/A^*$. This yields

$$A^* = \frac{\lambda + \sqrt{\lambda^2 - 4(K-1)J^2}}{2} \equiv A^*(\lambda; (K-1)J^2), \tag{11}$$

which along with (9) offer the equation for determining the marginal distribution, $q(H)$, in a self-consistent manner as

$$q(H) = \int \prod_{j=1}^{K-1} dH_j q(H_j) \left\langle \delta\left(H - \sum_{j=1}^{K-1} \frac{\mathcal{J}_j H_j}{A^*(\lambda; (K-1)J^2)}\right)\right\rangle_{\mathcal{J}}. \tag{12}$$

Tuning λ so that a non-trivial distribution, $q(H) \neq \delta(H)$, satisfies this equation yields the first eigenvalue, Λ. Paying attention to the first and the second order moments, $m_1 = \int dHq(H)H$ and $m_2 = \int dHq(H)H^2$, is sufficient for this purpose. This provides a generalized eigenvalue problem described by a couple of equations

$$m_1 = \frac{\Delta(K-1)J}{A^*(\lambda; (K-1)J^2)} m_1, \tag{13}$$

$$m_2 = \frac{(K-1)J^2}{(A^*(\lambda; (K-1)J^2))^2}(m_2 - m_1^2) + \left(\frac{\Delta(K-1)J}{A^*(\lambda; (K-1)J^2)} m_1\right)^2, \tag{14}$$

which requires the existence of a solution to $(m_1, m_2) \neq (0, 0)$. There are two possibilities that will satisfy this requirement. The first is characterized by $|m_1| > 0$, which in conjunction with (13) offers

$$1 = \frac{\Delta(K-1)J}{A^*(\lambda; (K-1)J^2)}. \tag{15}$$

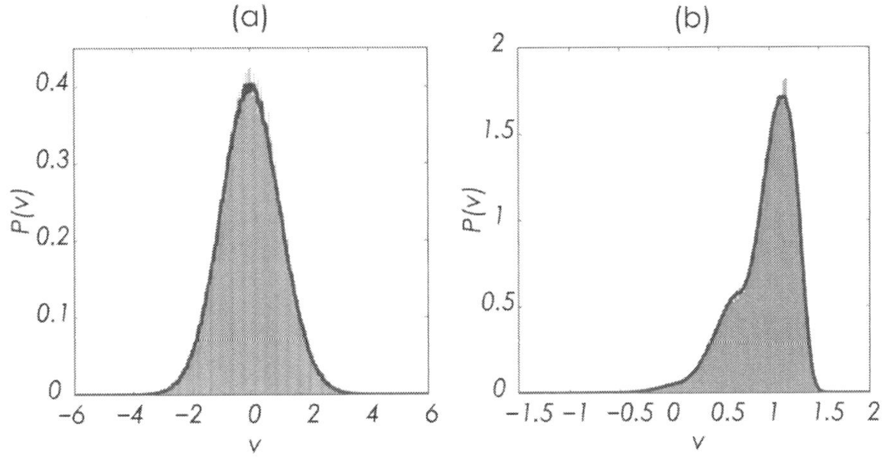

Figure 2. $P(v)$ for (a): $\Delta = 0.5$ and (b): $\Delta = 0.9$ of single degree model of $K = 4$. For both plots, the bars represent histograms obtained from 100 experiments of $N = 1024$ while the curves stand for theoretical predictions obtained from (12) and (10) by means of a Monte Carlo method of 10^6 populations. In handling the experimental data of (b), we employed conversion $\boldsymbol{v}^* \to -\boldsymbol{v}^*$ for data of $\sum_{i=1}^{N} v_i^* < 0$ in order to break the mirror symmetry between $\boldsymbol{v}^*$ and $-\boldsymbol{v}^*$, which is intrinsic in the eigenvalue problem.

The second is the case of $m_1 = 0$ and $m_2 > 0$, which along with (14) provide

$$1 = \frac{(K-1)J^2}{(A^*(\lambda; (K-1)J^2))^2}. \tag{16}$$

Solving (15) and (16) with respect to λ, we finally obtain an expression of the first eigenvalue,

$$\Lambda = \begin{cases} ((K-1)\Delta + 1/\Delta)\, J, & \Delta > \Delta_{\mathrm{c}}, \\ 2\sqrt{K-1}J, & \Delta < \Delta_{\mathrm{c}}, \end{cases} \tag{17}$$

where $\Delta_{\mathrm{c}} = 1/\sqrt{K-1}$. $\Delta > \Delta_{\mathrm{c}}$ and $\Delta < \Delta_{\mathrm{c}}$ correspond to the cases of $|m_1| > 0$ and $m_1 = 0$.

To justify the prediction of (17), we carried out numerical experiments for the systems of $K = 4$ and $J = 1$, whose results are in figures 1 (a) and (b). Figure 1 (a) has the plots of Λ versus Δ, which are in excellent agreement with (17) for relatively large Δ. However, there are non-negligible finite-size corrections for relatively small Δ, which makes it difficult to accurately detect the critical value, Δ_c. To overcome this difficulty, we turn to the profiles of $M = \left|N^{-1}\sum_{i=1}^{N} v_i^*\right|$, which are shown in figure 1 (b). The raw experimental data of M vary smoothly with Δ and do not exhibit any singularity at $\Delta = \Delta_{\mathrm{c}}$ either. Nevertheless, the data rescaled as $M \to N^{1/6}M$ and $\Delta \to N^{1/3}(\Delta - \Delta_{\mathrm{c}})/\Delta_{\mathrm{c}}$ support a scaling hypothesis, $M = N^{-1/6}F(N^{1/3}(\Delta - \Delta_{\mathrm{c}})/\Delta_{\mathrm{c}})$ (inset), which validates the transition at Δ_{c} assuming that $F(x)$ behaves as $O(x^{1/2})$ for $x \to +\infty$ and vanishes for $x \to -\infty$. A similar scaling hypothesis has been assumed in the examination of principal component analysis before [10].

Besides the eigenvalue, one can also assess the distributions, $P(v)$, of elements of the first eigenvector, $\boldsymbol{v}^*$, by numerically solving $q(H)$ with Monte Carlo (population dynamics) methods. The prediction with our approach is also in excellent agreement with the experimental results for both cases of $\Delta < \Delta_{\mathrm{c}}$ (figure 2 (a)) and $\Delta > \Delta_{\mathrm{c}}$ (figure 2 (b)) when $|\Delta - \Delta_{\mathrm{c}}|/|\Delta_{\mathrm{c}}|$ is sufficiently large.

4.2. Limit of large degrees and infinitesimal entries

The second example is offered by assuming that the average and the variance of $p(k)$ are provided as $\overline{k}$ and $O(\overline{k})$, respectively, and those of $p_J(J_{ij})$ are given as $\mu J/\overline{k}$ and $J^2/\overline{k}$ for sufficiently large $\overline{k}$ while $\mu > 0$ and $J > 0$ are certain finite constants. The assumption on the degree distribution, $p(k)$, covers the cases described by truncated Poissonian distributions of large mean values in which indices whose degrees are sufficiently larger than the means are eliminated from the basic graphs of Erdös-Rényi type [4]. Being combined with the assumption on $p_J(J_{ij})$, this, in conjunction with the law of large numbers, guarantees that the marginal distribution of A is provided as $q(A) = \delta(A - A^*(\lambda; J^2))$, where the functional form of $A^*(\lambda; J^2)$ is identical to that of (11). Inserting this into (12) and the resulting equations with respect to the first and the second order moments offer conditions of $1 = \mu J/A^*(\lambda; J^2)$ and $1 = J^2/(A^*(\lambda; J^2))^2$, respectively, which are counterparts of (15) and (16). This provides the first eigenvalue as

$$\Lambda = \begin{cases} (\mu + 1/\mu)J, & \mu > 1, \\ 2J, & \mu < 1. \end{cases} \tag{18}$$

$\mu > 1$ and $\mu < 1$ correspond to situations of $|m_1| > 0$ and $m_1 = 0$.

In the current case, the central limit theorem guarantees that $q(H)$ converges to a distribution of the Gaussian type. Consequently, this also makes it possible to analytically express the distribution, $P(v)$, of the elements of the first eigenvector in a Gaussian form. Under the constraint of $T = 1$, this yields

$$P(v) = \begin{cases} (2\pi(1 - (M(\mu))^2)^{-1/2} \exp\left(-(v - M(\mu))^2/(2(1 - (M(\mu))^2))\right), & \mu > 1, \\ (2\pi)^{-1/2} \exp\left(-v^2/2\right), & \mu < 1, \end{cases} \tag{19}$$

where $M(\mu) = \pm\sqrt{1 - 1/\mu^2}$.

4.3. Mixture of multiple degrees

The strategy we employed for analyzing the preceding two examples is summarized as follows. We first evaluate the marginal distribution, $q(A) = \int dHq(A, H)$, of the second order cavity fields in the form of a single delta distribution, $q(A) = \delta(A - A^*)$. This enables us to deal with A as a constant, A^*, in assessing the marginal distribution, $q(H) = \int dAq(A, H)$, of the first order cavity fields. Consequently, the original eigenvalue problem is reformulated as a generalized eigenvalue problem with respect to $q(H)$. This imposes certain conditions on the second order cavity field, A^*, which is a function of λ. The first eigenvalue is evaluated by solving the conditions with respect to λ.

Unfortunately, such clear decoupling of the first and the second order cavity fields cannot generally be exploited in handling the eigenvalue problem precisely. Even when $p(k)$ takes finite values for multiple k, the marginal distribution, $q(A)$, constitutes a closed equation with no necessity for considering the first order cavity field, H, as

$$q(A) = \sum_{k=1}^{k_{\max}} r(k) \int \prod_{j=1}^{k-1} dA_j q(A_j) \left\langle \delta\left(A - \lambda + \sum_{j=1}^{k-1} \frac{\mathcal{J}_j^2}{A_j}\right)\right\rangle_{\mathcal{J}}, \tag{20}$$

which, however, generally makes $q(A)$ a continuous distribution. This prevents us from handling A as a constant in assessing the marginal distribution, $q(H)$, and we eventually have to deal with the joint distribution, $q(A, H)$, directly. This requires much more effort in analysis than that in handling the marginal distributions, $q(A)$ and $q(H)$, separately.

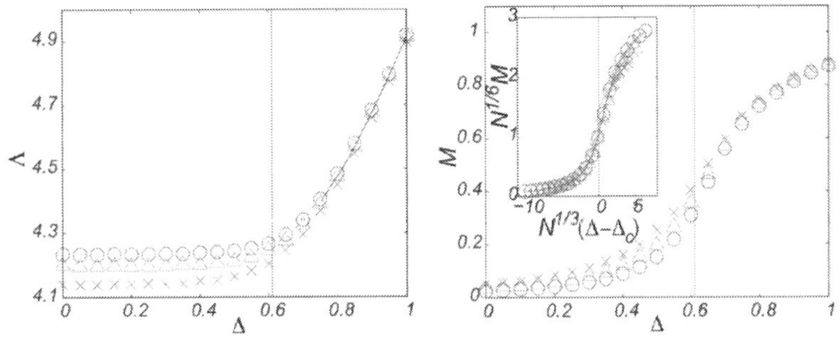

Figure 3. (a): First eigenvalue Λ versus Δ for mixture model of $p(k) = 0.9\delta_{k,4} + 0.1\delta_{k,8}$ and $J = 1$. The full curve represents the theoretical prediction assessed from (23) by means of a Monte Carlo method of 10^6 populations. Large statistical fluctuations due to a small portion of samples of $A \sim 0$ prevented us from accurately assessing the relation below $\Lambda \sim 4.4$, for which data have not been shown. The markers stand for data evaluated similarly for figure 1. (b): $M = |N^{-1}\sum_{i=1}^{N} v_i^*|$ versus Δ for the same experiments as (a). The scaling plots of $M \to N^{1/6}M$ and $\Delta \to N^{1/3}(\Delta - \Delta_c)/\Delta_c$ (inset) indicate $\Delta_c \simeq 0.611$, which is represented as vertical straight lines in (a) and (b).

To avoid this, let us move forward with our analysis, approximately, ignoring the correlations between A and H. Applying this approximate treatment to the assessment of the first and the second order moments of H by using (9) provides two conditions

$$1 = \left(\int d\mathcal{J} p_{\mathrm{J}}(\mathcal{J})\mathcal{J}\right)\left(\sum_{k=1}^{k_{\max}} r(k)(k-1)\right)\left(\int dA q(A) A^{-1}\right), \tag{21}$$

$$1 = \left(\int d\mathcal{J} p_{\mathrm{J}}(\mathcal{J})\mathcal{J}^2\right)\left(\sum_{k=1}^{k_{\max}} r(k)(k-1)\right)\left(\int dA q(A) A^{-2}\right), \tag{22}$$

where $q(A)$ is the solution to (20) for given λ. Equations (21) and (22) can be regarded as corresponding to generalizations of (15) and (16).

We examined the utility of (21) and (22) by applying them to cases of the bimodal degree distributions of $p(k) = (1-f)\delta_{k,4} + f\delta_{k,8}$ $(0 < f < 1)$ and $p_{\mathrm{J}}(J_{ij})$ of (1) with $J = 1$. For (1), equation (21) indicates that

$$\Delta = J^{-1}\left(\left(\sum_{k=1}^{k_{\max}} r(k)(k-1)\right)\left(\int dA q(A) A^{-1}\right)\right)^{-1}, \tag{23}$$

provides the relation between first eigenvalue Λ and Δ by dealing with $\lambda = \Lambda$ as a control parameter, as long as the right hand side is less than unity and greater than a certain critical value, Δ_c. Figure 3 (a) compares the theoretical prediction of (23) assessed with the Monte Carlo method and the results of numerical experiments for $f = 0.1$. Despite the fact that the treatment is not necessarily exact, the theoretical prediction agrees with the experimental results with excellent accuracy for relatively large Δ.

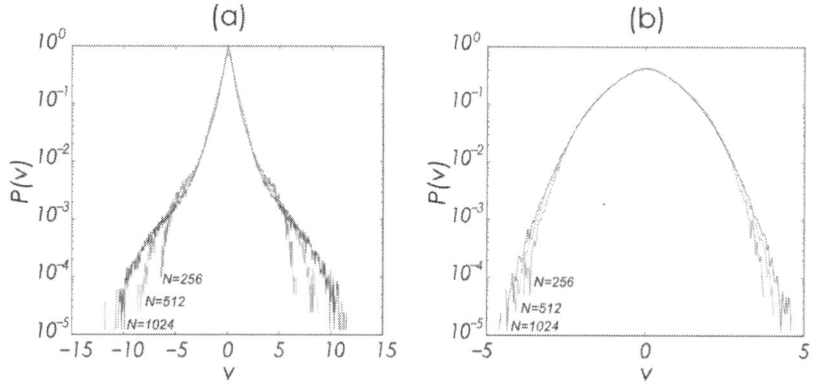

Figure 4. $P(v)$ of mixture model $p(k) = (1-f)\delta_{k,4} + f\delta_{k,8}$ for (a): $f = 0.1$ and (b): $f = 0.9$. In both plots, the curves were obtained from 2000 experiments setting $\Delta = 0$ for $N = 256$, 512 and 1024, respectively.

Rescaling $M = |N^{-1}\sum_{i=1}^{N} v_i^*|$ and $(\Delta - \Delta_c)/\Delta_c$ with $N^{1/6}$ and $N^{1/3}$, respectively, as has been assumed in the single degree model for the experimental data, indicates a critical value of $\Delta_c \simeq 0.611$ (figure 3 (b)). This should be characterized by satisfying (22) in the current approach. However, it is difficult to accurately determine Δ_c with the Monte Carlo method in practice because the right hand side of (22) greatly depends on the diverging contribution from a small portion of $A \sim 0$ at criticality, which is considerably sensitive to statistical fluctuations in the sampling. Another source of the difficulty is that (20) can make the support of $q(A)$ spread to a region of *negative* A for relatively small λ. The negative values of A imply that for certain indices i, (4) is minimized at $v_i = \pm\infty$, which may correspond to large elements of $\boldsymbol{v}^*$ that diverge as N tends to infinity. Such elements have been argued to be *defects* in several earlier studies before [12, 13, 15]. The experimental data indicate that the divergent behavior of the small fraction of the first eigenvector elements, which may be related to the defects, is more significant as f is relatively *smaller* (figures 4 (a) and (b)). Refining the current analysis so that it can accurately handle such divergent elements is currently under way.

5. Summary

In summary, we developed a methodology for analyzing the first eigenvalue problem of large symmetric random sparse matrices utilizing the cavity method of statistical mechanics. The scheme we developed makes it possible to assess the first eigenvalue and the distribution of elements of the first eigenvector on the basis of a functional equation concerning a joint distribution with respect to two kinds of cavity fields. Its validity was tested and confirmed by using two analytically solvable examples. Unfortunately, directly employing the approach we developed is technically difficult in general cases where the matrices are characterized by multiple degrees. However, we demonstrated that an approximate treatment that ignores certain correlations between the two kinds of cavity fields offers excellent capabilities for assessing the first eigenvalue when each entry of the matrices is i.i.d. following a distribution that has a sufficiently large positive mean value. However, singular behavior can be observed in the cavity-field distribution when the average of the matrix entries is not a sufficiently large positive value. Comparison with experimental results suggested that this may be related to the divergent elements of the first eigenvector, which have been argued in earlier studies.

In some problems, not only the first but also the second eigenvalue is of utility [23]. In addition to making refinements to the current study, analyzing the n-th eigenvalue ($n = 2, 3, \ldots$) by utilizing the cavity method may be an interesting project for future work.

Acknowledgments

This work was partially supported by Grants-in-Aids for Scientific Research on the Priority Areas "Deepening and Expansion of Statistical Mechanical Informatics" from the Ministry of Education, Culture, Sports, Science and Technology, Japan, No. 22300003 from the JSPS and the JSPS Global COE program, "Computationism as a Foundation for the Sciences" (YK and OW).

References

[1] Mardia K V, Kent J T and Bibby J M 1979 *Multivariate Analysis* (London: Academic Press)
[2] Langville A and Meyer C 2006 *Google's PageRank and Beyond: The Science of Search Engine Rankings* (Princeton: Princeton University Press)
[3] Alon N and Kahale N 1997 *SIAM J. Comput.* **26** 1733
[4] Coja-Oghlan A 2006 *Random Struct. Algorithms* **29** 351
[5] Born M, Heisenberg W and Jordan P 1925 *Zeit. Phys.* **35** 557
[6] Parisi G and Potters M 1995 *J. Phys. A: Math. Gen.* **28** 526
[7] Fischer K H and Hertz J A 1991 *Spin Glasses* (Cambridge: Cambridge University Press)
[8] Mehta M L 2004 *Random matrices, Third edition. Pure and Applied Mathematics 142* (Amsterdam: Elsevier/Academic Press)
[9] Marchenko V A and Pastur L A 1967 *Mat. Sb.* **72** 507
[10] Hoyle D C and Rattray M 2004 *Phys. Rev.* E **69** 026124
[11] Bray A J and Rodgers G J 1988 *Phys. Rev.* B **38** 11461
[12] Biroli G and Monasson R 1999 *J. Phys. A: Math. Gen.* **32** L255
[13] Semerjian G and Cugliandolo L F 2002 *J. Phys. A: Math. Gen.* **35** 4837
[14] Nagao T and Tanaka T 2007 *J. Phys. A: Math. Gen.* **40** 4973
[15] Kühn R 2008 *J. Phys. A: Math. Gen.* **41** 295002
[16] Rogers T, Rérez-Castillo I, Kühn R and Takeda K 2008 *Phys. Rev.* E **78** 031116
[17] Mézard M, Virasolo M A and Parisi G 1986 *Spin Glass Theory and Beyond* (Singapore: World Scientific)
[18] Dotzenko V S 2001 *Introduction to the Replica Theory of Disordered Statistical Systems* (Cambridge: Cambridge University Press)
[19] Steger A and Wormald N C 1999 *Combinatorics, Probability and Computing* **8** 377
[20] Erdös P and Rényi A 1959 *Publicationes Mathematicae* **6** 290
[21] Pearl J 1988 *Probabilistic Reasoning in Intelligent Systems: Networks of Plausible Inference* (San Francisco: Morgan Kaufmann)
[22] Albert R and Barabási A L 2002 *Rev. Mod. Phys.* **74** 47
[23] Hoory S, Linial N and Widgerson A 2006 *Bulletin (New series) of the American Mathematical Society* **43** 439

International Workshop on Statistical-Mechanical Informatics 2010 IOP Publishing
Journal of Physics: Conference Series **233** (2010) 012002 doi:10.1088/1742-6596/233/1/012002

Towards the generation of random bits at terahertz rates based on a chaotic semiconductor laser

Ido Kanter[1], Yaara Aviad[2], Igor Reidler[2], Elad Cohen[2] and Michael Rosenbluh[2]

[1] Minerva Center and Department of Physics, Bar-Ilan University, Ramat-Gan 52900, Israel
[2] Department of Physics, The Jack and Pearl Resnick Institute for Advanced Technology, Bar-Ilan University, Ramat-Gan, 52900 Israel

Abstract. Random bit generators (RBGs) are important in many aspects of statistical physics and crucial in Monte-Carlo simulations, stochastic modeling and quantum cryptography. The quality of a RBG is measured by the unpredictability of the bit string it produces and the speed at which the truly random bits can be generated. Deterministic algorithms generate pseudo-random numbers at high data rates as they are only limited by electronic hardware speed, but their unpredictability is limited by the very nature of their deterministic origin. It is widely accepted that the core of any true RBG must be an intrinsically non-deterministic physical process, e.g. measuring thermal noise from a resistor. Owing to low signal levels, such systems are highly susceptible to bias, introduced by amplification, and to small nonrandom external perturbations resulting in a limited generation rate, typically less than $100Mbit/s$. We present a physical random bit generator, based on a chaotic semiconductor laser, having delayed optical feedback, which operates reliably at rates up to $300Gbit/s$. The method uses a high derivative of the digitized chaotic laser intensity and generates the random sequence by retaining a number of the least significant bits of the high derivative value. The method is insensitive to laser operational parameters and eliminates the necessity for all external constraints such as incommensurate sampling rates and laser external cavity round trip time. The randomness of long bit strings is verified by standard statistical tests.

1. Introduction

Random bit generators (RBGs) constitute an important tool in statistical physics and are crucial in Monte-Carlo simulation [1], stochastic modeling [2], the generation of classical and quantum cryptographic keys [3–5]. The quality of a RBG is measured by the unpredictability of the bit string it produces and the speed at which the truly random bits can be generated. System complexity, cost, reliability and sensitivity to control parameters are also important factors for a successful RBG.

Deterministic algorithms generate pseudo-random numbers at high data rates as they are only limited by electronic hardware speed, but their unpredictability is limited by the very nature of their deterministic origin. When ultimate security is necessary one must turn to the only cipher which is theoretically unbreakable - the one-time pad [4]. It is widely accepted that the core of any true RBG must be an intrinsically non-deterministic physical process, e.g. measuring thermal noise from a resistor. Hence, it is no surprise that proposals for and implementations of RBGs range from measuring thermal noise from a resistor and shot noise from a Zener diode or a vacuum tube [6, 7], measuring radioactive decay from a radioactive source [7, 8], detection of a single photon arrival time, direction or polarization [9, 10] and sampling a stable high-frequency oscillator with an unstable low-frequency clock [11, 12]. Owing to low signal levels, such systems require extensive broadband amplification and are thus highly susceptible to bias introduced by the non-ideal amplifiers and small nonrandom external perturbations. Often, in order to obtain statistically passable sequences, post processing of the sequence on a digital computer is performed or a limited amplifier bandwidth is used, resulting in a limited generation rate, typically less than 100 Mbit/s. An intriguing possibility for a physical system for RBG is a semiconductor laser in the presence of external feedback, whose output consists of a large chaotic signal of pulses with a width less than 100 ps [12–14, 17]. The main challenge posed by such a source is that the external cavity which is responsible for the chaotic laser fluctuations has a photon round trip time associated with it, and the chaotic signal sequence is nearly identically repeated at this round trip time [14, 17]. The periodicity cannot be completely eliminated by increasing the length of the cavity to extremely long round trip times or introducing feedback from multiple external cavities with incommensurate feedback times. Though these help to reduce the correlation between the chaotic pulse sequence segments they do not eliminate them completely.

Recently, Uchida et. al. [15, 16] demonstrated a fast, 1.7 Gbits/s RBG based on chaotic lasers where the periodicity in the chaotic signal at the photon round trip frequency is eliminated by sampling the fluctuations of two independent lasers. They use **two single mode distributed feedback lasers** with external cavities with round trip times τ_1 and τ_2. An external clock with period τ_s, triggers the sampling and the one bit analog-to-digital conversion (ADC) of the two laser outputs. The parameters τ_1, τ_2, and τ_s are selected such that the ratios between the three possible pairs are **incommensurate** within the resolution of the experiment. The chaotic signals of each of the lasers are mapped at the clock frequency to a Boolean sequence by comparing with a threshold voltage for each signal, which is carefully tuned to achieve an unbiased sequence (equal overall probability of 0 s and 1 s). Finally an exclusive-OR (XOR) operation on the two Boolean sequences yields the random sequence.

2. The prototypical RBG

We use a single, off the shelf semiconductor laser, with absolutely no special requirements, operating in the coherence collapse regime due to feedback from an external cavity with photon round trip time τ [19]. Because of feedback the laser is chaotic, with a broadened, multimode lasing frequency spectrum and intensity fluctuating in time [14, 15]. Only **one incommensurate ratio** between the external cavity, τ, and an external clock rate, τ_s, is required. The detected laser output is sampled by an 8-bit, analog-to-digital converter (ADC) (Tektronix TDS6124C) and is used to generate a Boolean sequence in the following way: the difference between consecutive sampled 8-bit values is obtained and the m least significant bits (LSBs) of the difference value serve as the next m random bits of the sequence (Figure 1(a)). Our method is insensitive to variations of parameters such as the average laser power, and does not require the tuning or determination of a decision threshold value.

Figure 1 shows a schematic diagram of the RBG setup. The diode laser wavelength is near 656 nm and partial feedback is obtained from a reflector placed at a photon round trip time of 12.22 ns. The laser is operated moderately above the threshold current, $p_{op}/p_{th} = 1.55$, and the optical feedback strength is a few percent of the output intensity, though these parameters can be varied without affecting the RBG. The detection bandwidth (limited by the bias T) is about 40 GHz, which is sufficient to resolve the temporal dynamics of the laser output. The AC component of the detected signal (bias T 3dB low frequency cutoff=10KHz) is digitized by an 8-bit ADC triggered by a 2.5 GHz clock. The digital signal is stored and the difference between consecutive digital values is obtained using a software implementation. The $m \leq 8$ LSBs of the difference value are stored as the next m bits in a final RBG string. The rate of random number

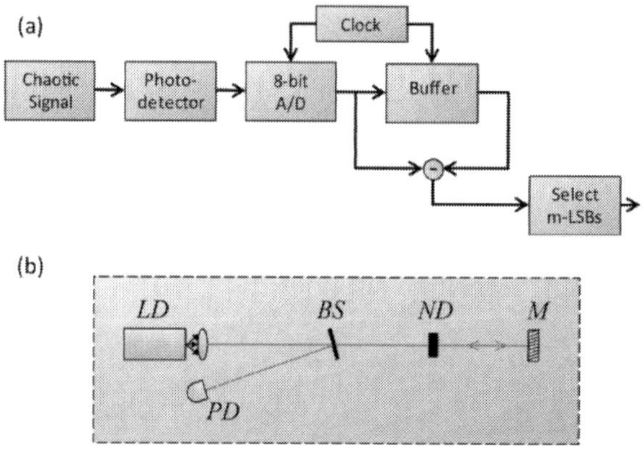

Figure 1. (a) A schematic diagram of the RBG. (b) Laser implementation, laser diode (LD), beam splitter (BS),neutral density filter (ND), mirror (M), high speed photo detector (PD).

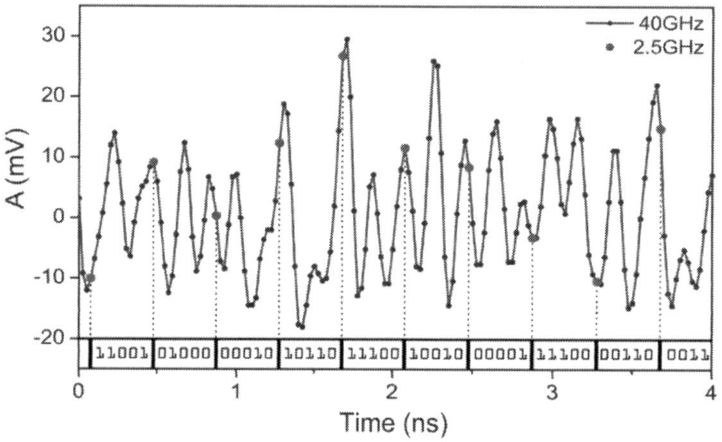

Figure 2. A 4 ns trace of laser intensity digitized at 40 GHz (blue dots connected by a line to guide the eye), and the sampling points at 2.5 GHz (red circles). At each sampling point, m-LSBs obtained from the difference between the current and the previous sampled point are generated and attached to the random bit stream. The generated bit stream for m=5 is depicted in the strip below the signal trace.

generation is therefore m times the ADC clock rate, and m can be varied up to a maximum value which depends on the resolution of the ADC. Figure 2 shows the AC component of the chaotic signal as recorded at a 40 GHz digitizing rate by the 8-bit ADC. The characteristic fluctuation time of the intensity is clearly shorter than the 2.5 GHz sampling rate used for the RBG, indicated by the red circles in the figure.

To test the randomness of the sequences we generated, standard statistical tests were applied to the sequences. We found that the bit sequences obtained from the differentiated chaotic laser intensity fluctuations using the 5 LSBs passed all of the NIST test [18] (performed for 1000 sequences of 1 Mbit length per sequence), thus allowing us to effectively generate random bits

Statistical test	P-value	Proportion	Result
Frequency	0.383827	0.9900	Success
Block frequency	0.591409	0.9890	Success
Cumulative sums	0.593478	0.9940	Success
Runs	0.869278	0.9930	Success
Longest run	0.980883	0.9890	Success
Rank	0.041709	0.9910	Success
Nonperiodic templates	0.007694	0.9910	Success
Overlapping templates	0.163513	0.9830	Success
Universal	0.670396	0.9870	Success
Approximate entropy	0.114040	0.9830	Success
Random excursions	0.133216	0.9919	Success
Random excursions variant	0.031213	0.9886	Success
Serial	0.272977	0.9870	Success
Linear complexity	0.208837	0.9850	Success

Table 1. Results of NIST Special Publication 800-22 statistical tests. For "success" using 1000 samples of 1 Mbit data and significance level= 0.01, the P value (uniformity of p values) should be larger than 0.0001 and the proportion should be greater than 0.9805608 [18]. For the tests which produce multiple Pvalues and proportions, the worst case is shown. Test results are shown for the 5-LSBs.

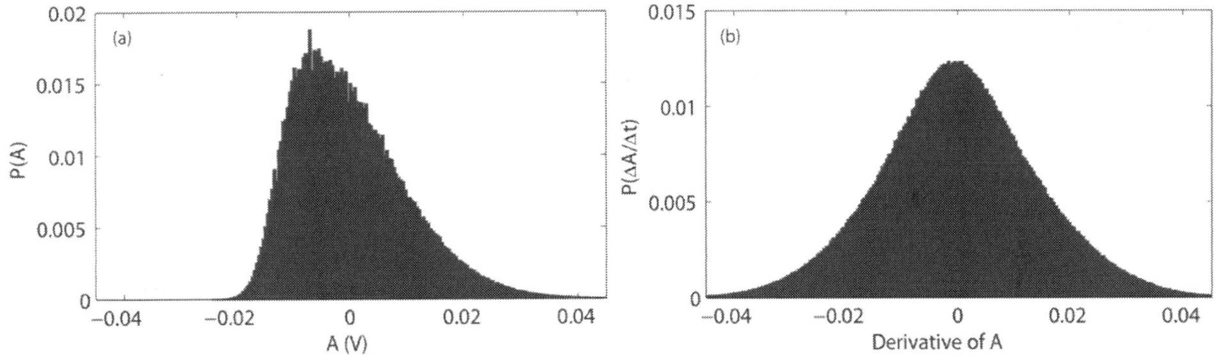

Figure 3. (a)Histogram of the laser intensity, independent of the sampling rate, obtained via an 8-bit ADC. The histogram is non symmetric and the small nonlinearity in the electronic measurement system results in population variance in some of the bins. (b) The first derivative of the time dependent laser intensity sampled at 2.5 GHz in the range [-255,255] (for clarity of the presentation only the occupation in the range [-150,150] is presented).

at a rate of 12.5Gbits/s (5 x 2.5Gbits/s). Typical results of the NIST tests are shown in Table 1 [19, 20].

3. First derivative of the chaotic signal and the number of used LSBs

A key element of our RBG method is the use of derivative of the signal rather than the signal itself. The derivative broadens the distribution of values used to generate the random sequence as the negative range of values becomes possible from -255 to 255. Its distribution is highly symmetric with mean value approaching zero (Figure 3(b)) and it enhances smaller changes in the chaotic signal to help eliminate correlations within the generated sequence. It is also responsible for the reduction of the population variance in some of the bins, caused by

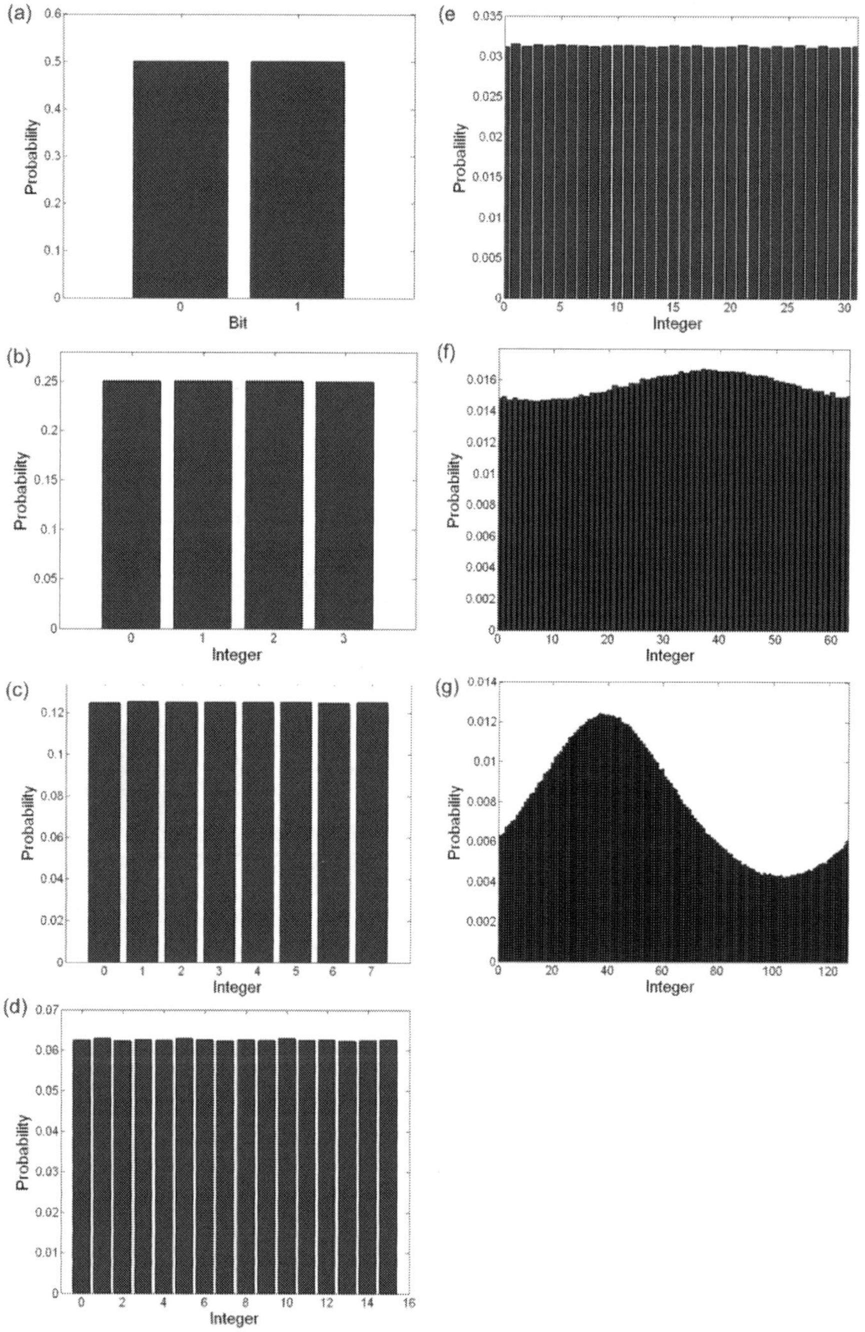

Figure 4. Uniformity of 2^m values (in integer representation) generated by the RBG algorithm retaining m-LSB data for cases (a) m=1; (b) m=2; (c) m=3; (d) m=4; (e) m=5; (f) m=6; (g) m=7.

nonlinearity in the digitizing electronics (Figure 3(a)).

An additional important element of our RBG method is the retention of only some of the

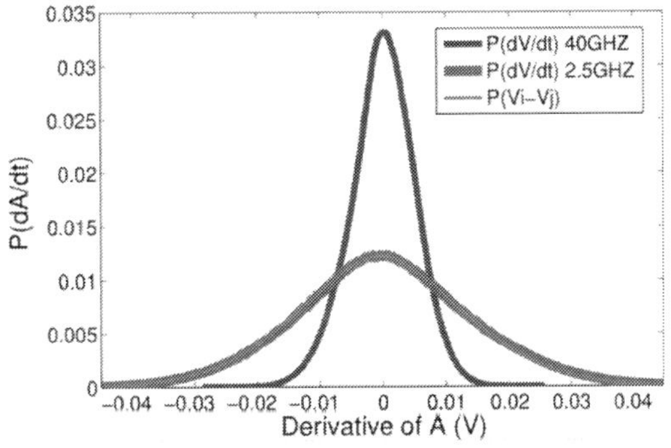

Figure 5. Histogram of the derivative of the laser intensity as obtained from the time series of laser intensities at 2.5 GHz sampling rate (green curve) and at 40GHz sampling rate (blue curve). The red line represents a histogram generated from the histogram of the waveform Fig. 4, whereby the temporal position of the values has been eliminated.

least significant bits (LSBs). Discarding the most significant bits enables to reduce correlations within the digitized sequence and to enhance uniformity of the generated sequence over short time windows. In our method, as opposed to the simple thresholding approach, m-LSBs divides the 256 possible levels for the derivatives calculated from the 8-bit level into groups of 2^{LSBs} where each group comprises of 2^{8-LSBs} different levels. For example, the first LSB, m=1, attributes 1/0 to the $2^7 = 128$ odd/even levels. The case of m=2 attributes for each level, k, among the 256 levels, a 2-bit number following mod(k,4) which can take the values 00/01/10/11 and for the general case of m-LSBs for each level k an m-bit is a signed following the binary representation of $mod(k, 2^m)$. This procedure of coarsening could affect the uniformity among the 2^m levels, as a result of the nonuniform shape and the width of the histogram of the derivative of the digitized waveform, Figure 3(b). It is obvious that a necessary condition for a RBG is that all 2^m windows of size m are equiprobable. Figure 4 demonstrates the uniformity of 2^m values (in integer representation) generated by the RBG algorithm by retaining m-LSB data, where m = 1, 2, ?, 7. The distribution of 5 LSBs or lower is uniform, whereas more LSBs show a non-uniform distribution. In this case, probability of some values is higher than others, and the produced sequence will show considerable bias .

4. High derivative of the chaotic signal

The chaotic laser intensity fluctuations in our experiments were digitized at 40 GHz sampling rate with an 8-bit resolution. After taking the derivative of the digitized values, the probability of obtaining a particular value of the derivative can be plotted as a histogram and is shown in Figure 5. The blue line is for the data digitized at 40 GHz rate while the green line is the histogram of values obtained at a digitizing rate of 2.5GHz. At a 40 GHz digitization rate, each intensity spike is sampled at a number of sample points, all lying on the same spike, as shown in Fig. 6. This will clearly lead in the thresholding approach to a correlation of consecutive values in the digitized signal and will disqualify the sequence from being random. Furthermore the approach based on the first derivative also fails, since for such a fast sampling rate (compared to the temporal structure of the signal) the derivative on the upward and downward slopes of the spike are almost constants (see red lines in Figure 6, representing local derivatives along a spike). Hence, successive first derivatives at 40 GHz sampling rate are strongly correlated.

One way to eliminate the correlations discussed above is to reduce the digitization rate of the signal. To estimate the maximal allowed sampling rate we construct the histogram

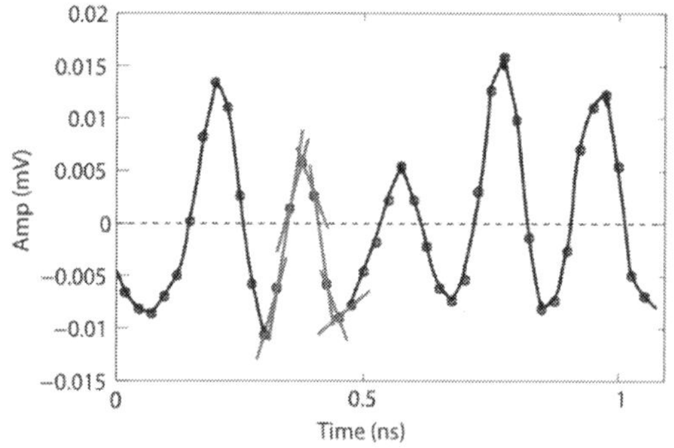

Figure 6. A 1 ns trace of laser intensity digitized at 40 GHz . The first derivative obtained from two consecutive sampling data points is demonstrated by the red lines for one spike.

of the values of the derivatives by two different procedures and look for their similarity [19, 21]. The first method is to use the original time series of the amplitudes and count the number of occurrences of a given derivative for the entire time sequence. This distribution is plotted for 2.5 GHz sampling rate (green histogram) and 40 GHz sampling rate (blue histogram) in figure 5. The second method to calculate the distribution of derivatives is to use only the distribution of amplitudes (e.g, histogram of figure 3(a)) from which all time dependence has been eliminated. More precisely, the histogram is calculated using the formula $P(n\bar{\Delta}) = \sum_{k,m} P(A_k)P(A_m)\delta(A_k - A_m - n\bar{\Delta})$, where $\bar{\Delta}$ is the amplitude value of the LSB of the 8-bit ADC and n is an integer ranging from -255 to +255.The resulting distribution of the derivative obtained from the second method is depicted by red curve of figure 5. Only if the assumption of the independence of amplitude on history is correct, will the two histograms be identical. The histograms obtained from data at 2.5 GHz sampling rate are nearly identical, implying that the correlation between two successive sampling data points is negligible, while for 40 GHz sampling rate remarkable temporal correlations are visible.

Another way to eliminate the correlations between closely lying values on a given intensity spike is to use higher derivatives of the digitized signal, $A(t)$. For instance, the first, second and third derivatives are given by $A(t) - A(t-1)$, $A(t) - 2A(t-1) + A(t-2)$, $A(t) - 3A(t-1) + 3A(t-2) - A(t-3)$, respectively.

The generation of the random bit stream consists of the following two steps [21]. In the first step the nth derivative is calculated (this uses $n + 1$ successive values of the recorded digitized waveform). In the second step the m least significant bits (LSBs) of the results of the nth derivative are appended to the bit sequence. The use of higher derivatives also enables the retention of an increasing number of available bits per sample point. This is because each derivative doubles the number of possible outcomes and, for the nth derivative, an original 8-bit value is converted to a number that has to be represented by 8+n bits. A schematic of the algorithm for n=3 is illustrated in Fig. 7. For the 16th derivative, even at 20GHz sampling rate and 15 LSBs retained, the random bit sequence passes the NIST Special Publication 800-22 statistical test suite using 1,000 sequences of 1 Mbit length [21]. The total speed of such a RBG is 300Gbits/s (15 x 20Gbits/s).

To demonstrate explicitly that the digitization rate (20 GHz) and the photon feedback time do not need to be incommensurate when using the high derivative method, the external cavity round trip time τ was carefully tuned to be commensurate with the sampling rate (equal to 10

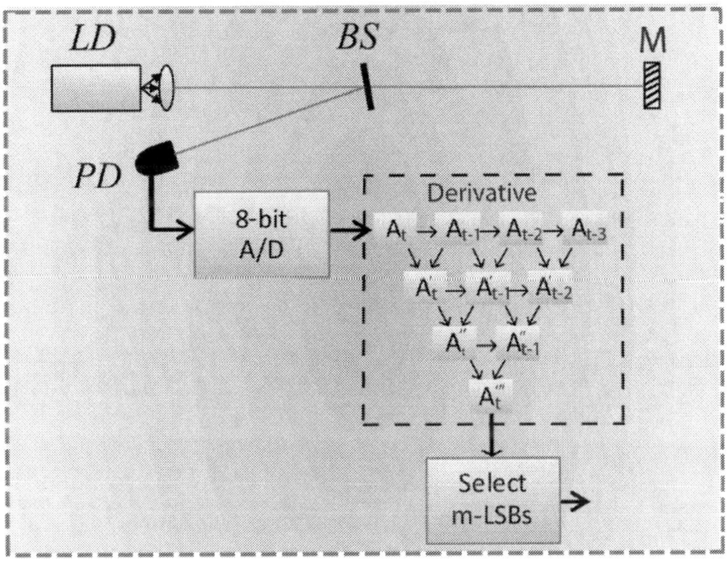

Figure 7. A schematic diagram of the RBG with high derivatives. The first part is the laser implementation, similar to Figure 1, where the external cavity round trip time was selected to be commensurate with the sampling rate, beam splitter (BS), high speed photo detector (PD), mirror (M). A schematic way to calculate the third derivative from a buffer of length four of the digitized waveform is depicted in the diagram.

ns). The feedback round trip time could be accurately adjusted by measuring the revival of the intensity autocorrelation at high integer multiples of τ. Thus, in our measurements, the ratio between the sampling rate and τ was as close to commensurate as experimentally achievable.

For a digitized waveform with a given temporal structure there are three tunable parameters (sampling rate, number of LSBs, order of the derivative) where the speed of the RBG is fixed by the product of the first two; (sampling rate) x (number of LSBs). We observed that the maximum allowable sampling rate grows quickly, reaching the experimental system's maximum of 20GHz, at the third and fifth derivative for 5 and 6 LSBs, respectively. Note that the bit string obtained from the first derivative does not pass the statistical tests using 5 or 6 LSBs, even at 0.5GHz for a **commensurate** round trip time of the external cavity (10ns) with the sampling rate. It is expected that data obtained with a higher sampling rate (e.g. 60 GHz) or with a higher resolution ADC (more than 8-bit ADC) the generation of random bits from a chaotic laser at terahertz rates will be achievable.

The nth derivative is extremely sensitive to small changes in a time window of $n+1$ sampling data points, so using a sufficiently high derivative and LSB truncation eliminates the quasi periodic structure and short range correlations present in the waveform. Figure 6 demonstrates this by showing that at 40 GHz sampling rate thresholding the waveform results in short range correlations, since in comparison to a random sequence there is a high probability for consecutive ones (zeros). The first derivative helps to suppress correlations, since the derivative can take a wide spectrum of values and especially it is positive/negative in the upward/downward slopes of a spike, independent if points are above/below the threshold. However, in a 40 GHz sampling rate a spike is sampled 6-8 times and the values of many consecutive first derivatives are very similar as depicted in figure 6. In contrast to the methods based on thresholding and first derivative, the nth derivative is extremely sensitive to small changes accumulated in a time window of n+1 sampling data points. Hence, using a sufficiently high derivative and LSBs truncation eliminate short range correlations by the amplification of small changes in the waveform over a window of n+1 sampling data points and in addition eliminates the quasi periodicity of the waveform .

References

[1] Metropolis N and Ulam S 1949 *J. Am. Statist. Assoc.* **44** 335
[2] Asmussen S and Glynn P W 2007 *Stochastic Simulation: Algorithms and Analysis* (New York: Springer-Verlag)
[3] Nielsen M A and Chuang I L 2000 *Quantum Computation and Quantum Information* (Cambridge: Cambridge University Press)
[4] Stinson D R 1995 *Cryptography : Theory and Practice* (Boca Raton, LA: CRC Press)
[5] Gallager R G 2008 *Principles of Digital Communication* (Cambridge: Cambridge University Press)
[6] Vincent C H 1971 *J. of Phys. E-Sci. Instrum.* **4** 825
[7] Maddocks R S Vincent C H Walker E W and Matthews S 1972 *J. of Phys. E-Sci. Instrum.* **5** 542
[8] Walker J 1995 HotBits: Genuine random numbers, generated by radioactive decay, http://www.fourmilab.ch/hotbits/
[9] Stefanov A Gisin N Guinnard O Guinnard L and Zbinden H 2000 *J. of Modern Optics* **47** 595
[10] Jennewein T Achleitner U Weihs G Weinfurter H and Zeilinger A 2000 *Rev. of Sci. Instrum.* **71** 1675-1680
[11] Rand Corporation. 1955 *A Million Random Digits With 100,000 Normal Deviates* (Glencoe Ill: Free Press)
[12] Fairfield R C Mortenson R L and Coulthart K B 1984 in Advances in Cryptology - Crypto '84 pp. 203-230 (New York: Springer-Verlag)
[13] Hales J Zhukov A Roy R and Dykman M I 2000 *Phys. Rev. Lett.* **85** 78
[14] Rosenbluh M et al. 2007 *Phys. Rev.* E **76** 046207
[15] Uchida A et al. 2008 *Nature Photonics* **2** 728
[16] Honjo T. et. al. 2009 *Optics Express* **17** 9053
[17] Murphy T E and Roy R 2008 *Nature Photonics* **2** 714
[18] NIST Statistical Tests Suit http://csrc.nist.gov/groups/ST/toolkit/rng/documentation_software.html
[19] Reidler I Aviad Y Rosenbluh M and Kanter I 2009 *Phys. Rev. Lett.* **103** 024102
[20] We have encountered a mistake in the FFT statistical test in [18]. Following our correspondence with NIST, we have been advised to disregard this test [18].
[21] Kanter I, Aviad Y., Reidler I., Cohen E. and Rosenbluh M. 2010 *Nature Photonics*, Published Online: 13 DECEMBER 2009 — DOI: 10.1038/NPHOTON.2009.235

Journal of Physics: Conference Series 233 (2010) 012003 doi:10.1088/1742-6596/233/1/012003

Correlation-based decimation in Constraint Satisfaction Problems

Saburo Higuchi[1,2] **and Marc Mézard**[1]

[1]Laboratoire de Physique Théorique et Modèles Statistiques, CNRS and Université Paris-Sud, Bât 100, 91405 Orsay Cedex, France

[2]Department of Applied Mathematics and Informatics, Ryukoku University, Otsu, Shiga, 520-2194, Japan

Abstract. We study hard constraint satisfaction problems using some decimation algorithms based on mean-field approximations. The message-passing approach is used to estimate, beside the usual one-variable marginals, the pair correlation functions. The identification of strongly correlated pairs allows to use a new decimation procedure, where the relative orientation of a pair of variables is fixed. We apply this novel decimation to locked occupation problems, a class of hard constraint satisfaction problems where the usual belief-propagation guided decimation performs poorly. The pair-decimation approach provides a significant improvement.

1. Introduction

Recent years have seen an important activity in the use of statistical physics concepts and methods to study discrete optimization problems (for a recent introduction, see [1]). The analysis of random constraint satisfaction problems (CSPs) has shown that the hardest problems are generated for a ratio of constraints-to-variable which is close to the critical ratio where a phase transition is found numerically, separating a "SAT" phase where almost all instances are satisfiable from an "UNSAT" phase where almost all instances are not satisfiable [2, 3, 4, 5]. Methods like the cavity or the replica method, developed in the statistical physics of disordered systems, have proved to be very efficient in locating the phase transition, both in satisfiability [6, 7, 8] and in colouring [9, 10, 11]. Another important result obtained with these methods is the existence of another phase transition, the "clustering transition", inside the SAT phase: this transition separates a phase where the space of solutions builds a large connected clusters from another SAT phase, close to the SAT-UNSAT threshold, where the space of solutions splits into ergodically separated groups – clusters [12, 6, 7]. Although the replica and cavity methods are not rigorous, some important aspects of their results have been confirmed rigorously: the probability that a random instance is SAT is known to have a sharper and sharper transition when the number of variables increases at a fixed ratio of constraints to variables [13], and the clustering property can be proven for random K-satisfiability formulas when K is large enough [14, 15].

Most interestingly, some of these mean-field based approaches have been turned into efficient algorithms which have the best performance for random satisfiability and colouring in their SAT phase close to the threshold, where the hardest instances are generated. These methods all aim at finding a satisfiable assignment (assuming that there is one). Their starting point is a 'Boltzmann'-type measure, the uniform measure over all satisfying assignment, and they are typically based on two main steps. The first one, which has been the topic of many studies, is the use of message-passing methods in order to get an estimate of the marginals of this measure. The archetype of these methods, belief propagation (BP), can be seen as an algorithmic version of the Bethe approximation, used on a given instance [16, 17, 18, 19]. When the system is in a glass phase, the survey propagation (SP) algorithm [7], an algorithmic version of the cavity method, gives very good performance. These message passing methods are used to estimate the marginals of each of the variables, with respect to the uniform measure over all solutions.

The second step makes use of these approximate estimates in order to find an assignment of variables which satisfies all constraints. If message passing would give the exact marginals, this second step would be exact, but in the most interesting problems the marginals obtained by BP or SP are approximate, so that a smart use of the information that they provide is not obvious. So far only two approaches have been used: decimation [7] and reinforcement [20]. Decimation consists in identifying from some criterion the most "polarized" variable (e.g. the one with the smallest entropy), and in fixing it to its most probable value. After this variable has been fixed, one obtains a new, smaller, CSP, to which one can apply recursively the whole procedure (BP-or SP- followed by identifying and fixing the most polarized variable). In reinforcement, one finds from the marginals the most probable value of each variable, and one adds, in the local measure of each variable, an extra bias in this preferred direction. The new CSP therefore has the same number of variables as the original one, but the local measure on each variable has been changed. One iterates this reinforcement procedure until the variables are infinitely polarized. If the algorithm is successful this returns a configuration of variables which satisfies all constraints. These two procedures, BP+decimation and BP+reinforcement, are remarkably efficient in random CSPs like K-satisfiability and graph colouring [10], and perceptron learning [21]. When one approaches the SAT-UNSAT threshold of these problems, a more elaborate version which uses the information on marginals from survey propagation (SP) is more effective [7, 6, 20], and at present the SP-based decimation and reinforcement methods are the most efficient incomplete SAT solvers for random K-satisfiability close to its threshold. Most of the work on these decimation and reinforcement procedures is numerical, as is also our present paper. On the analytical side, a first theoretical approach to the decimation problem has been done recently by [22, 23] who have analyzed the ideal decimation procedure, based on exact marginals.

There are two main reasons for exploring alternatives to the decimation and reinforcement procedures. First of all, it is clear that the estimates of the marginals obtained by message passing contain a lot of information, and it is likely that these two procedures do not exploit it fully. For instance one could think of using this information in order to bias intelligently the stochastic local search methods, or one could couple it to tree-search based solvers. On the second hand, some broad classes of constraint satisfaction problems have been found recently where these procedures perform rather poorly. These are the locked occupation problems (LOPs), a class of CSPs where the set of solution consists of isolated configurations, far away from each other [24, 25]. Apart from the XORSAT problem [26, 27, 28] which can be solved by Gaussian elimination, the random LOPs are very hard to solve in a broad region of the density of constraints, below their SAT-UNSAT transition. For these LOPs, it is known that SP is equivalent to BP. The BP+decimation method has been found to give rather poor results, and the BP+reinforcement, which works better, is still rather limited. One reason for this hardness is the fact that local marginals often convey little information on the solution.

This situation has motivated us to explore some extensions of the message-passing approaches, in which one estimates, beside local marginals, some correlation functions of the variables. Several possibilities to obtain information on the correlations from message-passing procedures have been explored recently [29, 30, 31, 32, 33, 34]. Here we use the susceptibility propagation initially introduced in [32]. We first study the general convergence properties of this method, and the reliability of its determination of correlations. We show that some of the hard LOPs that could not be solved by previous methods can now be solved by a mixture of the single-variable decimation with a new pair-decimation procedure which makes use of the knowledge of correlation. In the case of binary variables which we study here, this new procedure amounts to identifying a strongly correlated pair of variables, and fixing the relative orientation of the two variables.

The paper is organised as follows. In Sect. 2, we define the occupation problems. Sect. 3 introduces the message passing procedures, belief propagation and susceptibility propagation, and gives some simple basic properties of susceptibility propagation. Sect. 4 explains the new decimation procedure based on the estimate of correlations between pairs of variables. In Sect. 5, it is applied numerically to locked occupation problems and the accuracy of the method is examined: we measure the performance of the decimation process which makes use of the correlations obtained with this method. A short conclusion is contained in Sect. 6.

2. Occupation Problems

2.1. *Definition*

We consider a subclass of constraint satisfaction problems, the occupation problems [24, 25]. In these problems the elementary variables are binary, and the value of $x_i \in \{0,1\}$ is interpreted as an occupation number of a point $i \in \{1,\ldots,N\}$. These N occupation variables are related by M constraints $a \in \{1,\ldots,M\}$. The constraint a applies to the occupation variables of k_a points, $\partial a = \{i_{a,1},\ldots,i_{a,k_a}\}$, forming a subset of size k_a of $\{1,\ldots,N\}$. The constraint acts on the number of of occupied variables in this subset, $r_a = \sum_{i\in\partial a} x_i$. It is parametrized by a $(k+1)$-component "constraint-vector" $A_a = (A_a(0),\ldots,A_a(k_a))$ with binary entries: by definition, the constraint a is satisfied ($\psi_a = 1$) if and only if $A_a(r_a) = 1$.

Many well known CSPs belong to the category of occupation problems. For instance, k-XORSAT [26] is described by alternating constraint vectors $A = (0,1,0,1,0,1,...)$; 1-in-k satisfiability is described by $A = (0,1,0,0,0,...)$. We shall often denote a problem by the set of allowed values in the occupation-vector. For instance, 1-or-2-or-4 in 5 satisfiability corresponds to $A = (0,1,1,0,1,0)$.

A factor graph $G = (V;E)$ can be associated with an instance of an occupation problem [19]. Each variable and each constraint becomes a vertex of this graph (so that $|V| = N + M$), and an edge connects variable i and constraint a if and only if $i \in \partial a$. The neighborhood of the vertices are $\partial a = \{i \in V|(i,a) \in E\}$, $\partial i = \{a \in V|(i,a) \in E\}$. For a collection of variables in $S \subset V$, we shall write $\underline{x}_S = \{x_i|i \in S\}$. We also use the short-hand notation $\underline{x} = \underline{x}_V$. The factor graph representation of a locked occupation problem (1-in-4 satisfiability $A = (0,1,0,0,0)$) with a satisfying assignment is shown in Fig. 1.

An occupation problem is *locked* if the following three conditions are met [24, 25, 35]

- $A(0) = A(k) = 0$.
- $A(r)A(r+1) = 0$ for $r = 0,\ldots,k-1$.
- Each variable appears in at least two constraints: $|\partial i| \geq 2$

In a locked problem, each satisfiable assignment is an isolated point (one cannot obtain another assignment satisfying all constraints by changing a small number of variables). This feature seems to be at the origin of the difficulty to solve these problems with the usual message passing methods [24, 25].

2.2. *Random locked occupation problems*

We shall study random instances of locked occupation problems defined as follows: All the function nodes have degree $k_a = K$, and the constraint-vectors are all equal to the same vector A. The factor graph is uniformly chosen among the graphs where all function nodes have degree

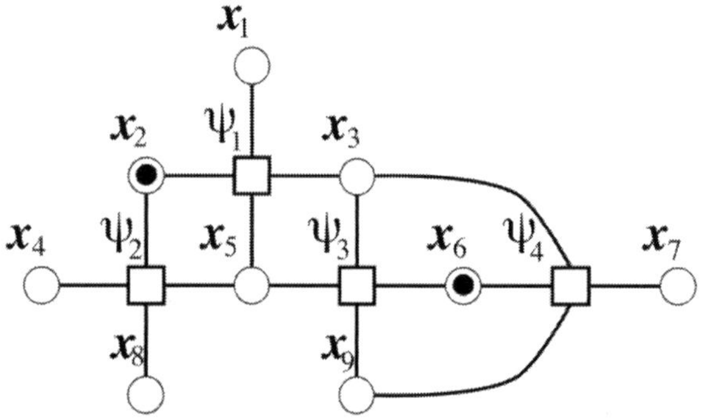

Figure 1. The factor graph representation of a small instance of 1-in-4 satisfiability ($A = (0, 1, 0, 0, 0)$). It has $N = 9$ variables (circles) and $M = 4$ constraints (squares) $\psi_1(x_1, x_2, x_3, x_5), \psi_2(x_2, x_4, x_5, x_8), \psi_3(x_3, x_5, x_6, x_9)$ and $\psi_4(x_3, x_6, x_7, x_9)$. This figure also shows a satisfying assignment in which x_2 and x_6 are occupied while other sites are not.

K and the variables have random degrees which are independent identically distributed variables chosen from the truncated Poisson degree distribution

$$q(\ell) = \begin{cases} 0 & (\ell = 0, 1) \\ \frac{e^{-c} c^{\ell}}{\ell!(1-(1+c)e^{-c})}. & (\ell \geq 2) \end{cases} \quad . \tag{1}$$

This ensures that all variable nodes have degree at least 2, making the problem locked; the average degree of a variable is

$$\bar{\ell} = \sum_{\ell=0}^{\infty} \ell q(\ell) = \frac{c(1 - e^{-c})}{(1 - (1 + c)e^{-c})}. \tag{2}$$

and the average number of variable nodes, $\overline{N}$, is given by $\overline{N} = KM/\bar{\ell}$.

The "thermodynamic limit" is taken by sending $M \to \infty$ at fixed $\bar{\ell}$, so that the density of constraints $M/\overline{N}$ is fixed. The phase diagram has been studied in [24]. When increasing $\bar{\ell}$, the probability that a satisfying assignment exists drops from 1 to 0 at the 'satisfiability threshold' ℓ_s. Between the 'clustering threshold' ℓ_d ([24] and the satisfiability threshold, the system is in a SAT phase, but the solutions are isolated points which are far away from each other. In this regime, although the satisfying assignments still exist with probability one, it is very difficult to find one by the algorithms known so far, because of the splitting of the set of solutions into clusters. This is the region which we will focus on most.

3. Message passing methods

3.1. Belief Propagation Update Rules

Consider an occupation problem described by a factor graph $G = (V, E)$ and a constraint-vector A. The uniform measure over all satisfying assignments (assuming that there is at least one such assignment), is defined as

$$p(\underline{x}) = \frac{1}{Z} \prod_{a=1}^{M} \psi_a(\underline{x}_{\partial a}) \tag{3}$$

where the function $\psi_a(\underline{x}_{\partial a})$ is equal to 1 if $A(\sum_{i \in \partial a} x_i) = 1$, and to 0 otherwise.

For later use, we introduce local 'external fields' h_ℓ^x ($x \in \{0,1\}, \ell \in V$), which will be sent to zero at the end, and consider a joint probability distribution

$$p(\underline{x}|\underline{h}^x) = \frac{1}{Z(\underline{h}^x)} \prod_{a=1}^{M} \psi_a(\underline{x}_{\partial a}) \times \prod_{\ell=1}^{N} \prod_{x} \mathrm{e}^{h_\ell^x \delta_{x_\ell, x}} . \tag{4}$$

This probability distribution is well defined as soon as there exists at least one ("SAT") configuration satisfying all the constraints. The constant $Z(\underline{h}^x)$ is a normalization factor. Our final aim is to extract solutions from the uniform measure $p(\underline{x}|\underline{0})$ over solutions satisfying all constraints (when there exists at least one solution).

The marginal distribution $p_i(x_i|\underline{h})$ can be estimated by the BP algorithm. The BP update rules for two families of messages, namely cavity fields and cavity biases, are given by [36, 1]

$$\nu_{i\to a}^{(t+1)}(x_i|\underline{h}^x) = \frac{1}{Z_{i\to a}^{(t)}(\underline{h}^x)} \prod_{b\in\partial i\setminus a} \hat{\nu}_{b\to i}^{(t)}(x_i|\underline{h}^x) \times \prod_{x} \mathrm{e}^{h_i^x \delta_{x_i,x}}, \tag{5}$$

$$\hat{\nu}_{a\to i}^{(t)}(x_i|\underline{h}^x) = \sum_{\underline{x}'_{\partial a}} \delta_{x_i,x'_i} \psi_a(\underline{x}'_{\partial a}) \prod_{\ell\in\partial a\setminus i} \nu_{\ell\to a}^{(t)}(x'_\ell|\underline{h}^x). \tag{6}$$

Here, we have decided to introduce a normalization factor $Z_{i\to a}^{(t)}(\underline{h}^x)$ for $\nu_{i\to a}^{(t)}(x_i|\underline{h}^x)$ and to avoid the normalization for $\hat{\nu}_{a\to i}^{(t)}(x_i|\underline{h}^x)$. This choice is perfectly valid for BP, and it helps to get relatively simple susceptibility propagation update rules (10)(11).

Assuming convergence to a fixed point, the BP estimate for the marginal distribution of variable i is:

$$p_i(x_i|\underline{h}^x) = \frac{1}{Z_i(\underline{h}^x)} \prod_{b\in\partial i} \hat{\nu}_{b\to i}^{(*)}(x_i|\underline{h}^x), \tag{7}$$

where $\hat{\nu}_{a\to i}^{(*)}(x_i|\underline{h}^x)$ is the fixed point of the BP iteration.

3.2. Susceptibility Propagation Update Rules

The 2-point connected correlation function at $\underline{h} = \underline{0}$ is obtained as

$$p_{ij}^{\mathrm{conn}}(x_i, x_j) \equiv p_{ij}(x_i, x_j) - p_i(x_i)p_j(x_j) = \left.\frac{\partial p_i(x_i|\underline{h}^x)}{\partial h_j^{x_j}}\right|_{\underline{h}=\underline{0}} . \tag{8}$$

To have a message-passing algorithm to calculate this quantity, we introduce the cavity susceptibility and its companion by

$$\nu_{i\to a,j}(x_i, x_j) = \left.\frac{\partial \nu_{i\to a}(x_i|\underline{h}^x)}{\partial h_j^{x_j}}\right|_{\underline{h}=\underline{0}} , \; ; \; \hat{\nu}_{a\to i,j}(x_i, x_j) = \left.\frac{\partial \hat{\nu}_{a\to i}(x_i|\underline{h}^x)}{\partial h_j^{x_j}}\right|_{\underline{h}=\underline{0}} . \tag{9}$$

Note that the roles of variables x_i and x_j are asymmetric; j can be an arbitrary variable while i is a neighbor of the constraint a.

The cavity susceptibility can be calculated by a message-passing method [29]. The susceptibility propagation update rules can be obtained by differentiating the belief propagation update rules (5) and (6) with respect to h_j^x. They read [32, 37]

$$\nu_{i\to a,j}^{(t+1)}(x_i, x_j) = \frac{1}{Z_{i\to a}^{(t)}(\underline{h}^x)} \prod_{b\in\partial i\setminus a} \left(\delta_{i,j}\delta_{x_i,x_j} + \sum_{b\in\partial i\setminus a} \frac{\hat{\nu}_{b\to i,j}^{(t)}(x_i, x_j)}{\hat{\nu}_{b\to i}^{(t)}(x_i)} + C_{i\to a,j}^{(t)}(x_j) \right), \tag{10}$$

$$\hat{\nu}_{a\to i,j}^{(t)}(x_i, x_j) = \sum_{\underline{x}'_{\partial a}} \delta_{x_i,x'_i} \psi_a(\underline{x}'_{\partial a}) \times \left(\prod_{\ell\in\partial a\setminus i} \nu_{\ell\to a}^{(t)}(x'_\ell) \right) \sum_{m\in\partial a\setminus i} \frac{\nu_{m\to a,j}^{(t)}(x'_m, x_j)}{\nu_{m\to a}^{(t)}(x'_m)}, \tag{11}$$

where

$$\nu^{(t)}_{i\to a}(x_i) = \nu^{(t)}_{i\to a}(x_i|\underline{h}^x = \underline{0}), \qquad \hat{\nu}^{(t)}_{a\to i}(x_i) = \hat{\nu}^{(t)}_{a\to i}(x_i|\underline{h}^x = \underline{0}). \tag{12}$$

The function $C^{(t)}_{i\to a,j}(x_j)$ originates from the derivative of $Z^{(t)}_{i\to a}$ and can be determined by requiring the normalization $\sum_{x_i} \nu^{(t)}_{i\to a,j}(x_i, x_j) = 0$.

Let us suppose that we have found a fixed point $(\nu^{(*)}_{i\to a}, \hat{\nu}^{(*)}_{a\to i}, \nu^{(*)}_{i\to a,j}, \hat{\nu}^{(*)}_{a\to i,j})$ of BP and the susceptibility propagation. By differentiating (7) with respect to the external fields, we can express the 2-point connected correlation function in terms of the messages at the fixed point as

$$p^{\text{conn}}_{ij}(x_i, x_j) = p_i(x_i)[\delta_{i,j}\delta_{x_i,x_j} + C_{ij}(x_j)] + \frac{1}{Z_i(\underline{0})} \sum_{b\in\partial i} \hat{\nu}^{(*)}_{b\to i,j}(x_i, x_j) \prod_{c\in\partial i\setminus b} \hat{\nu}^{(*)}_{c\to i}(x_i). \tag{13}$$

The constant $C_{ij}(x_j)$ is related to the derivative of $Z_i(\underline{h})$ and is conveniently fixed by the condition $\sum_{x_j} p^{\text{conn}}_{ij}(x_i, x_j) = 0$.

3.3. *Log-likelihood representation*

The rules (10,11) apply to all types of CSPs with discrete variables. When dealing with binary variables, it is helpful to rewrite the belief and susceptibility update equations in terms of log-likelihood variables. We introduce the cavity field and cavity bias in the log-likelihood representation $n_{i\to a}$ and $\hat{n}_{a\to i}$ as (we omit the time superscript (t) where it is obvious):

$$\nu_{i\to a}(x_i|\underline{h}) = A_{i\to a}\, e^{n_{i\to a}(\underline{h}) s_i} \quad ; \quad \hat{\nu}_{a\to i}(x_i|\underline{h}) = B_{a\to i}\, e^{\hat{n}_{a\to i}(\underline{h}) s_i}, \tag{14}$$

where s_i is the spin variable $s_i = 2x_i - 1 = \pm 1$ and the external fields in the two representations are related by $h_j = \frac{h^1_j - h^0_j}{2}$.

Naturally we define the cavity susceptibility in the log-likelihood representation as

$$\eta_{i\to a,j} = \left.\frac{\partial n_{i\to a}(\underline{h})}{\partial h_j}\right|_{\underline{h}=0} \quad ; \quad \hat{\eta}_{a\to i,j} = \left.\frac{\partial \hat{n}_{a\to i}(\underline{h})}{\partial h_j}\right|_{\underline{h}=0}. \tag{15}$$

The belief propagation update rules read

$$n^{(t+1)}_{i\to a} = \sum_{b\in\partial i\setminus a} \hat{n}^{(t)}_{b\to i} + h_i \quad ; \quad \hat{n}^{(t)}_{a\to i} = f_{a\to i}(\{n^{(t)}_{j\to a}\}_{j\in\partial a\setminus i}), \tag{16}$$

where

$$f_{a\to i}(\{n_{j\to a}\}_{j\in\partial a\setminus i}) = \frac{1}{2}\log\frac{F(+1)}{F(-1)} \quad ; \quad F(\sigma) = \sum_{\underline{s}_{\partial a}} \delta_{s_i,\sigma}\psi_a(\underline{s}_{\partial a}) \prod_{j\in\partial a\setminus i} e^{n_{j\to a}s_j}. \tag{17}$$

By differentiating both sides of (16), we obtain

$$\eta^{(t+1)}_{i\to a,j} = \sum_{b\in\partial i\setminus a} \hat{\eta}^{(t)}_{b\to i,j} + \delta_{i,j} \tag{18}$$

$$\hat{\eta}^{(t)}_{a\to i,j} = \sum_{m\in\partial a\setminus i} \frac{\partial f_{a\to i}(\{n^{(t)}_{j\to a}\}_{j\in\partial a\setminus i})}{\partial n_{m\to a}} \times \eta^{(t)}_{m\to a,j}. \tag{19}$$

Assuming that a solution $n_{j\to a}^{(t)}$ of the BP equations (16) is used, one sees that the susceptibility propagation update rule (18,19) is an inhomogeneous linear system in η and $\hat{\eta}$. The coefficient matrix takes the following form:

$$\frac{\partial f_{a\to i}(\{n_{j\to a}\}_{j\in\partial a\setminus i})}{\partial n_{m\to a}} = \frac{\langle s_m s_i\rangle - \langle s_m\rangle\langle s_i\rangle}{1-\langle s_i\rangle^2}. \tag{20}$$

Here the brackets in $\langle s_i\rangle$ and $\langle s_m s_i\rangle$ for $i,m\in\partial a$ mean the expectation value with respect to the joint probability distribution for variables that are neighbors of a constraint; it can be obtained solely from beliefs [1, Sec.14.2.3].

The magnetization and the pair correlation between arbitrary sites are given in terms of the fixed-point messages by

$$\langle s_i\rangle = \tanh\left(\sum_{b\in\partial i}\hat{n}_{b\to i}^{(*)}\right), \tag{21}$$

$$\langle s_i s_j\rangle_{\mathrm{conn}} \equiv \langle s_i s_j\rangle - \langle s_i\rangle\langle s_j\rangle = \left[1-\tanh^2\left(\sum_{b\in\partial i}\hat{n}_{b\to i}^{(*)}\right)\right]\times\left[\sum_{c\in\partial i}\hat{\eta}_{c\to i,j}^{(*)}+\delta_{i,j}\right]. \tag{22}$$

Although the symmetry between the sites is not manifest in the right-hand side of (22), it is in fact symmetric when the graph is a tree and the fixed point gives the exact magnetization.

3.4. Basic properties of susceptibility propagation

In order to study the structure of susceptibility propagation update rules (18,19), we construct a kMN-component column vector

$$\mathbf{y}^{(t)} = (\eta_{i\to a,j}^{(t)}, \hat{\eta}_{a\to i,j}^{(t)})^{\mathrm{t}}_{(i,a)\in E, j\in V}. \tag{23}$$

Then the fixed point condition associated with (18,19) can be written as a linear equation

$$\mathbf{y}^{(*)} = \mathbf{M}\mathbf{y}^{(*)} + \mathbf{b}, \tag{24}$$

with the inhomogeneous term

$$\mathbf{b} = (\delta_{i,j}, 0)^{\mathrm{t}}_{(i,a)\in E, j\in V}. \tag{25}$$

The coefficient matrix is block-diagonal in j:

$$\mathbf{M}_{(iaj),(i'a'j')} = \delta_{j,j'}\mathcal{M}_{(ia),(i'a')}, \tag{26}$$

$$\mathcal{M} = \begin{pmatrix} 0 & \mathbf{1}(a'\in\partial i\setminus a)\delta_{i,i'} \\ \mathbf{1}(i'\in\partial a\setminus i)\delta_{a,a'}\frac{\partial f_{a\to i}(\{n_{j\to a}^{(*)}\}_{j\in\partial a\setminus i})}{\partial n_{i'\to a'}} & 0 \end{pmatrix}, \tag{27}$$

where the block $\mathcal{M}$ is independent of the block index j.

Thus we obtain the unique fixed point

$$\mathbf{y}^{(*)} = (\mathbf{1}-\mathbf{M})^{-1}\mathbf{b} \tag{28}$$

if $(\mathbf{1}-\mathbf{M})$ is invertible, or equivalently, if $(\mathbf{1}-\mathcal{M})$ is invertible.

The susceptibility propagation update rules (18,19) can be regarded as an iterative method to solve the linear equation (28). It converges to a value irrespective of the initial vector if all the eigenvalues of $\mathcal{M}$ have moduli smaller than unity. Because the block $\mathbf{M}$ does not depend on

j, the existence of the fixed points and convergence to them are solely determined by $\mathcal{M}$ and do not depend on j.

When the factor graph is a tree, even in presence of the external fields $\underline{h}^x$, the exact marginals are obtained by (7) on a fixed point $\nu^{(*)}_{i\to a}, \hat{\nu}^{(*)}_{i\to a}$ [19]. Therefore, by differentiation with respect to $\underline{h}^x$, there exists a susceptibility fixed-point which gives the exact 2-point correlation function. In the examples which we have considered, the iteration of susceptibility propagation converges to this fixed-point. On the other hand, if the graph has more than one loop, there is no guarantee either that the fixed point exists or the iteration leads to that fixed point.

A simple test of these statements is obtained by studying the the 1-in-2 satisfiability problem, which is nothing but the anti-ferromagnetic Ising model at zero temperature, or a version of the XORSAT problem [26]. If one considers this problem on a chain of length N, it is easy to check that the belief and susceptibility propagation equations have a unique fixed point, which gives the correct results for the magnetization and for the pair correlation ($\langle s_i \rangle = 0$ and $\langle s_i s_j \rangle - \langle s_i \rangle \langle s_j \rangle = (-1)^{i-j}$).

Consider now the same problem on the simplest graph with one loop, a ring. The BP iteration now has a continuous family of fixed points, characterized by 'right moving' messages which alternate between the value A_+ and $-A_+$, and 'left moving' messages which alternate between the value A_- and $-A_-$, where $A_\pm$ are two constants [1]. As a consequence of the existence of this family of fixed points, $(1 - \mathcal{M})$ is not invertible; in fact it has an eigenvector with zero eigenvalue, $\mathbf{y}_0 = (\mathbf{1}, -\mathbf{1})$ where $\mathbf{1}$ corresponds to the η-block and $-\mathbf{1}$ corresponds to the $\hat{\eta}$ block. In agreement with the existence of this dangerous eigenvector, one finds that the susceptibility propagation update rule does not converge. As the susceptibility messages are updated, $\eta_{i\to i+\frac{1}{2},j}$ picks up the constant shift $\delta_{i,j} = 1$. This effect is accumulated as the messages go around the ring, and the consequence is that the messages diverge as $t \to \infty$.

In summary, for 1-in-2 satisfiability on a ring, the belief propagation can converge to a family of solutions for the magnetization among which only one solution is exact. On the other hand, the susceptibility propagation update does not have a fixed point, it diverges. In the simple case of a ring, this behaviour can be cured by using the finite temperature version of the BP and susceptibility propagation update equations. But in general there is no guarantee of convergence of loopy BP and loopy susceptibility propagation, and when they converge the quality of their results cannot be assessed a priori. Fig. 2 gives an example of analysis of a small instance of 1-in-4 satisfiability, giving an idea of the errors made by susceptibility propagation on small factor graphs. On the other hand, as for standard BP, one may hope that the method becomes better for large instances when the factor graph is locally tree-like.

4. Correlation based decimation

4.1. Single-variable and pair decimation

As we mentioned in the introduction, decimation consists in finding a strongly polarized variable and fixing it to its most probable value. One way to measure the degree of polarization is to compute the marginal $p_i(x)$ of a variable x_i, and the corresponding entropy $S_i = -\sum_x p_i(x) \log p_i(x)$. At each step one chooses the variable with the smallest entropy and fixes it: one gets a new occupation problem which has one variable less, and one iterates the procedure.

Assuming that the susceptibility propagation provides us with the good estimate for the 2-point connected correlation, we can think of a decimation procedure which acts on a pair of variable instead of a single variable. Let x_i and x_j be variables. If one defines a random variable $y_{ij} = \mathbf{1}(x_i = x_j)$, one can compute the probability $p_{ij}(y_{ij})$ of y_{ij}, and the corresponding pair entropy $S_{ij} = -\sum_y p_{ij}(y) \log p_{ij}(y)$, once one knows the 2-variable marginal $p_{ij}(x_i, x_j) = p_i(x_i)p_j(x_j) + p^{\rm conn}_{ij}(x_i, x_j)$. In pair decimation, one identifies the pair (i, j) with the smallest entropy S_{ij}, and one fixes the "pair relation" as either $x_i = x_j$ or $x_i + x_j = 1$,

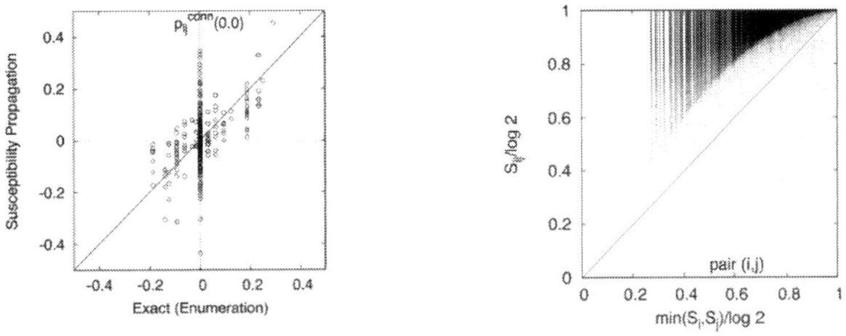

Figure 2. Left: Comparison between the 2-point connected correlation function calculated exactly and that estimated with susceptibility propagation. A 1-in-4 satisfiability instance on a randomly generated factor graph with $N = 27$ variables and $M = 16$ constraints with Poisson degree distribution with average degree $\overline{\ell} = 2.4856$. Right: Comparison between the minimum entropy $\min(S_i, S_j)$ (where S_i and S_j are the entropies of x_i, x_j)and S_{ij}, that of $y_{ij} = \mathbf{1}(x_i = x_j)$. For each pair i, j, we plot $S_{ij}/\log 2$ versus $\min(S_i, S_j)/\log 2$. The points in the bottom-right half part correspond to correlated pairs. The instance is 1-in-4 satisfiability on a random factor graph with $N = 1618$ variables and $M = 1000$ factors with the truncated Poisson degree distribution with average degree $\overline{\ell} = 2.4856$.

depending on which event is the most probable according to the measured correlation. This results in a reduced smaller CSP, which is still an occupation problem.

In both cases, it is convenient, when one fixes a variable (or the relation between two variables), to find out if this has direct logical implications on other variables. These implications, called "unit clause propagation" in the context of satisfiability, can be found efficiently by the "warning propagation" algorithm, a min-sum message passing procedure described in [28, 22].

The efficiency of the new pair decimation process depends on whether one can find a pair with less entropy than the single variable with the smallest entropy. It is easy to see that, in the absence of correlations, namely if $p_{ij}(x_i, x_j) = p_i(x_i)p_j(x_j)$, then the entropy of y_{ij} is larger than the one of x_i or x_j. So the whole procedure relies on being able to detect correlations. Fig. 2 shows that strongly correlated pairs can be found, making it possible to use pair decimation efficiently.

4.2. Algorithm

Our algorithm has two main parts, the message passing (belief + susceptibility propagation) part and the decimation part. The basic message-passing algorithm that we use is described by the following pseudocode:

`Input:` Factor graph, constraint-vector, convergence criterion, initial messages

`Output:` Estimate for 2-point connected correlation functions (or ERROR-NOT-CONVERGED)

- Initialize messages
- Repeat until all messages have converged to a fixed point (within a given resolution):
  - Update cavity fields and cavity biases $\nu^{(t)}_{i\to a}(x_i)$ and $\hat{\nu}^{(t)}_{a\to i}(x_i)$ with (5)
  - Update cavity susceptibilities $\nu^{(t)}_{i\to a,j}(x_i, x_j)$ and $\hat{\nu}^{(t)}_{a\to i,j}(x_i, x_j)$ with (10)(11) with the help of $\nu^{(t)}_{i\to a}(x_i)$ and $\hat{\nu}^{(t)}_{a\to i}(x_i)$ obtained above
- Compute 1-variable marginals $p_i(x_i)$ and entropies S_i from the fixed-point messages $\hat{\nu}^{(*)}_{a\to i}(x_i)$ by (7)

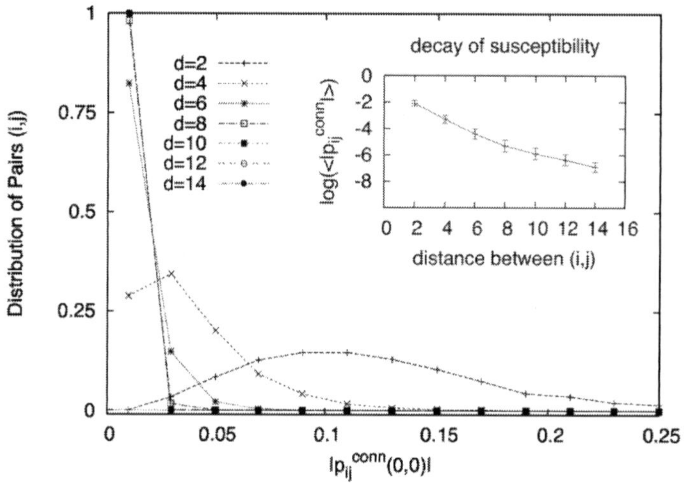

Figure 3. This graph shows how the 2-point connected correlation $p_{ij}^{\rm conn}(x_i, x_j)$ decays as the distance d between x_i and x_j increases. At each distance d, the distribution of $|p_{ij}^{\rm conn}(0,0)|$ is plotted. In the inset, logarithm of the average of that quantity is plotted against the distance. The instance is 1-in-4 satisfiability on a random factor graph with $N = 1618$ variables and $M = 1000$ factors with the truncated Poisson degree distribution with average degree $\bar{\ell} = 2.4856$.

- Compute 2-point connected correlation functions $p_{ij}^{\rm conn}(x_i, x_j)$ and entropies S_{ij} from the fixed-point messages $\hat{\nu}^{(*)}_{a\to i}(x_i)$ and $\hat{\nu}^{(*)}_{a\to i,a}(x_i, x_j)$ by (13)

This algorithm requires a memory proportional to kMN, and each step of iteration requires a computation of $\mathcal{O}(N^2)$ for fixed k.

One accelerates it by observing that the correlations tend to decay rather fast with the distance between variables on the factor graph. Fig. 3 shows the distribution of magnitude of 2-point connected correlation function computed with susceptibility propagation for all pairs of points in a graph for a fixed distance between the points. One observes a broad dispersion of correlations, and an approximate exponential decay with the distance. Here we measure the distance d with the convention that each edge connecting a variable to a constraint is of length 1. Because of this exponential decay, it is possible to use in some cases approximate versions of susceptibility propagation which are faster and use less memory. This is done by truncating to zero the cavity susceptibilities $\nu_{i\to a,j}, \hat{\nu}_{a\to i,j}$ beyond some prescribed distance $\mathrm{dist}(a,j) > d$ or $\mathrm{dist}(i,j) > d$ and keeping only the correlation functions between pairs of variables not far from each other. This truncation provides us with an efficient practical method to compute the 2-variable correlations between variables with $d \geq 4$.

The decimation algorithm that we have used is described by the following pseudo-code:

`Input:` Factor graph, constraint-vector, convergence criterion, initial messages, threshold entropy $S_{\rm th}$.

`Output:` A satisfying assignment (or FAIL-NOT-FOUND)

- While graph has more than R variables:
  - Run the message passing algorithm (belief + susceptibility propagation)
  - Find the variable i with the smallest entropy S_i
  - Find the pair ij with the smallest entropy S_{ij}
  - if ($S_i < S_{\rm th}$ or $S_i < S_{ij}$), fix variable i

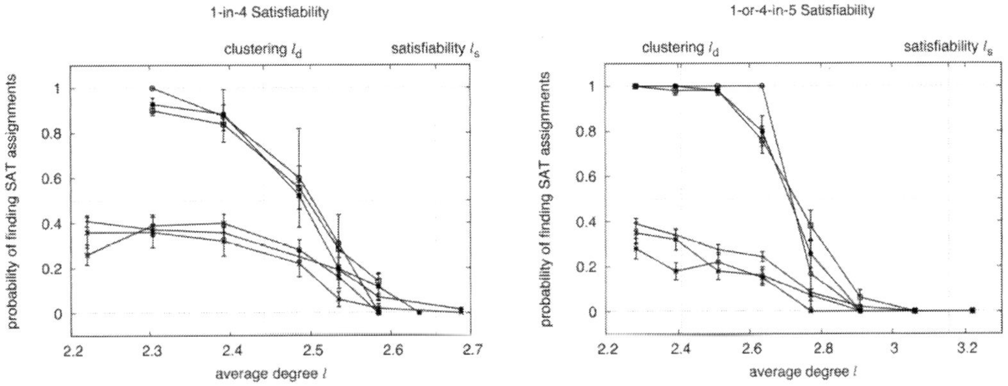

Figure 4. Success probability of pair decimation process for 1-in-4 satisfiability $A = (0, 1, 0, 0, 0)$ (left) and 1-or-4-in-5 satisfiability $A = (0, 1, 0, 0, 1, 0)$ (right) on a random factor graph with M constraints and average degree $\bar{\ell}$ (right), plotted versus $\bar{\ell}$. For comparison, the performance of simple belief-guided decimation process is shown. The vertical lines show the clustering and satisfiability thresholds.

  - else, fix the pair relation ij
  - Clean the graph :
    * Fix the value of isolated variables
    * Do warning propagation
    * Find degree 2 constraints, and if they enforce that $y_{ij} = 0$ or that $y_{ij} = 1$, fix the pair relation
- When the number of variables is equal to or smaller than R: perform an exhaustive search for satisfying assignments. If found
  - Then return the satisfying assignment
  - Else return FAIL-NOT-FOUND

The threshold entropy $S_{\rm th} = 0$ is an ad-hoc parameter. For the optimal reduction of the entropy within a decimation step, it is reasonable to set $S_{\rm th} = 0$. However, we have found that $S_{\rm th} > 0$ performs better for finding a satisfying assignment. The optimal value of $S_{\rm th}$ depends on the type of locked occupation model and the average degree. This behaviour can be understood from the fact that the estimation of 1-variable marginals is more precise than the 2-variable ones within given computational resource. Thus it is advantageous to respect the former if it is decisively small.

5. Numerical results

We have run some simulations of random LOPs using the above algorithm.

The convergence criterion for message passing was set to 10^{-2}. The threshold for exhaustive search has been fixed to $R = 16$, and the values of S_{th} were typically 0.20. The performance of this algorithm is shown for 1-in-4 satisfiability $A = (0, 1, 0, 0, 0)$ and 1-or-4-in-5 satisfiability $A = (0, 1, 0, 0, 1, 0)$ in Fig. 4. For 1-in-4 satisfiability, data with randomization is presented: instead of fixing the most polarized variable or pair, we fix a variable or pair randomly chosen among a fixed number (here we adopt 8) of most polarized variables/pairs. The figure also shows the two important thresholds for these problems, ℓ_d and ℓ_s, which are values of the average degree (a measure of the density of constraints) separating qualitatively distinct phases. In both LOPs the performance of the new algorithm using pair decimation is clearly improved compared to the simple belief-guided decimation employed in [24]. Especially for 1-or-4-in-5, the present algorithm works well above the clustering threshold, a region of $\bar{\ell}$ where all known algorithms are reported to perform poorly [24].

6. Conclusion and Discussion

We have shown how to find satisfying assignments for locked occupation problems in their hard (clustered) phase by using a new pair decimation technique based on the measurement of correlation among variables. This improves significantly upon the conventional decimation method which is guided by 1-variable marginals only. The intuitive understanding of this behaviour is that, since flipping a variable in an LOP typically forces another variable far away to be flipped, the performance of the algorithm is improved when one estimates the correlations.

Acknowledgment

S.H. was supported by Ryukoku University Research Fellowship (2008).

References

[1] Mézard M and Montanari A 2009 *Information, Physics and Computation* (New York: Oxford University Press)
[2] Cheeseman P, Kanefsky B and Taylor W M 1991 *Proc. 12th Int. Joint Conf. on Artificial Intelligence* ed J M and R R (San Mateo, CA: Morgan Kaufmann) pp 331–7
[3] Mitchell D G, Selman B and Levesque H J 1992 *Proc. 10th Nat. Conf. on Artificial Intelligence* (Menlo Park, CA: AAAI Press) pp 459–65
[4] Kirkpatrick S and Selman B 1994 *Science* **264** 1297–301
[5] Selman B, Mitchell D G and Levesque H J 1996 *Artificial Intelligence* **81** 17–29
[6] Mézard M, Parisi G and Zecchina R 2002 *Science* **297** 812–5
[7] Mézard M and Zecchina R 2002 *Phys. Rev.* E **66** 56126
[8] Mertens S, Mézard M and Zecchina R 2006 *Random Structure and Algorithms* **28** 340–73
[9] Mulet R, Pagnani A, Weigt M and Zecchina R 2002 *Phys. Rev. Lett.* **89** 268701
[10] Krzakala F, Montanari A, Ricci-Tersenghi F, Semerjian G and Zdeborová L 2007 *Proc. of the National Academy of Sciences* **104** 10318–23
[11] Zdeborová L and Krzakala F 2007 *Phys. Rev.* E **76** 031131
[12] Biroli G, Monasson R and Weigt M 2000 *Eur. Phys. J.* B **14** 551–68
[13] Friedgut E 1999 *J. Amer. Math. Soc.* **12** 1017–54
[14] Mézard M, Mora T and Zecchina R 2005 *Phys. Rev. Lett.* **94** 197205
[15] Achlioptas D and Ricci-Tersenghi F 2006 *Proc. of 38th annual ACM symposium on Theory of Computing* (New York: ACM) pp 130–9
[16] Bethe H A 1935 *Proc. Roy. Soc. London* A **150** 552–75
[17] Gallager R 1962 *IRE Transactions on Information Theory* **8** 21–8
[18] Pearl J 1982 *Proc. of American Association of Artificial Intelligence National Conference on AI* (Pittsburgh: AAAI Press) pp 133–6
[19] Kschischang F, Frey B and Loeliger H 2001 *IEEE Transactions on Information Theory* **47** 498–519
[20] Chavas J, Furtlehner C, Mézard M and Zecchina R 2005 *J. Stat. Mech.* P11016
[21] Braunstein A and Zecchina R 2006 *Phys. Rev. Lett.* **96** 030201
[22] Montanari A, Ricci-Tersenghi F and Semerjian G 2007 Preprint arXiv:0709.1667
[23] Ricci-Tersenghi F and Semerjian G 2009 *J. Stat. Mech.* **9** P09001
[24] Zdeborová L and Mézard M 2008 *Phys. Rev. Lett.* **101** 078702
[25] Zdeborová L and Mézard M 2008 *J. Stat. Mech.* P12004
[26] Creignou N and Daude H 1999 *Discrete Applied Mathematics* **96** 41–53
[27] Cocco S, Dubois O, Mandler J and Monasson R 2003 *Phys. Rev. Lett.* **90** 047205
[28] Mézard M, Ricci-Tersenghi F and Zecchina R 2003 *Journal of Statistical Physics* **111** 505–33
[29] Montanari A and Rizzo T 2005 *J. Stat. Mech.* **10** P10011
[30] Chertkov M and Chernyak V 2006 *Phys. Rev.* E **73** 065102
[31] Parisi G and Slanina F 2006 *J. Stat. Mech.* L02003
[32] Mézard M and Mora T 2009 *J. of Physiology-Paris* **103** 107–13 preprint arXiv:0803.3061
[33] Tanaka K 2003 *IEICE Transactions on Information and Systems* **E86-D** 1228–42
[34] Welling M and Teh Y W 2004 *Neural Computation* **16** 197–211

[35] Zdeborová L 2008 *Statistical Physics of Hard Optimization Problems* Ph.D. thesis Université Paris-Sud preprint arxiv:0806.4112

[36] Yedidia J, Freeman W and Weiss Y 2003 *Understanding belief propagation and its generalizations* (San Francisco: Morgan Kaufmann) pp 239–236

[37] Mora T 2007 *Géométrie et inférence dans l'optimisation et en théorie de l'information* Ph.D. thesis Université Paris-Sud http://tel.archives-ouvertes.fr/tel-00175221/en/

An extended ensemble Monte Carlo study of a lattice glass model

Koji Hukushima and Shin-ichi Sasa
Department of pure and applied Sciences, The University of Tokyo, Tokyo, 153-8902, Japan
E-mail: hukusima@phys.c.u-tokyo.ac.jp

Abstract. The thermodynamic transition to a glass phase from a liquid phase is studied in a simple lattice glass model both on a random graph and in two dimensions by employing an extended ensemble Monte Carlo method, which enables us to explore its equilibrium behavior in a high density regime. The results for the random-graph case support the scenario of an equilibrium glass transition as predicted from one-step replica symmetry breaking schemes via the cavity method. Numerical evidence for the glass transition is also found for the system in two dimensions. The equation of state measured indicates a discontinuous change in the compressibility at the transition point and the overlap distribution function has a sharp double peak structure in the glassy phase. They suggest that this singularity originates from a one-step replica symmetry breaking.

1. Introduction

Drastic changes in various properties of a liquid are observed at a laboratory glass transition point T_{g}, whose value is determined by a few conventional criteria [1]. The understanding of glass transitions has been one of challenging problems in physics [2]. However, as understood from a fact that T_{g} depends on the cooling rate, the consideration of dynamical processes seems inevitable, which causes difficulties of experimental and theoretical studies. Nevertheless, if there exists an idealized structural glass transition, which is defined as a thermodynamic singularity of an assembly of molecules without translational symmetry breaking, the nature of a laboratory glass transition might be elucidated on the basis of simple laws discovered in the idealized situation.

The study along with such a motivation has a long history. In addition to insightful phenomenological considerations [3–6], the analysis within statistical mechanics has provided quantitative results for the idealized structural glass transition [7–10]. To the present, it has become widely recognized that an idealized glass transition, if it exists, corresponds to a one-step replica symmetry breaking (1RSB). This conjecture, together with other phenomenological considerations, is called *a random-first order transition scenario* [5]. (See also [1, 11] as reviews.)

However, despite these progresses, statistical mechanics studies are restricted to models with infinite range interaction or on a random graph; otherwise, an approximation valid for those models is applied to the analysis of a finite-dimensional system. As far as we know, there are no direct evidences for the existence of the idealized structural glass transition in a finite-dimensional system. There are two reasons. First, in general, the crystallization is preferred in finite-dimensional systems. The other reason is in the operational impossibility of accessing equilibrium states, which is due to the emergence of a dynamical singularity before the thermodynamic transition is observed.

The main achievement in this paper is to present clear numerical evidences for the existence of an idealized glass transition in a two-dimensional lattice model. Concretely, among lattice models exhibiting dynamical behavior similar to that observed near glass transition points [9, 12–14], we study a model proposed by Biroli and Mézard (BM) [9]. Note however that the model defined on a cubic lattice does not exhibit an idealized glass transition, but the first-order transition to a crystal in the sense of thermodynamics. This corresponds to the first difficulty mentioned above. In order to overcome it, we consider a more frustrated model by introducing a hexagonal lattice. We then employ an extended ensemble Monte Carlo (MC) method [15] recently developed so as to reveal equilibrium properties behind dynamical singularities. The multicanonical method [16] and the exchange MC (EMC) method [17] are typical examples of the extended ensemble methods, in which the ensemble to be simulated is modified or extended from the original canonical distribution. In this paper we use the EMC method that has been applied so far to various glassy systems. With these two ideas, we shall propose our claim.

The outline of the paper is the following. In section 2 we define the BM model that is studied in this paper and explain some details of our simulation methods. In section 3 we present our numerical results for the BM model on a regular random graph and discuss the efficiency of our method. Section 4 is concerned with a possible equilibrium phase transition in a two dimensional model. Finally, section 5 is devoted to our conclusion and discussions.

2. Model and Monte Carlo algorithm

In this section, we define the model that is studied in the present work. Some details on our simulation methods are also given.

We concentrate on a simple lattice glass model, proposed by Biroli-Mézard [9], on a lattice with a fixed connectivity k. On each site i, we define a binary variable $\sigma_i = \{0, 1\}$, representing whether the i-th site is occupied by a particle or not. A particle configuration denoted by $\boldsymbol{\sigma} = \{\sigma_i | i = 1, \cdots, N\}$ is restricted by a hard constraint that each site cannot have more than l among its k neighbors occupied. We thus introduce an indicator function $C(\boldsymbol{\sigma})$ which is 1 whenever a configuration $\boldsymbol{\sigma}$ satisfies the constraint condition, and zero otherwise. Explicitly, for $l = 1$ and $k = 3$, $C(\boldsymbol{\sigma}) = \prod_i^N (1 - \sigma_i D_i)$ with $D_i = \sigma_{i_1}\sigma_{i_2} + \sigma_{i_2}\sigma_{i_3} + \sigma_{i_3}\sigma_{i_1} - 2\sigma_{i_1}\sigma_{i_2}\sigma_{i_3}$, where i_1, i_2 and i_3 are three nearest neighbors of the ith site. This model can be regarded as a constraint-satisfaction problem in computer science. In particular, the BM model for $l = 0$ on a random graph is equivalent to the vertex-cover problem recently studied by the replica theory [18] and the cavity method [19]. Using this function, the grand-canonical distribution function $P(\boldsymbol{\sigma}; \mu)$ of the BM model is given by

$$P(\boldsymbol{\sigma}; \mu) = \frac{1}{Z(\mu)} C(\boldsymbol{\sigma}) e^{\mu \sum_{i=1}^N \sigma_i}, \tag{1}$$

where $Z(\mu) = \sum_{\boldsymbol{\sigma}} C(\boldsymbol{\sigma}) e^{\mu \sum_{i=1}^N \sigma_i}$ and μ is a chemical potential governing the number of occupied sites. We denote the grand-canonical average by

$$\langle f(\boldsymbol{\sigma}) \rangle = \frac{1}{Z(\mu)} \sum_{\boldsymbol{\sigma}} f(\boldsymbol{\sigma}) C(\boldsymbol{\sigma}) e^{\mu \sum_{i=1}^N \sigma_i}, \tag{2}$$

and calculate the average occupation density as

$$\rho(\mu) = \langle \hat{\rho} \rangle = \frac{1}{N} \frac{\partial}{\partial \mu} \ln Z(\mu), \tag{3}$$

where $\hat{\rho} = \frac{1}{N}\sum_{i=1}^{N}\sigma_i$. The average occupation density can be obtained by an MC simulation discussed below. Integrating $\rho(\mu)$ from $-\infty$ to μ, we obtain $Z(\mu)$ and then all the thermodynamic quantities such as entropy.

The BM model on a regular random graph has been studied by the cavity method [20] with one-step replica-symmetry-breaking ansatz [9, 10, 21], which is conjectured to provide an exact solution in equilibrium for a class of mean-field like models on sparse random graphs. Here we briefly review some results by the cavity method [10, 21]. As the chemical potential or the average density increases, the models exhibit two distinct transitions at $\mu_{\rm d}$ and $\mu_{\rm s}$. At $\mu = \mu_{\rm d}$, which is termed the clustering or dynamical transition, the phase space is divided into exponentially many subsets and the equilibrium time diverges, while no singularity is found in the thermodynamic functions. This transition can be detected by the emergence of a non-trivial and metastable glass solution in the cavity method with 1RSB ansatz. Further, a static transition corresponding to a Kauzmann transition occurs at $\mu = \mu_{\rm s} > \mu_{\rm d}$, where the 1RSB solution is stable and the thermodynamic functions exhibit a singularity: e.g., the average density $\rho(\mu)$ is continuous but not smooth at $\mu = \mu_{\rm s}$.

Another important aspect from the equilibrium studies is the dynamical behavior. At $\mu = \mu_{\rm d}$ the system gets trapped into one of the exponentially many subsets. Thus, when the chemical potential is increased up to some value greater than $\mu_{\rm d}$ with a sweep rate, the average density does not follow the equilibrium line and depends on the increasing rate of μ. Such a behavior has been frequently observed in glassy systems. It is remarkably suggested that in some models the average quantities hardly reach the equilibrium ones even when an annealing process is conducted for MC simulations, which is based on a local update, with infinitely slow rate of μ [22]. This means that it is quite difficult to obtain the equilibrium distribution above $\mu_{\rm d}$ in numerical simulations.

In order to obtain the thermodynamic functions as a function of μ, a typical grand-canonical Monte Carlo algorithm with heat-bath transition probability has been used as a local update [23]. For lattices with a fixed connectivity that include a regular random graph and regular lattices in finite dimensions, it is useful to implement a multi-spin coding technique originally developed in an Ising-model simulation [24, 25]. In the method, occupation variables at the same site of the lattice with different values of μ are stored in a computer word. For an update of the occupation variables with different μ, one prepares a large integer vector with length $N_{\rm d}$, in which each bit of the element takes 1 with probability according to the heat-bath transition probability and 0 otherwise. Generating a random integer number r from uniform distribution in the range from 1 to $N_{\rm d}$, the chosen r-th element of the vector represents the transition probability with different μ in its bit code. Then, the variables of 64 systems are able to be updated simultaneously with a single random-number sequence in a 64-bit computer.

In addition to the local update, we employ an exchange MC method [17] so as to reveal equilibrium properties behind dynamical inaccessibility mentioned above. We prepare M-copies of the system with different particle baths and denote the particle configurations and the chemical potentials by $\boldsymbol{\sigma}^{(k)}$ and $\mu^{(k)}$, respectively, with $k = 1, \cdots, M$. For convenience, we take $\mu^{(k)} < \mu^{(k+1)}$ in order. The equilibrium distribution of the total system is expressed as the product of the grand-canonical distributions: $P(\boldsymbol{\sigma}^{(1)}, \cdots, \boldsymbol{\sigma}^{(M)}) = \prod_{k=1}^{M} P(\boldsymbol{\sigma}^{(k)}; \mu^{(k)})$. Each configuration $\boldsymbol{\sigma}^{(k)}$ is updated simultaneously and independently by the conventional grand-canonical MC method. In our simulations, M is set to be 64. This is efficiently performed by the multi-spin coding method explained above. Then, we introduce an exchange between two neighboring configurations $\boldsymbol{\sigma}^{(k)}$ and $\boldsymbol{\sigma}^{(k+1)}$ so as to satisfy the detailed balance condition with respect to the whole distribution. By these processes, each configuration can effectively move and wander in the system with a wide range of μ. The equilibration in the system at lower μ is achieved more easily. Thus, as a result of the exchange process, the relaxation to equilibrium at large chemical potentials is accelerated by mixing of the fast relaxation at lower

μ. Actually, a temperature-exchange version of the exchange MC method has been applied to various statistical-mechanical models with many barriers in phase space such as spin glasses. We thus expect that the grand-canonical distribution can be efficiently generated in the BM model by using the method. In the present work, we prepared $\sigma_i^{(k)} = 0$ for all i and k as an initial condition. A unit of time consists of N local updates for each $\boldsymbol{\sigma}^{(k)}$ of the M replicated configurations and one exchange update for every two adjacent chemical potentials. We set $\mu_1 = -0.3215$ and $\mu_M = 10$. The intermediate chemical potentials between them are determined by a simple iteration method [26] in order that acceptance probability of the exchange process is roughly constant. Discarding the data until a MC step $t = t_{\mathrm{w}}$, we estimate the expectation value of observables from sampled configurations in the time interval $[t_{\mathrm{w}}, 2t_{\mathrm{w}}]$. We choose the value of t_{w} so that the result is independent of the choice. That is, the statistical averages presented below are defined in equilibrium state, which makes a sharp contrast to standard studies in glass transitions.

3. MC results for BM model on a regular random graph

In this section we show the results of our MC simulations for the BM model defined on a regular random graph and compare them with the results by the cavity method. Our purposes in this section are twofold; one is to confirm the validity of the 1RSB cavity method by direct numerical simulations for finite size system and the other is to show the efficiency of the exchange MC in the BM model. In particular, equilibration check beyond μ_{d} is crucial for us to search for the equilibrium glass transition of the BM model in two dimensions, which will be discussed in the next section.

We here present our numerical results for the BM model with $k = 3$ and $l = 1$. Figure 1 (left) shows the average density as a function of the chemical potential μ with several annealing rates in the standard Glauber dynamics with time-dependent μ. As shown in the figure, the rate dependence hardly vanishes above the expected transitions, that is a typical phenomena found in glassy systems. We also show the result obtained by the EMC method, that is independent of the discarding initial MC steps. This strongly indicates that the result by the annealing method is still far from equilibrium while the EMC method is efficient for sampling from the equilibrium distribution.

It is of interest to study the extent to which the average density $\rho(\mu)$ for a finite calculation time is close to the equilibrium value $\rho_N^{(\mathrm{eq})}$. We fix $\mu = 10$ which is well above the expected static transition. Figure 1 (right) presents the difference $\rho_N^{(\mathrm{eq})} - \rho(\mu)$ as a function of calculation time (CTIME) for the MC annealing method and the EMC method, where CTIME is defined as the time interval for taking averages in the system with $\mu = 10$. The equilibrium value $\rho_N^{(\mathrm{eq})}$ for each size N is estimated by the EMC method. The thermodynamic-limit value extracted from the data is consistent with the value of random close packing $\rho_{\mathrm{RCP}} = 0.575\,74$ calculated by the cavity method [10]. At relatively short CTIME, the difference in the annealing method is smaller than that in the EMC method, but it is reversed for longer CTIME. This also demonstrates the efficiency of the EMC method over the annealing method in a practical manner. Interestingly, the long-time limit of the difference in the annealing method seems to take a finite value in the large-size limit, which is very close to the value predicted by the isocomplexity approximation [10, 22].

Next we show the equilibrium thermodynamic behavior of the BM model obtained by the EMC method. The most primitive quantity is the average density as a function of μ, shown in figure 1. This function is expected to exhibit a singularity in the thermodynamic limit at the static transition point μ_{s} but it is rounded by the finite-size effect. In order to detect a thermodynamic singularity clearly, the density fluctuation or susceptibility $\chi(\mu) = N[\langle(\hat{\rho} - \langle\hat{\rho}\rangle)^2\rangle]$ with $[\cdots]$ being the average with respect to the random-graph ensemble is measured, which is

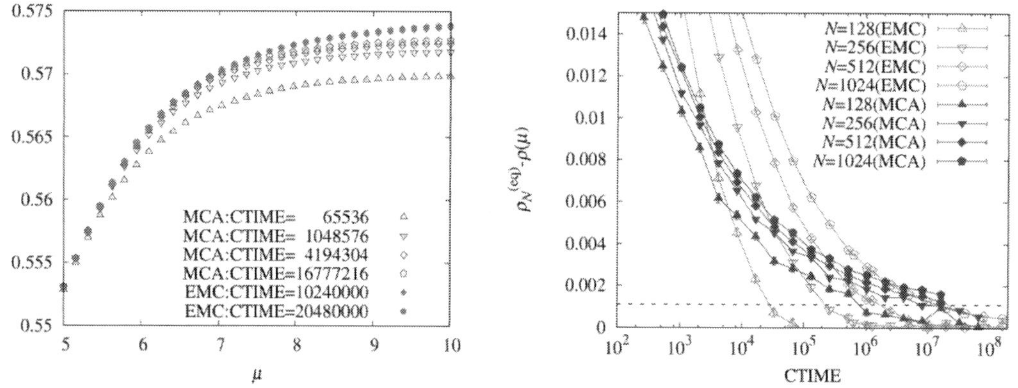

Figure 1. Left: Average density ρ as a function of μ for the BM model on a random graph with $N = 512$. Open marks are obtained by the MC annealing (MCA) method in which the data at each μ is averaged over caluculation time (CTIME) shown in the legend of figure. The filled marks are obtained by the EMC method with the same definition of CTIME.
Right: CTIME dependence of the difference between the average density at $\mu = 10$ and the equilibrium value of the BM model for different system sizes by the MC annealing (MCA) method (closed symbols) and the EMC method (open). The dotted line is the result by the isocomplexity approximation [22].

equal to $d\rho/d\mu$ through the fluctuation-response relation. Note also that the compressibility is given by χ/ρ^2. In analogy with spin-glass study, we define an overlap variable

$$\hat{q} = \frac{1}{N}\sum_{i=1}^{N}(\sigma_i - \hat{\rho})(\sigma'_i - \hat{\rho}'), \tag{4}$$

where $\boldsymbol{\sigma}$ and $\boldsymbol{\sigma}'$ are configurations of two independent systems touched to the particle baths with the same μ. We carry out an independent MC simulation for each of them and estimate the overlap distribution $P(q) = [\langle\delta(q - \hat{q})\rangle]$ by calculating a histogram of $\hat{q}$. We show our numerical result for $P(q)$ in figure 2. It is found that at μ below some value μ^* $P(q)$ has a single peak and it splits into two peaks, one at $q = 0$ and the other at a finite q, above μ^*. This is a clear evidence that the system exhibits the 1RSB transition.

In order to determine the static glass transition point μ_s precisely, we need to analyze the finite-size rounding of the singularity. We discuss the moment of the distribution $P(q)$ for finite N. The overlap susceptibility $\chi_q(\mu)$ is related to the second moment of $P(q)$; explicitly it is defined as

$$\chi_q(\mu) = N\mu\left\langle(\hat{q} - \langle\hat{q}\rangle)^2\right\rangle + N\mu\left\langle\hat{\rho}\hat{\rho}'\hat{q}\right\rangle, \tag{5}$$

which is equivalent to a response function with respect to the coupling $-\epsilon\sum_{i=1}^{N}\sigma_i\sigma'_i$. The emergence of the expectation value of $\hat{q}$ implies the divergence of the overlap susceptibility as $N \to \infty$. Thus, the transition is detected as a value at which χ_q diverges firstly as μ increases. The measured χ_q has a tendency to diverge at a finite μ as N increases (not shown). Instead, it is convenient to measure the Binder parameter g_μ [27] defined by

$$g_\mu = \frac{1}{2}\left(3 - \frac{\langle\hat{q}^4\rangle}{\langle\hat{q}^2\rangle^2}\right). \tag{6}$$

This function is used in the context of finite-size scaling analysis of second-order phase transitions. The coefficients in the expression are defined such that in the thermodynamic

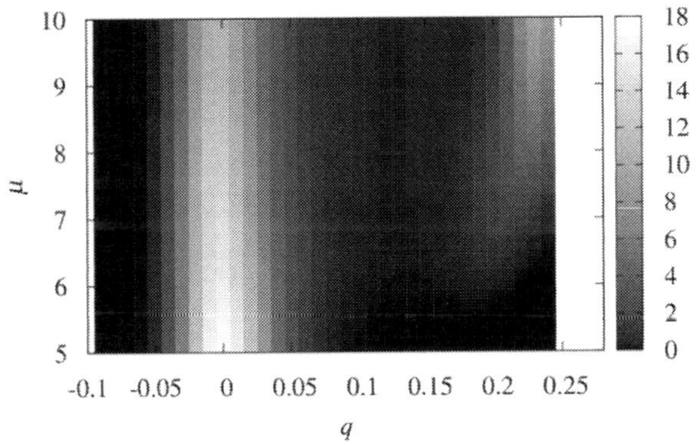

Figure 2. Gray-scale representation of the overlap distribution $P(q)$ for the BM model on a regular random graph with $N = 512$.

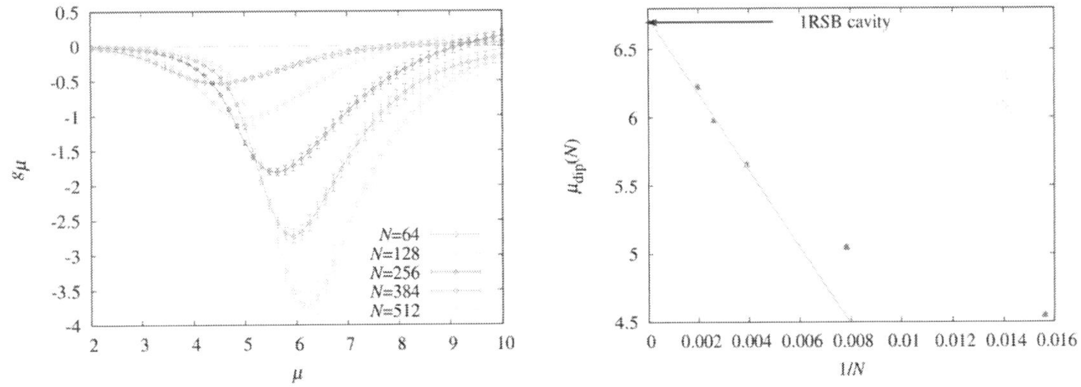

Figure 3. Left: Binder parameter of overlap q as a function of μ for several sizes N of the BM model on a regular random graph with connectivity $k = 3$. Right: (inverse) size dependence of a dip value of the chemical potential at which the Binder parameter has a minimum. The straight line stands for a fitting line for largest three sizes and the transition point predicted by the cavity method is indicated by an arrow.

limit g_μ is zero in the disordered phase and non-zero in an ordered phase. In the scaling limit, g_μ for different sizes N is expected to intersect in a common point at the transition point in a standard second-order transition. Figure 3 shows the Binder parameter obtained as a function of μ. As shown in the figure, g_μ does not have a common intersection but exhibits a clear negative dip at $\mu = \mu_{\rm dip}$. Recall that $g_\mu \to -\infty$ at $\mu = \mu_{\rm s}$ when a 1RSB transition occurs [28], in contrast to a conventional transition. Therefore, the shape of g_μ is another evidence of the 1RSB. Furthermore, $\mu_{\rm dip}$ provides us with an estimated value of the transition point $\mu_{\rm s} \simeq 6.8$, which is consistent with the cavity method.

4. Phase transition of the BM model in two dimensions

We here turn our attention to the BM model in two dimensions. In this section we study a hexagonal lattice where the number of neighboring sites, $k = 3$, is the smallest in two-dimensional regular lattices. We express the hexagonal lattice Λ by two integers i_1 and i_2, where $1 \le i_1 \le L$ and $1 \le i_2 \le L$ with L being a linear dimension. For each site $i = (i_1, i_2) \in \Lambda$, the set of three nearest neighbor site $B(i)$ are defined as $B(i) = \{(i_1 - 1, i_2), (i_1 + 1, i_2), (i_1, i_2 + ip)\}$ where $ip = 1$

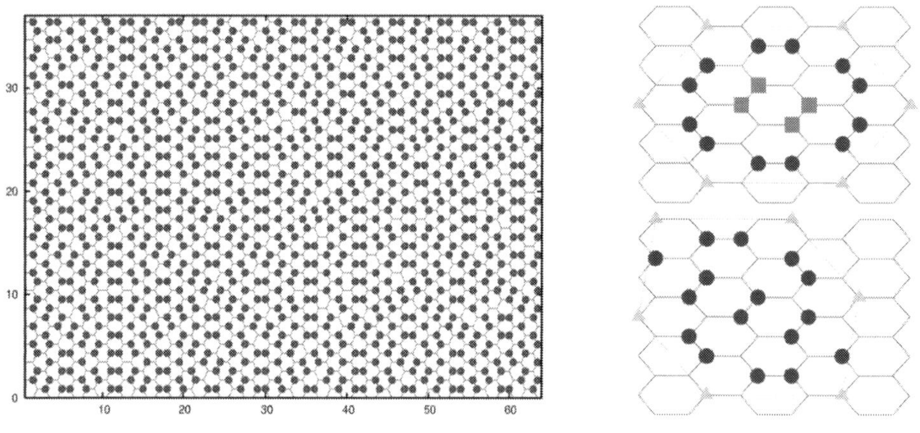

Figure 4. Left: typical equilibrium configuration for $\mu = 7.0$ and $L = 42$ under free boundary conditions. Right: elementary hexagonal regions constituting densest packing. Note that four square particles in the up pattern have three different configurations. The down pattern is obtained by shearing the left pattern in one direction. Depending on the direction, there are three types of skewed hexagonal region.

if $i_1 + i_2$ is odd and $ip = -1$ otherwise. The total number of sites N is equal to L^2. In our simulations, we investigated different system sizes up to $L = 50$.

We start with the left panel of figure 4, in which we display a typical particle configuration of the system with $\mu = 7$ and $L = 42$ under free boundary conditions. In the argument below, we wish to convince readers that the configuration is obtained just after the one-step replica symmetry breaking occurs. Before presenting the numerical evidences for our claim, we first consider densest packing.

We conjecture that the densest packing in an infinite space is obtained by patching hexagonal regions shown in the right-up in figure 4. These patterns provide the maximum density 9/16. On the other hand, densest packing in systems with a finite size are highly non-trivial. The exceptional cases are that L is a multiple of 8 under periodic boundary conditions, in which periodic patterns consisting of symmetric hexagonal regions (right-up in figure 4) provides the density 9/16. A densest packing for the other cases of finite sizes is generated by inserting skewed hexagonal regions (right-down in figure 4) into the symmetric hexagonal pattern in a complicated manner. For this reason, the qualitative behavior of quantities we measured depends on the system size and the boundary condition, up to $L = 50$ which is the current limitation of our computation. In the present paper, we focus on the system under free boundary conditions, and the results for the system under periodic boundary conditions, in which more complicated size dependence is observed than that presented below, will be reported in another paper.

Next, we briefly mention how particle configurations change as μ is increased. For sufficiently small μ, there is no feature in configurations. Increasing μ, one can find several types of local configuration that yields a density larger than 0.5. With larger μ, the two types of the hexagonal regions on the right-hand side of figure 4 come to dominate a large part of the system in a heterogeneous way. Such a transition reminds us of a one-step replica symmetry breaking. However, it is not easy to analyze configurations by counting the number of local configurations with large density. Instead, following the analysis in the previous section, we shall present statistical quantities supporting the existence of the transition.

In the inset of figure 5, the average density is shown as a function μ and strong rate dependence is observed in the annealing method, as shown in figure 1. In figure 5, we plot the density

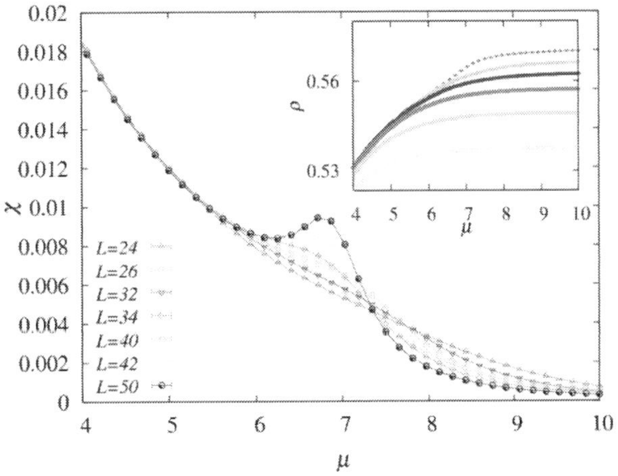

Figure 5. Intensity of density fluctuation χ as a function μ for different sizes. (Inset): Average density ρ as a function of μ for $L = 50$. The value of μ is changed from $\mu = 0$ to $\mu = 10$ during a time interval $[0, t_{\rm a}]$. The five curves correspond to the data by MC annealings with $t_{\rm a} = 10^3$, 10^4, 10^5, 10^6, and 10^7, respectively, from below. The circle symbols (top) represent the equilibrium result obtained by the exchange MC method.

fluctuation χ, obtained by EMC, as a function of μ for the system with several values of L. The size dependence observed in the regime $\mu \geq 6.0$ indicates the anomalous behavior.

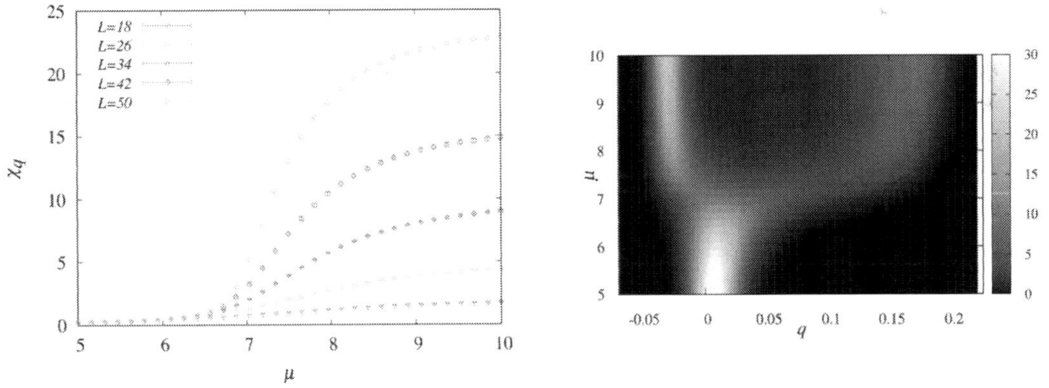

Figure 6. Left: Overlap susceptibility χ_q as a function μ for several values of L. Right: gray-scale representation of the distribution function of q for the system with $L = 42$.

In figure 6, the numerical data of $\chi_q(\mu)$ are displayed for the system with several values of L. This result also suggests that the symmetry breaking occurs in a region $\mu \geq 6.5$. As a further evidence for the existence of the replica symmetry breaking, the distribution function $P(q)$ is displayed in the right panel of figure 6. This picture indicates the two peaks of $P(q)$ when $\mu \geq 6.5$, while the distribution function of $\hat{\rho}$ has a single peak (not shown). Since the peaks appear with a finite separation, this indicates that a one-step replica symmetry breaking occurs, although a possibility of a continuous transition cannot be completely ruled out, in contrast with the sharp discontinuous transition observed in the BM model on a random graph. Figure 7

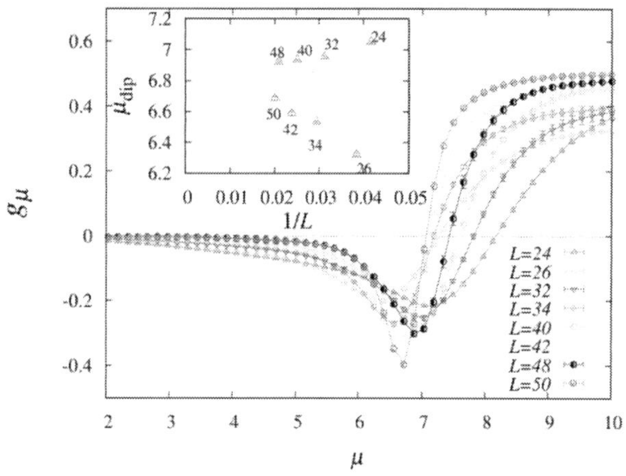

Figure 7. Binder parameter g_μ of q as a function of μ for several values of L. The inset represents $1/L$ dependence of the dip position of g_μ.

presents the Binder parameter of the overlap, which has a clear negative dip at $\mu = \mu_{\rm dip}$. This is quite similar to that observed in the previous section. As seen in the inset of figure 7, $\mu_{\rm dip}$ approaches the transition point from above for $L = 8m$ and below for $8m + 2$ with integer m. This result implies that the transition indeed occurs in the thermodynamic limit and the transition point is estimated as $\mu_{\rm s} \simeq 6.8$, which is unexpectedly close to that on the random graph with $k = 3$.

5. Conclusion and discussions

In this paper we have presented the first equilibrium Monte Carlo study based on the extended ensemble method for the Biroli-Mézard lattice glass models both on the regular random graph and on the hexagonal lattice in two dimensions. According to the cavity analysis, it was predicted that the models on the random graph exhibit both a dynamical transition at $\mu_{\rm d}$ and a static transition at $\mu_{\rm s} > \mu_{\rm d}$. The issues to be addressed hence are to verify the theoretical predictions from MC simulations with finite sizes. In particular, we focus our attention on the static glass transition. To clarify them is not only of interest for a better understanding of the equilibrium properties of the BM model but also useful for interpreting numerical results for glassy models in general. We have established that our MC results support the theoretical predictions qualitatively, where the static transition has been estimated from the analysis of the Binder parameter. Further, it has been found that a direct observation of the overlap distribution $P(q)$ is consistent with the 1RSB scenario. The alternative view of these results is that the exchange MC method employed turns out to be an efficient tool for studying equilibrium glassy behavior beyond the dynamical singularity.

Using the exchange MC method, we also have found that the equilibrium transition occurs at a finite value of μ in the BM model on the hexagonal lattice. At the transition point $\mu_{\rm s}$, the system exhibits a discontinuous change of the density fluctuation and divergent behavior of the overlap susceptibility. With the help of knowledge from the random-graph model, it is strongly suggested that this singularity originates from a 1RSB. As we already discussed, when μ is slightly larger than $\mu_{\rm s}$, an amorphous packing of the two types of hexagonal regions is observed. On the other hand, we conjecture that a crystal consisting of symmetric hexagonal regions is formed in the infinite-size system with $\mu = \infty$, because the number of configurations for the crystal with the density $9/16$ is much larger than that for the other mixed patterns with the same density. We thus need to understand the competition between the amorphous packing and the crystal packing near $\mu = \mu_{\rm s}$. Let $\rho_{\rm s}$ be the density corresponding to $\mu_{\rm s}$ in the infinite size limit. The crystal order might sustain for the configurations generated by removing particles randomly from the perfect crystal with the ratio $1 - 16\rho_s/9$. However, vacants generated in this process can be an endpoint of an array of skewed hexagonal regions, with keeping the density constant, and the arrays form a random network. By estimating the number of the random networks, we will be able to consider the competition problem theoretically. Furthermore, to this point, we cannot deny a possibility that the transition we observe is described by a complicated order parameter. This possibility will be studied in future.

At any rate, our numerical results will stimulate further studies. One example is to extract numerically a correlation length associated with the 1RSB. More importantly, the connection of the thermodynamic singularity with the nature of laboratory glass transition points, dynamical singularities and non-equilibrium properties will be discussed quantitatively through numerical experiments.

Acknowledgments

The authors thank G. Biroli, F. Krzakala and H. Tasaki for fruitful discussions on statistical mechanics of this model. The numerical calculations were partly performed using the facilities of the Supercomputer Center, Institute for Solid State Physics, University of Tokyo. This work was supported by a grant from the Ministry of Education, Science, Sports and Culture of Japan, Nos. 18079004 (KH), 19540394 (SS), and 21015005 (SS).

References

[1] Cavagna A 2009 *Physics Reports* **476** 51–124
[2] Anderson P W 1995 *Science* **267** 1615–6
[3] Gibbs J H and DiMarzio E A 1958 *J. Chem. Phys.* **28** 373–83
[4] Adam G and Gibbs J H 1965 *J. Chem. Phys.* **43** 139–46
[5] Kirkpatrick T R, Thirumalai D and Wolynes P G 1989 *Phys. Rev.* A **40** 1045–54
[6] Bouchaud J P and Biroli G 2004 *J. Chem. Phys.* **121** 7347–54
[7] Mézard M and Parisi G 1999 *Phys. Rev. Lett.* **82** 747–50
[8] Monasson R 1995 *Phys. Rev. Lett.* **75** 2847–50
[9] Biroli G and Mézard M 2002 *Phys. Rev. Lett.* **88** 025501
[10] Krzakala F, Tarzia M and Zdeborová L 2008 *Phys. Rev. Lett.* **101** 165702
[11] Parisi G and Zamponi F 2010 *Rev. Mod. Phys.* **81** 789–845
[12] Ciamarra M P, Tarzia M, de Candia A and Coniglio A 2003 *Phys. Rev.* E **67** 057105
[13] Weigt M and Hartmann A K 2003 *Europhys. Lett.* **62** 533–9
[14] Jack R L, Garrahan J P and Sherrington D 2005 *Phys. Rev.* E **71** 036112
[15] Iba Y 2001 *Int. J. Mod. Phys.* C **12** 623–56
[16] Berg B A and Neuhaus T 1991 *Phys. Lett.* B **267** 249–53
[17] Hukushima K and Nemoto K 1996 *J. Phys. Soc. Jpn.* **65** 1604–8
[18] Weigt M and Hartmann A K 2001 *Phys. Rev.* E **63** 056127
[19] Zhou H 2003 *Eur. Phys. J.* B **32** 265–70
[20] Mézard M and Parisi G 2001 *Eur. Phys. J.* B **20** 217–33
[21] Rivoire O, Biroli G, Martin O C and Mézard M 2004 *Eur. Phys. J.* B **37** 55–78
[22] Montanari A and Ricci-Tersenghi F 2004 *Phys. Rev.* B **70** 134406
[23] Landau D P and Binder K 2000 *A Guide to Monte Carlo Simulations in Statistical Physics* (Cambridge: Cambridge University Press)
[24] Bhanot G, Duke D and Salvador R 1986 *Phys. Rev.* B **33** 7841–4
[25] Bhanot G, Duke D and Salvador R 1986 *J. Stat. Phys.* **44** 985–1002
[26] Hukushima K 1999 *Phys. Rev.* E **60** 3606–13
[27] Binder K 1981 *Z. Phys.* B **45** 61–9
[28] Hukushima K and Kawamura H 2000 *Phys. Rev.* E **62** 3360–5

Statistical inference using weak chaos and infinite memory

Max Welling and Yutian Chen

Donald Bren School of Information and Computer Science, University of California Irvine CA 92697-3425 USA

E-mail: welling@ics.uci.edu, yutian.chen@uci.edu

Abstract. We describe a class of deterministic weakly chaotic dynamical systems with infinite memory. These "herding systems" combine learning and inference into one algorithm, where moments or data-items are converted directly into an arbitrarily long sequence of pseudo-samples. This sequence has infinite range correlations and as such is highly structured. We show that its information content, as measured by sub-extensive entropy, can grow as fast as $K \log T$, which is faster than the usual $\frac{1}{2}K \log T$ for exchangeable sequences generated by random posterior sampling from a Bayesian model. In one dimension we prove that herding sequences are equivalent to Sturmian sequences which have complexity exactly $\log(T+1)$. More generally, we advocate the application of the rich theoretical framework around nonlinear dynamical systems, chaos theory and fractal geometry to statistical learning.

1. Introduction

There are two theoretical frameworks for statistical inference: the frequentist and the Bayesian paradigm. A frequentist assumes a true objective value for some parameter and tries to estimate its value from from samples. Except for the simplest models, estimation usually involves an iterative procedure where the value of the parameter is estimated with increasing precision. In information theoretic terms, this means that more and more information from the data is accumulated in more decimal places of the estimate. With a finite data-set, this process should stop at some scale because there is not enough information in the data that can be transferred into the decimal places of the parameter. If we continue anyway, we will overfit to the dataset at hand.

In a Bayesian setting we entertain a posterior distribution over parameters, the width of which determines the amount of information it encodes. In Bayesian estimation, the width automatically adapts itself to the amount of available information in the data. In both cases, the information contained in our parameter (distribution) can be estimated to leading order as $H(N) = \frac{1}{2}K \log(N)$ where K is the number of parameters and N is the number of data-cases. Note that the learnable information in the data grows sub-linearly (or sub-extensively).

The learning process itself can be viewed as a dynamical system. For a frequentist this means a convergent series of parameter estimates $w_1, w_2, \ldots$. For a Bayesian running a MCMC procedure this means a stochastic process converging on some equilibrium distribution. In this paper we introduce a third possibility especially suitable for Markov random field models for which maximum likelihood (ML) learning and even more so Bayesian posterior inference is highly intractable. The idea is to model the correlations in data by a deterministic weakly chaotic nonlinear dynamical system which was inspired by taking a zero temperature limit of the ML gradient update rules. Unlike ML learning however, the weights never converge but trace out a quasi-periodic trajectory on an attractor set which is often found to be of fractal dimension.

In this paper we study this dynamical system using the tools developed for chaotic systems. We estimate its topological entropy and find the characteristic behavior of weak chaos: $H(T) = K\log(T)$. We then relate this quantity to the sub-extensive entropies of the frequentist and Bayesian methods and find to our surprise that is can be higher by a factor of two. This remarkable result is explained by the fact that herding generates highly structured sequences with infinite range negative auto-correlations. This infinite memory effect renders the herding samples more efficient in encoding the information present in the data and can have implications for designing better MCMC methods.

2. Herding Dynamics

We will first introduce the herding system in its simplest form and show its equivalence to some well-studied theories in mathematics. Next, the system will be generalized and its relevance to machine learning explained in more detail.

2.1. A single Neuron Model

Consider a single (artificial) neuron, which can take on two distinct states: either it fires ($s = 1$) or it doesn't fire ($s = 0$).[1] Assume, we want this neuron to represent an irrational number, $p \in [0, 1]$, by firing at a rate that is asymptotically equal to p on average: $\lim_{T\to\infty} \frac{1}{T}\sum_{t=1}^{T} s_t = p$. One way to achieve that is to introduce a "synaptic strength", w, which keeps track of the current error. Assume that at time $t = 0$ we start with some arbitrary initial value w_0. By using the following updates for w and s we achieve the averaging property above:

$$s_{t+1} = \mathbb{I}[w_t > \alpha], \tag{1}$$
$$w_{t+1} = w_t + \gamma(p - s_{t+1}) \tag{2}$$

where $\mathbb{I}[\cdot]$ is the indicator function (equal to 1 if the argument is true and 0 if not). These updates are easily interpreted: at each iteration the synaptic strength increases by an amount γp. When the synaptic strength rises above a certain threshold, α, the neuron fires. If it fires its synaptic strength is depressed by a factor γ. The actual values of α and γ will only have some effect on the transient behavior of the resulting sequence $s_1, s_2, ...$ (in fact, for two different herding systems α, γ and α', γ' you can find initial conditions w_0 and w_0' such that the resulting sequences are identical). In Figures 1 we plot the weights (a) and the states (b) resulting from herding with a 1-neuron model. One may think of the synaptic strength as an error potential that keeps track of the total error so far.

It is perhaps interesting to note that by setting $p = \phi$ with ϕ the golden mean $\phi = \frac{1}{2}(\sqrt{5}-1)$ and initializing the weights at $w_0 = 2\phi - 1$, we exactly generate the "Rabbit Sequence": a well studied Sturmian sequence which is intimately related with Fibonacci numbers[2]). For a proof, please see Appendix A.

We can view this as a dynamical system on the state space s, i.e. $s_{t+1} = F_t(s_{[1:t]}; w_0)$. We note that the map depends on the entire history, $\{s_1, ..s_t\}$. This same information is equivalently

[1] Physicists may want to think of (classical) spin systems.

[2] Imagine two types of rabbits: young rabbits (0) and adult rabbits (1). At each new generation the young rabbits grow up ($0 \to 1$) and old rabbits produce offspring ($1 \to 10$). Recursively applying these rules we produce the rabbit sequence: $0 \to 1 \to 10 \to 101 \to 10110 \to 10110101$ etc. The total number of terms of these sequences and incidentally also the total number of 1's (lagged by one iteration) constitutes the Fibonacci sequence: $1, 1, 2, 3, 5, 8, ...$.

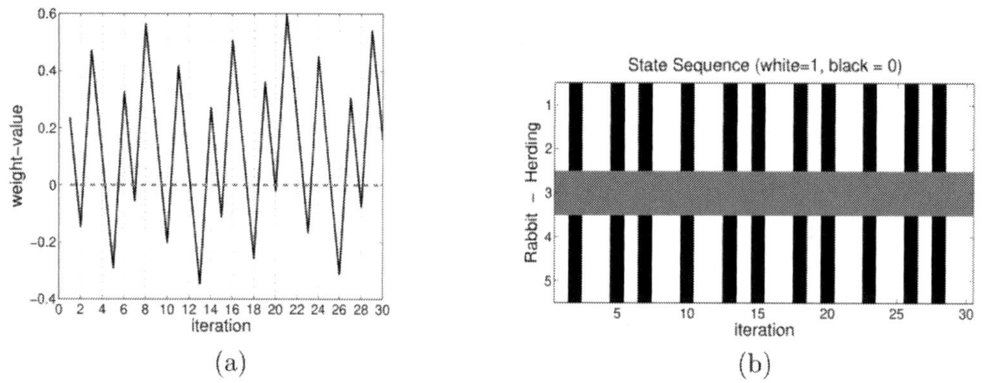

Figure 1. (a) Sequence of weight values for the "Fibonacci neuron" based on herding dynamics. Note that the state results by checking if the weight value is larger than 0 (in which case $s_t = 1$) or smaller than 0 (in which case $s_t = 0$). By initializing the weights at $w_0 = 2\phi - 1$ and using $p = \phi$, with ϕ the golden mean, we obtain the Rabbit sequence (see main text). (b) Top stripes show the first 30 iterates of the sequence obtained with herding. For comparison we also show the Rabbit sequence below it (white indicates 1 and black indicates 0). Note that these two sequences are identical as we show in Appendix A.

stored in the current value of the parameter w_t. However, marginalizing over w, the system has infinite memory. One can also compute the complexity of the state sequences, which is defined as the total number possible sequences of length T. This number turns out to be exactly $T+1$, which is the absolute bare minimum for sequences that are not eventually periodic. These facts imply that our neuron model generates Sturmian sequences for irrational values of p which are precisely defined to be the non-eventually periodic sequences of minimal complexity (Lu & Wang, 2005). (For a proof, please see Appendix A.)

2.2. A Hopfield Network Model

Now let's generalize the one neuron model to a network of N neurons (a Hopfield network (Hopfield, 1982)). Each neuron can be in a firing or non-firing state ($s_i = 0/1$) and with each neuron we associate a dynamic weight, w_i, as before but we also introduce dynamic weights (or synapses) between two neurons, w_{ij} . The update equations for this entire network become:

$$s_{i,t+1} = \mathbb{I}\left[\sum_{j\backslash i} w_{ijt}s_{jt} + w_{it} > \alpha_i\right] \tag{3}$$

$$w_{i,t+1} = w_{it} + \gamma_i(p_i - s_{i,t+1}) \tag{4}$$

$$w_{ij,t+1} = w_{ijt} + \gamma_{ij}(p_{ij} - s_{i,t+1}s_{j,t+1}). \tag{5}$$

The first equation sums up all the synaptic weights from all neighboring, spiking neurons and adds its own weight. It emits a spike when the total is larger than a threshold. This equation is typically run until all states, s, have stabilized. After that, we perform updates on the synaptic strengths as indicated. Once again, synapses grow linearly in strength but may get instantaneously depressed when the neurons associated with that synapse fire. It is not hard to show that this type of dynamics guarantees that averages over trajectories will converge to their associated population averages:

$$\lim_{T\to\infty} \frac{1}{T}\sum_{t=1}^{T} s_{it} = p_i, \tag{6}$$

$$\lim_{T\to\infty} \frac{1}{T}\sum_{t=1}^{T} s_{it}s_{jt} = p_{ij}. \tag{7}$$

Here, $\{p_i\}$ and $\{p_{ij}\}$ represent the probability of a single neuron firing and the probability of a pair of neurons firing respectively. These values should be mutually consistent and can for instance be obtained as averages over data-cases.

To guarantee that the weights in herding dynamics remain contained in a bounded region of space (i.e. do not run away to infinity), we need to find the exact maximizing argument of the Hopfield energy at every iteration. This is arguably hard in most cases. However, we claim (empirically for now) that we can replace exact maximization with approximate local maximization, initializing at the final state of the previous iteration, and still obtain excellent results. This approximate version is in fact the algorithm described above.

One way to understand dynamical synapses is as bookkeeping devices: they keep track of the number of times a state was sampled relative to other states. When a state is under-sampled it will attract the system to this state and when it is over-sampled it will repel the system away from this state. In this sense, the dynamics is similar to the weakly chaotic system of (Aihara, 1994) with destabilizing attractors.

2.3. General Features

We now generalize one step further and extend herding dynamics to the case of general features. In the examples described above we have been using the features $f_i(s_i) = s_i, f_{ij}(s_i s_j) = s_i s_j$, but there is no reason to restrict ourselves to that. In the following we will use general features denoted with $f_\alpha(S_\alpha)$, where the sub-indices fulfill the dual role of labeling both feature functions and (possibly overlapping) subsets of variables (e.g. $f_5(S_5) = f_5(s_1, s_4, s_8)$). We start by writing an energy function as follows: $E(S,w) = -\sum_k w_\alpha f_\alpha(S_\alpha)$. Herding dynamics is now defined through the repeated application of the following update equations:

$$S_{t+1} = \arg\min_S E(S, w_t), \tag{8}$$

$$w_{\alpha,t+1} = w_{\alpha,t} + < f_\alpha >_P - f_\alpha(S_{\alpha,t+1}), \tag{9}$$

where P is the probability distribution from which the data was sampled (the empirical distribution can be used instead). For tractability, we can again replace the minimization by a local minimization initialized at the final state of the last iteration.

This system shares similar properties with the Hopfield model described above. In particular we can show that under certain mild conditions the weights stay contained in a finite region of parameter space and the average of the features over the samples S_t will converge to the population average.

2.4. Hidden Variables

The models described above have the property that the generated state sequence reproduces the average of the sufficient statistics of the data. This can sometimes be restrictive in terms of the capacity to model arbitrary higher order statistical dependencies. A well-known method to model higher order dependencies is to introduce hidden variables, i.e. variables that are not directly observed in the data-sequence. The main question is whether we can extend herding dynamics to incorporate hidden variables as well.

The answer turns out to be affirmative. The only thing we have to change is the term $< f_\alpha >_P$ in the herding updates. Instead, it has to be replaced with the term $< f(Z^*, X) >_{P(X)}$, where $S = (Z, X)$ and Z^* is the hidden state that minimizes the energy,

$$Z^* = \arg\min_Z E(X, w, Z) \tag{10}$$

In practice we use for $P(X)$ the empirical distribution. This implies that at every iteration for every data-case separately we impute the most likely value for Z by minimizing the energy, and then average over all data-cases. (For more details see (Welling, 2009a).)

Conceptually, the situation has changed relative to the fully observed model of the previous sections. In particular, in those cases we only needed to store a small number of sufficient statistics and we could subsequently delete the actual dataset. In other words, the model was simply given in terms of these average sufficient statistics and we provided a way to probe this model through the generation of samples. For hidden variables we seem to actually have to store all the data-cases for our herding dynamics to run. We note that this might still be a useful exercise because the Z variables can often be interpreted as more abstract and semantically meaningful representations of the input data. These representations are often useful for subsequent processing such as classification or retrieval tasks. Moreover, the visible subset of pseudo-samples samples (X_t) can be used to interpolate between the data-samples. However, it is sometimes desirable to decouple the data altogether by replacing the data dependent terms $< f >_P$ with parameterized functions (i.e. a "model"). We have indeed performed a preliminary study and found that these functions can be effectively learned from the data. We have shown that the herding system that emerges from that process can be used to compress data and predict attributes of new data (Chen & Welling, 2010). In Figure 2 (c) we show a herding sequence trained on the digit "2" from the USPS digits dataset.

3. Learning in the Zero Temperature Limit

Consider the Kullback-Leibler divergence between the target distribution P_0 and the model distribution P_w (where w stands for parameters of the model),

$$KL[P_0||P_w] = -\sum_S P_0(S) \log P_w(S) - H[P_0] \tag{11}$$

The goal of learning would be to adjust the parameters w in such a way as to make this KL-divergence as small as possible.

In the following we will consider a Hopfield network (or Ising model) for definiteness, but the same conclusions hold for the more general systems described above. For these models we may write,

$$\begin{aligned} KL[P_0||P_w] &= < E(S;w) >_{P_0} + \log(Z) - H[P_0] & (12) \\ &= (< E(S;w) >_{P_0} - < E(S;w) >_{P_w}) - (H[P_0] - H[P_w]) & (13) \end{aligned}$$

with

$$E(S;w) = -(\sum_{ij} w_{ij} s_i s_j + \sum_i w_i s_i) \tag{14}$$

This represents a difference of two free energies, ΔF or "contrastive free energy" (Welling & Sutton, 2005).

Next, we introduce a temperature by substituting $E \rightarrow E/T$ for the energy. We will consider the quantity,

$$T\Delta F = \Delta E - T\Delta H \tag{15}$$

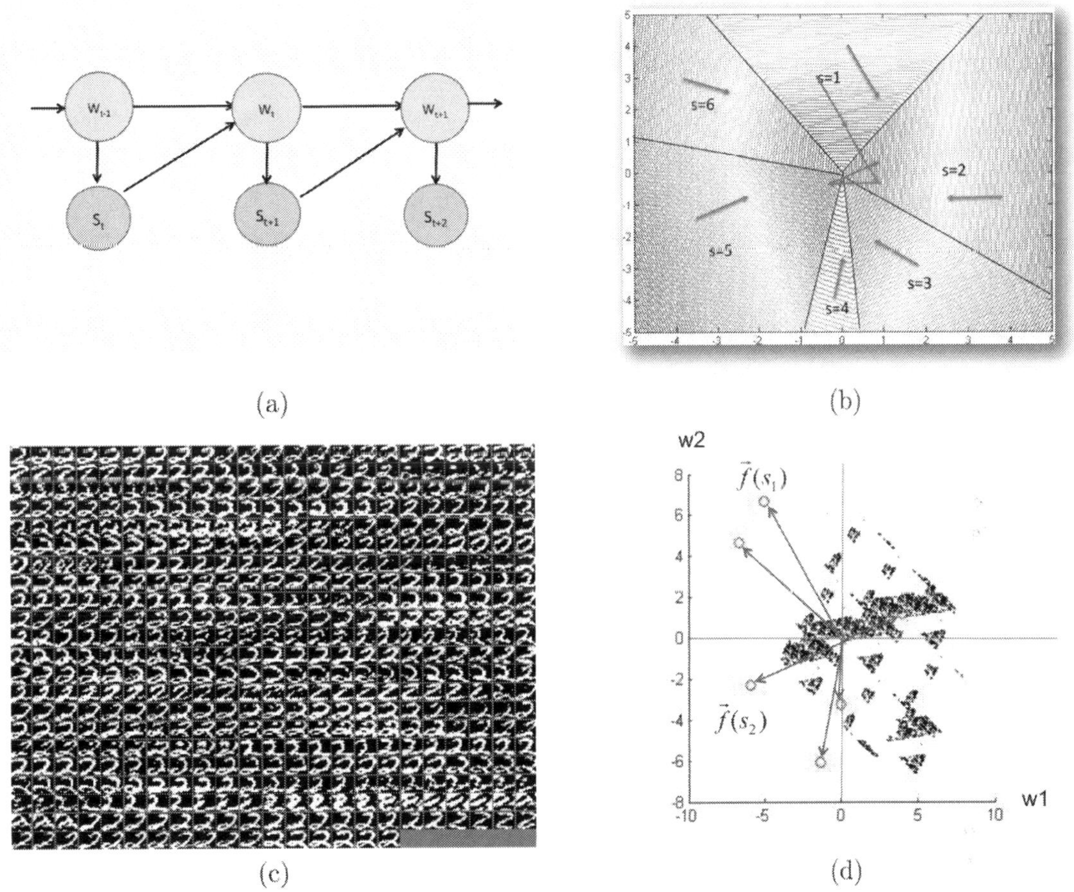

Figure 2. (a) Herding as a nonlinear dynamical system over the joint space (w, s). (b) Cones in parameter space $\{w_1, w_2\}$ that correspond to the discrete states $s_1, ...s_6$. Arrows indicate the translation vectors associated with the cones. (c) Sequence of "two's" generated using herding. (d) Fractal attractor set for herding with two parameters. The circles represent the feature-vectors evaluated the states $s_1, ...s_6$. Hausdorff dimension for this example is between 0 and 1.

and take the limit $T \to 0$. The result of this limit is then that all entropy terms vanish and averaging over P_w turns into maximization,

$$\ell_0 \doteq < E(S; w) >_{P_0} - \max_S E(S; w) \tag{16}$$

This function has a number of interesting properties (see (Welling, 2009b; Welling, 2009a)), namely: A) it is concave with a maximum at the origin, B) it is piecewise linear where the flat faces are formed by cones starting at the origin, C) it is scale free in that $\ell_0(\lambda \mathbf{w}) = \lambda \ell_0(\mathbf{w})$.

Using this function, we now propose to generate samples by alternating two steps: I) *maximize* ℓ_0 over $\mathbf{w}$ by taking a gradient step $\mathbf{w}_{t+1} = \mathbf{w}_t + \nabla_w \ell_0$ (It turns out that due to the scale-free property (C) the stepsize will not matter for the sequence S_t and will only scale $\mathbf{w}$-space by a constant factor.), II) *minimize* the energy $E(S, w)$ over S to compute the second term of

Eqn. 16. This is like performing gradient descent with a finite stepsize on a surface that looks like a "Tipi" tent. Because it will continuously overshoot the tip of the tent the sequence of weights will never converge to a fixed point. Instead we find an attractor set with (often) fractal dimension (see Figure 2 (d)). The sequence itself can be characterized as "weakly chaotic" as we will further explain in the next sections.

4. A Simple Example

For illustrative purposes only, we will now assume a finite discrete state space, $S = \{y_1, .., y_K\}$ and choose features: $f_k(S, y_k) = \delta(S, y_k)$, $k = 1..K$. Every feature is also associated with a parameter, i.e the energy is given as:

$$E(S; w) = -\sum_k w_k \delta(S, y_k) \tag{17}$$

Hence, every degree of freedom has its own feature and the model is unconstrained. We will denote the probabilities $P(S = y_k) = p_k$. In this situation the herding equations become,

$$S_{t+1} = \arg\max_S \sum_k w_{kt} \delta(S, y_k) \tag{18}$$

$$w_{k,t+1} = w_{k,t} + p_k - \delta(S_{t+1}, y_k) \tag{19}$$

In this case it is not hard to see that this will produce a sequence of states, S_t, where each symbol y_k will appear asymptotically with exactly the correct probability p_k. In figures 3 (a) and (b) we have compared the speed with which this sampling procedure converges onto the correct probabilities with drawing independent samples from the (correct) distribution. Interestingly, the above deterministic dynamical system has much better convergence properties.

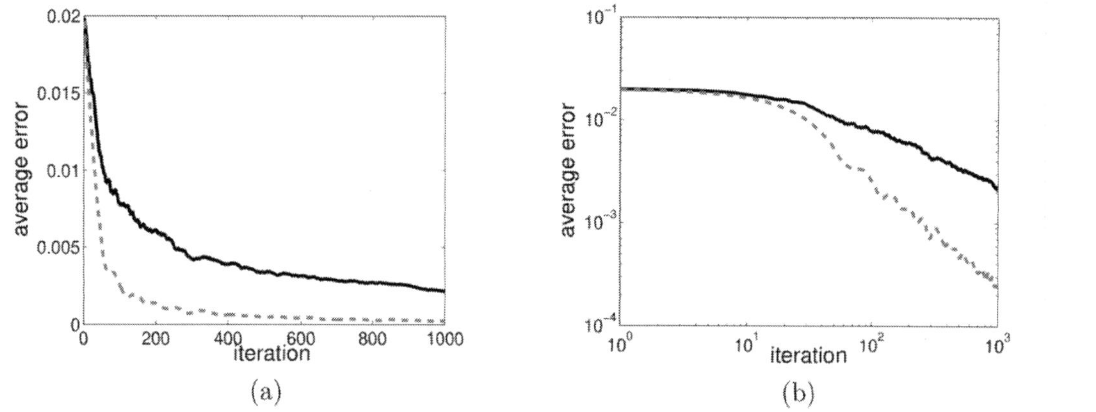

Figure 3. 100 probabilities were uniformly drawn between [0, 1] and subsequently normalized. We then drew 1000 samples either independently from the correct distribution or using the method described in the main text (herding). We measured the mean absolute difference between the true probabilities and the estimated probabilities computed as averages over samples. The black solid curve represents IID sampling while the red dashed curve represent herding. Left figure (a) shows the errors as a function of iteration while the right plot (b) shows the same on a log-log plot. Clearly, the error converges much faster for herding both initially and asymptotically.

5. Herding as a Nonlinear Dynamical System

Herding can be viewed as a nonlinear dynamical system in K dimensions where K is the total number of parameters involved (Figure 2 (a)). In the previous section we learned that herding may be viewed as a series of translations where each discrete state s corresponds to one translation (i.e. $\rho(s) =< f >_P - f(s)$). Parameter space is partitioned into cones emanating from the origin (Figure 2 (b)). If the current location of the weights is inside cone i, then one applies the translation corresponding to that cone and moves along ρ_i to the next point. This system is an example of what is known as a piecewise translation (or piecewise isometry more generally) (Goetz, 2000).

It is clear that this system has zero Lyapunov exponents everywhere (except perhaps on the boundaries between cones but since this is a measure zero set we will ignore these). One can also show that that the evolution of the weights will remain bounded inside some finite ball (Welling, 2009b; Welling, 2009a) so the evolution will converge on some attractor set. Moreover, the dynamics is non-periodic in the typical case (more formally, the translation vectors must form an incommensurate (possibly over-complete) basis set; for a proof see Appendix B). It can often be observed that this attractor has fractal dimension (see Figure 2 (d) for an example). All these facts clearly point to the idea that herding is on the edge between full chaos (with positive Lyapunov exponents) and regular periodic behavior (with negative Lyapunov exponents). In fact, herding is an example of "weak chaos", which is usually defined through its (topological) entropy. We will study the entropy of herding in the next sections.

Herding is a dissipative system in the sense that there are many locations where a point has more than one pre-image. At these places it becomes obviously impossible to determine $\mathbf{w}_{t-1}$ from $\mathbf{w}_t$. In other words, we can not run the system backwards in time, i.e. it is *time irreversible*. We interpret these regions as "entropy sinks", i.e. regions where entropy is destroyed. These entropy sinks are full dimensional (i.e. of dimension K). On the other hand, entropy is also generated on the boundaries between the cones. We call these $K-1$ dimensional areas "entropy sources". It is here that a region gets split into two (or more) regions and the points in one half become completely uncorrelated with points in the other half.[3].

Note that in breaking a region in two we do not expand the volume of our space at a microscopic level. However at the sinks, regions overlap and we do indeed loose volume there. So, we can conclude that the volume will only decrease as time moves on. Whether the volume will become zero or reach a pedestal remains an open question. We conjecture that (under the right conditions) the volume will shrink to zero leading to fractal attractors[4].

Although the $K-1$-dimensional entropy sources do not generate K-dimensional volume at a microscopic level, they do generate volume at a coarse grained level by pushing apart the two halves of the region that it splits. It is this process which is usually referred to when we speak of the second law of thermodynamics, namely that the entropy of a system always increases. Starting with a small (infinitely dense) region we will indeed find that at a coarse grained level it will expand and converge on a stable attractor set. This despite the fact that it will loose K-dimensional volume. Again this picture suggests a fractal structure of the attractor set.

6. Topological Entropy

In this section we will estimate the entropy production rate of herding. This will inform us further of the properties of this system and how it processes information. We will first focus on topological entropy.

[3] Amusingly, herding therefore satisfies its own version of the "holographic principle" where entropy is generated on two-dimensional surfaces (such as black-hole horizons) and entropy gets destroyed in the three dimensional bulk of the universe.

[4] We have numerically verified this statement for some two dimensional cases.

From Figure 2 (b) we see that the sequence $s_1, s_2, ...$ can be interpreted as the symbolic system of the continuous dynamical system defined for the parameters $\mathbf{w}$. A sequence of symbols (states) is sometimes referred to as an "itinerary". Every time $\mathbf{w}$ falls inside a cone we record its label which equals the state s. The topological entropy for the symbolic system can be easily defined by counting the total number of subsequences of length T, which we will call $M(T)$. One may think of this as a dynamical language where the subsequences are called "words". The topological entropy is defined as,

$$h = \lim_{T\to\infty} h(T) = \lim_{T\to\infty} \frac{\log M(T)}{T} \tag{20}$$

It was rigorously proven in (Goetz, 2000) that $M(T)$ grows polynomially in T for general piecewise isometries, which implies that the topological entropy vanishes for herding. It is however interesting to study the growth of $M(T)$ as a function of T to get a sense of how chaotic its dynamics is.

To count the number of subsequences of length T, we can study the T-step herding map that results from applying herding T steps at a time. The original cones are now further subdivided into smaller convex polygons, each one labeled with the sequence $s_1, s_2, ..., s_T$ that the points inside the polygon will follow during the following T steps. Thus as we increase T, the number of these polygons will increase and it is exactly the number of those polygons which partition our parameter space that is equal to the number of possible subsequences. We first claim that every polygon, however small, will break up into smaller sub-pieces after a finite amount of time. This is proven in Appendix C. In fact, because we can show that in general every pair of points will break up as well, we infer that the diameter of the polygons must shrink. A partition with this property is called a *generating partition*.

Thus, to estimate the rate with which the number of sequences grow, we need to estimate how quickly a partition breaks up into smaller pieces. Define $c(t)$ to be the total number of cells (polygons) at time t. Recall that the cells split when they overlap with a cone-boundary. The probability that that happens is proportional to the diameter of the cell, which scales as $d(t) \sim c(t)^{-\frac{1}{K}}$ where K is the number of dimensions (which is equal to the number of herding parameters). This follows because $c(t) \approx (D/d(t))^K$ with D the size of the initial cell. At time t the probability of splitting a particular cell is $\pi \sim d(t)/L$ where L is the diameter of the volume over which the cells are distributed. Therefore, the change in the expected number of cells between time t and $t + \mathrm{d}t$ equals the number of cells at time t times the probability of a cell splitting: $\mathrm{d}c(t) = c(t)\pi(t)\mathrm{d}t$. Which then gives,

$$\mathrm{d}c(t) = c(t)\pi(t)\mathrm{d}t \sim c(t)d(t)\mathrm{d}t \sim c(t)^{1-\frac{1}{K}}\mathrm{d}t \tag{21}$$

which leads to the solution: $c(t) \sim t^K$. It has been rigorously proven that the growth rate must have an exponent less or equal than K (Goetz, 2000), but we conjecture based on the above argument and some preliminary numerical simulations that it is actual equal to K for herding in the typical case (a.k.a. with an incommensurate translation basis, see Appendix B). Thus, if we accept this conjecture, then the entropy $H(T) = Th(T)$ of a sequence of length T (for T large enough) is given by,

$$H(T) = K\log(T) \qquad \alpha \in (0,1] \tag{22}$$

This result is interesting, because it implies that there may be more *learnable* information in a herding sequence than in a sequence of IID samples. We will return to this issue in section 8.

7. Dynamical Tsallis Entropy

Dynamical entropies are defined by first dividing space into small cells and initializing a large number of particles in 1 such cell. We subsequently measure (or estimate) the total accumulated probability mass in each cell of our partition. The definition for discrete time systems is,

$$h_{\mathrm{Dyn}} = \lim_{\ell \to 0} \lim_{T \to \infty} \lim_{N \to \infty} \frac{H(T) - H(0)}{T} \tag{23}$$

where N is the number of particles in the simulation, ℓ is the diameter of the cells in the partition and $H(T)$ counts the total entropy of the ensemble at time T. If we use the usual Shannon entropy $H_{\mathrm{Shannon}}(t) = -\sum_i p_i(t) \log p_i(t)$ then the Kolmogorov-Sinai entropy follows. (Note that if we *count* the number of occupied cells instead of using the Shannon entropy we retrieve the topological entropy). From the previous section we know however, that $h_{\mathrm{KS}} = 0$ because $H_{\mathrm{KS}}(T)$ would grow logarithmically in T.

There is however an alternative definition of entropy specifically designed for systems with powerlaw growth in their complexity called the Tsallis entropy (Tsallis et al., 2006) defined as,

$$H^q_{\mathrm{Tsallis}} = \frac{1 - \sum_i p_i^q}{q - 1} \tag{24}$$

The Tsallis entropy converges to the well known Shannon entropy as $q \to 1$. By demanding that the Tsallis entropy grows *linearly* in time (Tsallis et al., 2006; van Beijeren, 2004) we can compute the exponent q. For that we recall the result: $c(t) \sim t^K$. Thus, because the number of occupied cells grows as t^K, each cell itself will contain $p_{\mathrm{occupied}} \sim t^{-K}$ probability mass. Therefore:

$$\sum_i p_i^q \sim t^K (t^{-K})^q \propto t \quad \Rightarrow \quad q = \frac{K-1}{K} \tag{25}$$

This result is interesting because it tells us that as we increase the dimensionality, K, of the system, we will approach the Shannon entropy. Therefore, very high dimensional systems are expected to behave increasingly like random systems and will be more accurately described by a Gibbs distribution. This conclusion is indeed what we observe in learning systems with a very high number of variables.

8. Learning with Long Range Order

Consider the task of finding a binary sequence (a.k.a. a code) such that the mean of that sequence represents an irrational number p, i.e. $\frac{1}{T}\sum_{t=1}^{T} s_i = p$. The squared error is given as,

$$\mathrm{error}^2 = (\frac{1}{T}\sum_{t=1}^{T} s_t - p)^2 \tag{26}$$

We now wish to find the sequence that has minimal error *at every iteration t*. It is not hard to show that at time t the optimal decision is given by,

$$s_t = \mathbb{I}\left[(p - \frac{1}{2}) - (\sum_{t=1}^{T-1} s_t - (t-1)p) > 0\right] \tag{27}$$

This sequence results from herding with an initial weight $w_0 = p - \frac{1}{2}$. (Note that this is different from the initialization of the Rabbit sequence by a factor of 2.)

Alternatively, one can generate IID samples from a Bernoulli distribution $P(s) = p^s(1-p)^{1-s}$. While independent samples from a Bernoulli distribution also converge to p, they do so at a slower rate (see Figure 3). This phenomenon can be understood by the fact that a herding sequence has infinite range auto-correlations while IID samples have zero auto-correlations. The long range auto-covariances of herding are depicted in Figure 4. We observe strong negative correlations between time steps which are responsible for the improved convergence rate. Note that we have become used to the positive auto-correlations of MCMC sampling methods which have the opposite effect of decreasing the convergence rate to the mean. It's interesting that a deterministic dynamical system with long range order manages to improve the convergence rate.

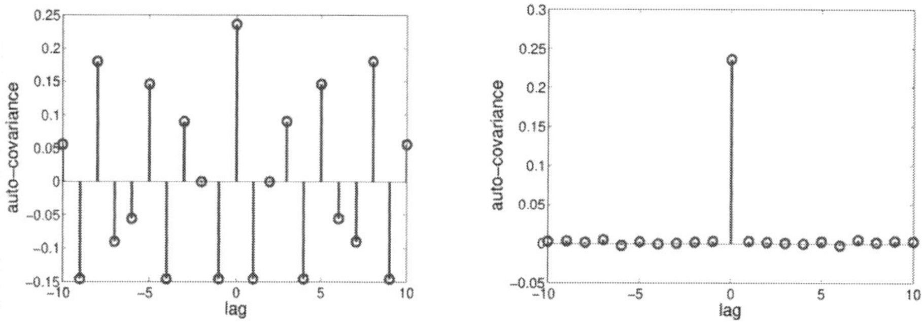

Figure 4. Autocovariance function for herding (left) and IID sampling (right) for the two-state single neuron problem of section 2.1 with $p = \phi$ the golden mean. We observe that the long range order with negative autocorrelation is causing the faster convergence to the (golden) mean.

Recall that in section 6 we estimated the dominant contribution to the entropy of a herding sequence to be $H_{\text{top}}(T) \sim K \log(T)$ with K the number of parameters. We can also compute the entropy of a *posterior* sample from a Bernoulli distribution where we will use a Beta-prior on the parameter p. This results in an exchangeable sequence of samples (no longer IID due to the fact we integrate over the p which induces dependencies). The dominant part of that sequence grows linearly in T, i.e. $H_{\text{ext}} = hT$. As explained in (Bialek et al., 2001) however, this extensive contribution will not contribute the predictive information defined as the mutual information between the future and past sequence. The authors argue that it is the sub-extensive part of the entropy that should be interpreted as the amount of predictive information about the future. In fact the authors go one step further in stating that sub-extensive entropy is a good measure of the *complexity* of the sequence where both regular and random sequences have low complexity. The dominant contribution of the sub-extensive part for posterior sampling can easily be computed as $H_{sub}(T) = \frac{1}{2}K \log T$ (with $K = 1$ for the Beta-Bernoulli distribution). This sub-extensive entropy is equal the minimum description length as it measures the number of bits required to encode the learnable (predictable) part of the sequence. It can also be derived as the *bits-back* term (Hinton & Zemel, 1994) in the Bayesian evidence but now applied to the parameters instead of the hidden variables.

Comparing to the entropy of herding[5] we can draw two important conclusions: I) The entropy of herding does not have an extensive component due to its long range order. Thus, all the information in a herding sequence is useful for prediction and no bits are wasted on unpredictable randomness. II) The entropy of herding can grow faster by a factor of two than the sub-extensive part of the entropy of the Bayesian evidence. From this we conclude that the samples of herding can be interpreted as a filtered version of samples obtained from sampling from the corresponding Bayesian model (a Beta-Bernoulli distribution in this case) where the useful signal has been separated from a noisy background. (The term "noisy" is being used in its more general interpretation as useless for prediction.)[6].

[5] We do not expect that the topological entropy and the Kolmogorov-Sinai entropy will differ significantly in their leading order behavior. The KS-entropy can be identified with the H_{sub}.

[6] It is interesting to note that the authors of (Bialek et al., 2001) comment on the equivalence of information

Herding becomes more practical, if we use it for higher dimensional learning problems like the ones discussed in sections 2.2,2.3 and 2.4. However, for these problems the number of possible states is exponentially larger than the number of parameters. We can therefore not hope to apply herding to all degrees of freedom of the problem (this would mean that we define one feature per state as in section 3). In this case it becomes important to think about the behavior of the moments which are not forced to converge to their true values. This will namely reflect our *inductive bias*. The usual approach without hidden variables is that of maximizing the entropy subject to the enforced constraints. This makes intuitive sense because we remain maximally ignorant about the unconstrained degrees of freedom. We have strong indications that herding does not follow a maximum entropy distribution. We are currently looking into alternative variational principles perhaps based on the Tsallis framework. The conclusion is thus that herding sacrifices the maximum entropy to some extend in order to achieve faster convergence on estimating some sufficient statistics relative to say posterior sampling from a hierarchical Bayesian model.

9. Conclusions

The information flow for herding is similar to that of Bayesian posterior sampling: We take in data and convert that into a sequence of parameters (sampled from the posterior $P(w|\text{data})$) and given these weights we can produce samples from the model $P(x|w)$. The overall effect of this is samples from $P(x|\text{data})$. However, Bayesian sampling is stochastic while herding is deterministic. This has interesting consequences: firstly, herding is more tractable than posterior sampling, particularly for Markov random field models. Secondly, herding has no extensive entropy leading to highly structured samples. This structure causes the sequence of samples to potentially carry more useful (predictable/learnable) information than posterior random samples. We believe that this observation underlies the success of the herding in real world learning problems as demonstrated in (Welling, 2009b) and (Welling, 2009a). In large state spaces, when the number of states is much higher than the number of moments, herding is no longer maximizing entropy for the unconstrained degrees of freedom. Elucidating the effective distribution from which herding is sampling is important because it relates to the inductive bias of the modeler.

Herding in Hopfield networks shows strong similarities to the relatively recently discovered dynamical weights in real neurons (Maass & Zador, 1998; Tsodyks et al., 1998; Pantic et al., 2002; Pfister et al., 2009). There are clear differences between herding and the phenomenon of depression in neurons, most markedly the fact that real neurons depress when the pre-synaptic neuron fires while "herding neurons" depress when both pre- and post-synaptic neurons fire. Yet, it can be argued in both cases that the depression is causing the dynamics to be on the edge between regular and chaotic behavior (Levina et al., 2007) which is often associated with high complexity and efficient information processing. Our results clearly underscore these conclusions and we believe that a good deal of progress is yet to be made in building networks of artificial neurons with depressing synapses. In this context the work of (Tieleman & Hinton, 2009) should also be mentioned where fast weights fulfil a similar role.

In summary, we believe that we have identified a problem that has components in physics, mathematics, cognitive science, neuroscience, statistics and machine learning and it could greatly benefit from the combined wisdom of these disciplines.

10. Aknowledgements

We thank Romain Thibaux, Anton Goredetski and Richard Palais for discussions and suggestions.

generated through parameter tying or through other types of infinite range correlations in the data.

Appendix A. All Sturmian sequences can be generated by herding

We wish to proof that any Sturmian sequence can be generated using herding with irrational probabilities p. We first recall the definition of Sturmian sequences. A sequence is Sturmian if it can be written as,

$$r_k = \lfloor (k+1)\gamma + \alpha \rfloor - \lfloor k\gamma + \alpha \rfloor, \qquad k = 1, \ldots \tag{A.1}$$

with $0 < \gamma < 1$ an irrational real and α real. Rewrite the second term as,

$$\lfloor k\gamma + \alpha \rfloor = \left\lceil \sum_{i=1}^{k-1} r_i \right\rceil + \lfloor \gamma + \alpha \rfloor. \tag{A.2}$$

Next we note that r_k can only take on values 0 or 1, and that we can equivalently write

$$r_k = \mathbb{I}\left[(k+1)\gamma + \alpha - \lfloor k\gamma + \alpha \rfloor > 1\right], \qquad k = 1, \ldots \tag{A.3}$$

with $\mathbb{I}[\cdot]$ a indicator function. Combining with A.2 we get the condition,

$$r_k = \mathbb{I}\left[(k+1)\gamma + \alpha - \sum_{i=1}^{k-1} r_i - \lfloor \gamma + \alpha \rfloor > 1\right], \qquad k = 1, \ldots \tag{A.4}$$

Lets consider a herding sequence with initial weight w_0. A herding sequence is given as,

$$s_k = \mathbb{I}\left[w_{k-1} > 0\right] \tag{A.5}$$

$$w_k = w_{k-1} + \gamma - s_k \tag{A.6}$$

Hence, we find that,

$$w_k = w_0 + k\gamma - \sum_{i=1}^{k} s_i \tag{A.7}$$

$$s_k = \mathbb{I}\left[w_0 + (k-1)\gamma - \sum_{i=1}^{k-1} s_i > 0\right] \tag{A.8}$$

where in the second line we used the first line. If we now compare this with Eqn.A.4 then we find that we need to identify,

$$w_0 = 2\gamma + \alpha - \lfloor \gamma + \alpha \rfloor - 1 \tag{A.9}$$

We conclude that for every Sturmian sequence there is a herding algorithm. However, there are more herding algorithms than Sturmian sequences because we can find initializations for which there is no corresponding for γ, α (note that γ) is fixed by the requirement that $p = \gamma$). However, the difference is only a transient effect and solely due to initialization. Finally, the Rabbit sequence (connected to the Fibonacci sequence) is obtained by choosing $p = \gamma = \phi$, $\alpha = 0$, and $w_0 = 2\phi - 1$ (with ϕ the golden mean).

Appendix B. Herding is non-periodic

We define the translation vector for every cone i to be $\boldsymbol{\varrho}_i$. It is given by

$$\boldsymbol{\varrho}_i = \langle \mathbf{f} \rangle_p - \mathbf{f}(s_i^*) \tag{B.1}$$

where s_i^* is the maximizing state in that cell. We call the collection of translation vectors a translation basis. Note that we assume that the number of basis vectors is at least equal to the number of dimensions. Herding will shift the current weight vector $\mathbf{w}$ over $-\boldsymbol{\varrho}_i$ for some (optimal) i.

Proposition: If there is no solution to the Eqn. $\sum_i n_i \boldsymbol{\varrho}_i = 0$ for $n_i \in \mathcal{N}$, $\forall k$, i.e. the translation vectors form an incommensurate basis, then every orbit is non-periodic.

Proof: Assume some orbit starting at $\mathbf{w}_0$ is periodic with period T. Then $\sum_{t=1}^{T} \delta\mathbf{w}_t = 0$, with $\delta\mathbf{w}_t = \mathbf{w}_t - \mathbf{w}_{t-1}$. However, by definition, $\delta\mathbf{w}_t = \boldsymbol{\varrho}_{\iota[t]}$ where $\iota[t]$ is the index of the translation vector picked at time t. Thus, if we denote with n_i the number of times cell i was visited in the finite time sequence under consideration, i.e. $n_i = \sum_{t=1}^{T} \mathbb{I}[\iota[t] = i]$ then we find that the following equation must hold for a periodic orbit: $\sum_{t=1}^{T} n_i \boldsymbol{\varrho}_i = 0$, which is excluded by the premise of the theorem. Contradiction.

Appendix C. Cells always split

Proposition: For herding with an incommensurate basis we have that any region, however small, will eventually split.

Proof: Consider a small region, R_0. Assume that it will never split. Then, because we know that any point will remain inside a compact region U, there is some finite time τ such that the region obtained after τ_2 iterations, R_{τ_2}, will overlap with some region generated at an earlier time R_{τ_1}. (The visited region will grow linearly in time with only a finite total volume available to move around in.) Call $R_{\tau_1} = R_1$ and $R_{\tau_2} = R_2$. Because the map is non-periodic (Appendix B), when the intersection happens, every element in the region R_2 will be translated by a vector $\mathbf{v} = \sum_i n_i \boldsymbol{\varrho}_i$ for some (finite) set of integers $\{n_i\}$. The conjoined region $R_{12} = R_1 \cup R_2$ must inherit the property that it will never split. However, after another $\tau_2 - \tau_1$ iterations, the regions R_1 has shifted again to R_2 and R_2 has shifted *over the same vector* $\mathbf{v}$ to R_3 (this follows because the region R_{12} as a whole cannot split and so R_1 and R_2 must follow the same sequence of translation vectors). Repeating this argument shows that the union of all these regions $R_{12..k}$ will grow *linearly*. Hence, since it cannot outgrow the compact region U, it will have to overlap with an edge at some finite time which causes the region to $R_{12..k}$ to split for some k. This constitutes a contradiction.

References

Aihara, K. (1994). Chaos in neural response and dynamical neural network models: toward a new generation of analog computing. In M. Yamaguti (Ed.), *Towards the harnessing of chaos*, 83–98. Elsevier, Science Publishers B.V.,Amsterdom.

Bialek, W., Nemenman, I., & Tishby, N. (2001). Predictability, complexity and learning. *Neural Computation, 13*, 2409–2463.

Chen, Y., & Welling, M. (2010). Parametric herding. *Proceedings of the Conference on Uncertainty in AI.*

Goetz, A. (2000). Dynamics of piecewise isometries. *Illinois Journal of Math, 44:3*, 465–478.

Hinton, G., & Zemel, R. (1994). Autoencoders, minimum description length, and helmholtz free energy. *Neural Information Processing Systems.*

Hopfield, J. (1982). Neural networks and physical systems with emergent collective computational abilities. *Proceedings of the National Academy of Sciences, 79*, 2554–2558.

Levina, A., Herrmann, J., & Geisel, T. (2007). Dynamical synapses causing self-organized criticality in neural networks. *Nature Physics, 3*, 857 – 860.

Lu, K., & Wang, J. (2005). Construction of sturmian sequences. *J. Phys. A: Math. Gen., 38*, 2891–2897.

Maass, W., & Zador, A. M. (1998). Dynamic stochastic synapses as computational units. *Advances in Neural Information Processing Systems* (pp. 903–917). MIT Press.

Pantic, L., Torres, J., Kappen, H., & Gielen, C. (2002). Associative memory with dynamic synapses. *Neural Computation, 14*, 2903–2923.

Pfister, J., Dayan, P., & Lengyel, M. (2009). Know thy neighbour: A normative theory of synaptic depression. *Advances in Neural Information Processing Systems – NIPS.*

Tieleman, T., & Hinton, G. (2009). Using Fast Weights to Improve Persistent Contrastive Divergence. *Proceedings of the International Conference on Machine Learning.*

Tsallis, C., Gell-Mann, M., & Sato, Y. (2006). Extensivity and entropy production. *Europhys. News, 36*, 186–189.

Tsodyks, M., Pawelzik, K., & Markram, H. (1998). Neural networks with dynamic synapses. *Neural Computation, 10*, 821–835.

van Beijeren, H. (2004). Generalized dynamical entropies in weakly chaotic systems. *Physica D, 193*, 90–95.

Welling, M. (2009a). Herding dynamic weights for partially observed random field models. *Proc. of the Conf. on Uncertainty in Artificial Intelligence.* Montreal, Quebec, CAN.

Welling, M. (2009b). Herding dynamical weights to learn. *Proceedings of the 21st International Conference on Machine Learning.* Montreal, Quebec, CAN.

Welling, M., & Sutton, C. (2005). Learning markov random fields using contrastive free energies. *International Workshop on Artificial Intelligence and Statistics* (pp. 397–404).

Journal of Physics: Conference Series 233 (2010) 012006 doi:10.1088/1742-6596/233/1/012006

Combining binary machines for multi-class: Statistical model and parameter estimation

Shiro Ikeda
The Institute of Statistical Mathematics, Tokyo, 190-8562, Japan
E-mail: shiro@ism.ac.jp

Abstract. Combining binary machines for multi-class classification problems is a popular idea, and many related methods have been proposed. When each binary machine reports a hard decision (a binary classifier), one of the most popular methods is to use the error correcting output codes (ECOC), while when each machine reports a soft decision (a binary predictor), another interesting idea is to use the Bradley-Terry (BT) model. In this paper, these methods are reviewed from a statistical viewpoint. As a result, a common framework will be given and natural extensions are derived for both of the ECOC and the BT model approaches.

1. Introduction

Since the proposal of Dietterich and Bakiri [1], the idea of combining binary classifiers for multi-class classifications has become popular.

The error correcting output codes (ECOC) approach [1, 2] is a simple method to combine hard decisions of binary classifiers with a predetermined code matrix. In order to combine soft decisions of binary predictors, the ECOC approach can be used, but Hastie and Tibshirani proposed a different method [3–5] which utilizes the Bradley-Terry (BT) model [6]. In this paper, we study the multi-class classification problem from a statistical viewpoint. As a result, a unified framework is given and parameter estimation algorithms for these two models are derived.

The parameter estimation of the ECOC approach is called the "code matrix" design [7]. Our proposal is to connect it with the multinomial logistic regression (MLR), and to use the elastic-net type sparse estimation method [8] which provides a new method for the code matrix design. The BT model approach is slightly complicated, but it naturally estimates the class probability. The estimated probability is correct if each binary predictor is accurate, otherwise it is not. We propose a method to re-estimate the binary predictors to make the combined statistical model better.

2. Preliminary

Consider a K-class classification problem ($K > 2$). Let $\boldsymbol{x} \in \mathcal{X}$ be the feature vector and $y \in \{1, \cdots, K\}$ be the class label. Our problem is to create a machine which predicts the class label y of a new vector $\boldsymbol{x}$ when N observations $(\boldsymbol{x}_1, y_1), \cdots, (\boldsymbol{x}_N, y_N)$ are available for training. We first define binary machines.

2.1. *Binary machines*

Let us define R pairs of disjoint sets C_r^+ and C_r^- as

$$C_r, C_r^+, C_r^- \subset \{1, \cdots, K\},\ C_r^+, C_r^- \neq \emptyset,\ C_r^+ \cap C_r^- = \emptyset, \quad C_r = C_r^+ \cup C_r^-, \quad r = 1, \cdots, R.$$

Each binary machine classifies the data point into one of the two class label sets C_r^+ and C_r^-. If $|C_r^+| = |C_r^-| = 1$ it is a "1-vs.-1" machine and if $|C_r^+| = 1$ and $|C_r^-| = K - 1$ it is a "1-vs.-the rest" machine. Two types of binary machines are defined as follows.

binary classifier: A classifier $f_r(\boldsymbol{x})$ reports $+1$ if it predicts $y \in C_r^+$ or -1 if it predicts $y \in C_r^-$. It reports one of $\{-1, +1\}$ for all $\boldsymbol{x}$ even if $y \notin C_r$,

$$f_r(\boldsymbol{x}) \in \{-1, +1\}, \quad \forall \boldsymbol{x} \in \mathcal{X}.$$

binary predictor: A predictor $q_r(\boldsymbol{x})$ reports an estimate of $p(y \in C_r^+ | y \in C_r, \boldsymbol{x})$. From the definition, $1 - q_r(\boldsymbol{x})$ is the estimate of $p(y \in C_r^- | y \in C_r, \boldsymbol{x})$. The following property is assumed

$$0 < q_r(\boldsymbol{x}) < 1, \quad \forall \boldsymbol{x} \in \mathcal{X}.$$

In the following, we consider combining binary machines for multi-class problems.

3. ECOC approach

3.1. *ECOC approach*

The ECOC approach is a popular method to combine binary classifiers. The code matrix $M \in \{-1, 0, +1\}^{K \times R}$ plays an important role, which is defined as follows [2],

$$m_{kr} = \begin{cases} +1, & k \in C_r^+ \\ -1, & k \in C_r^- \\ 0, & k \notin C_r \end{cases}. \tag{1}$$

When $\boldsymbol{x}$ is given, the outputs of $f_r(\boldsymbol{x})$ $(r = 1, \cdots, R)$ are computed and y is estimated as follows,

$$\hat{y} = \underset{k}{\operatorname{argmax}}\ v_k(\boldsymbol{x}), \quad v_k(\boldsymbol{x}) = \sum_{r=1}^{R} m_{kr} f_r(\boldsymbol{x}). \tag{2}$$

3.2. *Stochastic model*

Let $g(\cdot)$ be a positive and monotonically increasing function, and consider the problem

$$\hat{y} = \underset{k}{\operatorname{argmax}}\ v'_k(\boldsymbol{x}), \ v'_k(\boldsymbol{x}) = g\Big(\beta \sum_{r=1}^{R} m_{kr} f_r(\boldsymbol{x})\Big), \tag{3}$$

where $\beta > 0$. From the monotonicity of $g(\cdot)$, $\hat{y}$ in (2) and (3) are identical. We further define a probability function as

$$p(y = k|\boldsymbol{x}) = \frac{g(\beta \sum_{r=1}^{R} m_{kr} f_r(\boldsymbol{x}))}{\sum_{l=1}^{K} g(\beta \sum_{r=1}^{R} m_{lr} f_r(\boldsymbol{x}))}.$$

Note that $\hat{y} = \text{argmax}_y\, p(y|\boldsymbol{x})$ is still the same. One natural choice of $g(\cdot)$ is $\exp(\cdot)$, that is,

$$p(y|\boldsymbol{x}) = \frac{\exp\left(\delta_k(y)\beta \sum_{r=1}^{R} m_{kr} f_r(\boldsymbol{x})\right)}{\sum_{l=1}^{K} \exp\left(\beta \sum_{r=1}^{R} m_{lr} f_r(\boldsymbol{x})\right)}, \quad \delta_k(y) = \begin{cases} 1 & y = k \\ 0 & y \neq k \end{cases}. \tag{4}$$

This is equivalent to the well-known multinomial logistic regression (MLR). The general definition is found in [9, 10]

$$p(y|\boldsymbol{x};\boldsymbol{\theta}) = \frac{\exp\left(\sum_k \delta_k(y) \sum_{r=0}^{R} \theta_{kr} f_r(\boldsymbol{x})\right)}{\sum_l \exp\left(\sum_{r=0}^{R} \theta_{lr} f_r(\boldsymbol{x})\right)}, \quad \theta_{kr} \in \Re \ \ (k = 1, \cdots, K,\ r = 0, \cdots, R), \tag{5}$$

where $f_0(\boldsymbol{x}) = 1$ is a constant function. MLR is an exponential family. Strictly speaking, $\{\theta_{kr}\}$ and $\{\delta_k(y) f_r(\boldsymbol{x})\}$ are redundant since $\sum_{k=1}^{K} \delta_k(y) = 1$ holds, and they can be reduced from $K(R+1)$ to $(K-1)(R+1)$. But we keep the redundancy in this paper, and we call $\{\theta_{kr}\}$ and $\{\delta_k(y) f_r(\boldsymbol{x})\}$ the natural parameters and the sufficient statistics, respectively.

The model in (5) becomes equivalent to (4) by setting $\theta_{k0} = 0$ and $\theta_{kr} = \beta m_{kr}$. However, the natural parameters $\{\theta_{kr}\}$ can take any real value. This motivates us to estimate the parameters from data. Similarly a sufficient statistic $\delta_k(y) f_r(\boldsymbol{x})$ is restrictive since it only takes $\{-1, +1\}$ if $y = k$ and 0 otherwise. It can be generalized to any real valued function or even to reproducing kernel Hilbert space (RKHS) [9, 10]. This is a promising direction of extensions, but we stay with combining binary classifiers.

A similar extension has been discussed [7], and we propose a different parameter estimation method in the next subsection.

3.3. Parameter estimation

This special choice of the parameters $\theta_{k0} = 0$, $\theta_{kr} = m_{kr}$ connects the ECOC approach with MLR. This is justified under the assumption of the error correction codes. We review it in the following.

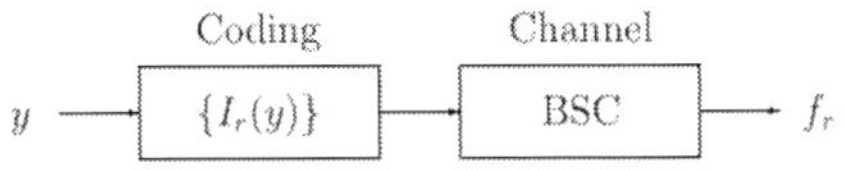

Figure 1. The idea of ECOC.

First, a "code" $I_r(y)$ of y is defined as

$$I_r(y) = \begin{cases} +1, & y \in C_r^+ \\ -1, & y \in C_r^- \\ -1 \text{ or } +1 \text{ with prob. } 1/2, & y \notin C_r \end{cases}.$$

Let us assume a binary symmetric channel (BSC), where each $I_r(y)$ is independently flipped to the other sign with probability $0 < \alpha < 1/2$ and observed as f_r. The probability of f_r is

$$p(f_r|y) = \begin{cases} \dfrac{\exp(\beta I_r(y) f_r)}{2\cosh(\beta)} & y \in C_r \\ 1/2 & y \notin C_r \end{cases}, \beta = \frac{1}{2} \ln \frac{1-\alpha}{\alpha}.$$

This can be rewritten with m_{kr} defined in (1), as

$$p(f_r|y) = \frac{\exp\big(\beta\delta_k(y)m_{kr}f_r\big)}{\sum_{f_r}\exp\big(\beta\delta_k(y)m_{kr}f_r\big)}.$$

Since f_r is independent, the joint distribution is

$$p(f_1,\cdots,f_R|y) = \frac{\exp\big(\beta\delta_k(y)\sum_r m_{kr}f_r\big)}{\sum_{f_1,\cdots,f_R}\exp\big(\beta\delta_k(y)\sum_r m_{kr}f_r\big)}.$$

By assuming $p(y) = 1/K$,

$$p(y|f_1,\cdots,f_R) = \frac{\exp\big(\beta\delta_k(y)\sum_r m_{kr}f_r\big)}{\sum_k \exp\big(\beta\sum_r m_{kr}f_r\big)},$$

which is the same as (4).

However, when $\boldsymbol{x}$ is given, f_r is not a random variable but a function of $\boldsymbol{x}$. They are not independent and there is no need to fix θ_{kr} equal to m_{kr}. Instead of fixing it, let us estimate θ_{kr} from data. One of the popular estimation methods is the maximum likelihood estimation (MLE), which is defined as follows,

$$\hat{\boldsymbol{\theta}} = \underset{\boldsymbol{\theta}}{\operatorname{argmax}}\, L(\boldsymbol{\theta}), \quad L(\boldsymbol{\theta}) = \sum_t \ln p(y_t|\boldsymbol{x}_t;\boldsymbol{\theta}), \tag{6}$$

where $p(y_t|\boldsymbol{x}_t;\boldsymbol{\theta})$ is defined in (5)[1]. Unfortunately the direct maximization of $L(\boldsymbol{\theta})$ does not work from two reasons. One is that the log-likelihood of MLR can be made arbitrarily large if the data are separable. The other reason is the redundancy. We keep the redundancy to have the sparse parameterization, since one of the important properties of the code matrix M is the sparsity. We will come back to this point later.

In order to solve these problems, we consider the following regularized ML,

$$\hat{\boldsymbol{\theta}} = \underset{\boldsymbol{\theta}}{\operatorname{argmax}}\Big[L(\boldsymbol{\theta}) - \lambda_1\sum_{kr}|\theta_{kr}| - \frac{1}{2}\lambda_2\sum_{kr}\theta_{kr}^2\Big], \quad \lambda_1,\lambda_2 > 0. \tag{7}$$

This cost function shares its idea with the elastic net [8]. By minimizing the cost function in (7), we can obtain sparse natural parameters.

Note that it is necessary to keep λ_2 positive to reduce all the redundancy. This can be checked as follows. Let $\lambda_2 = 0$, ϵ be a small real number, and $\theta'_{rk} = \theta_{rk} + \epsilon$ $(k = 1,\cdots,K)$ for some r. From $\sum_k \delta_k(y) = 1$, $L(\boldsymbol{\theta})$ in (6) does not change by replacing θ_{rk} with θ'_{rk}. Also it is easy to check that $\sum_k|\theta_{kr}| = \sum_k|\theta'_{kr}|$ if the numbers of positive and negative θ_{rk} $(k = 1,\cdots,K)$ are the same, and (7) does not change.

Our proposal is different from the extension in [7], because their cost function is not based on the log-likelihood function. However, their cost function can be optimized with QP. It is not possible to apply QP in order to optimize (7), but an efficient iterative algorithm can be derived by modifying the results in [10], which is developed by a bound optimization approach. We do not go into the details, but note that major load of the computation in each iteration is the inversion of a $KR \times KR$ matrix and convergence is practically fast enough.

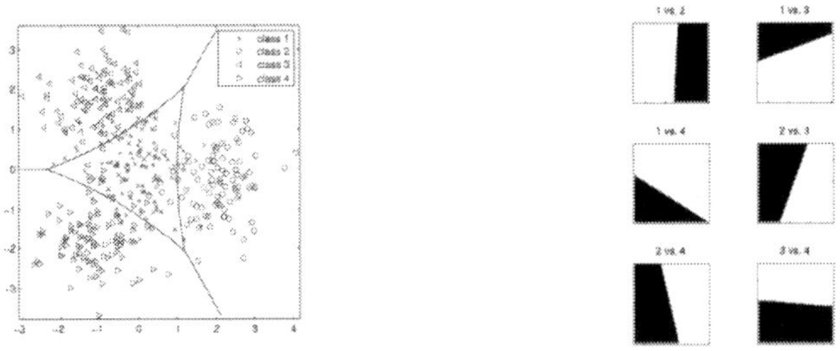

Figure 2. Left: 4 class data with optimal boundary computed from the true distribution, right: 1-vs.-1 classifiers.

3.4. Numerical simulation

We show how the parameter estimation method works with a numerical simulation. Four-class data, where each class follows a 2-dimensional Gaussian distribution are prepared (figure 2, left). We consider combining 1-vs.-1 classifiers with ECOC. The classifiers are prepared as the hard decisions of binary logistic regressions which are estimated from data.

When they are combined with the code matrix, the decision boundaries are shown in figure 3, left. And as a result of the learning algorithm we obtain the decision boundaries in figure 3, right. The decision boundaries are combinations of line segments of binary classifiers. Some boundaries in figure 3 are miss-specified, while the estimated boundaries become clear.

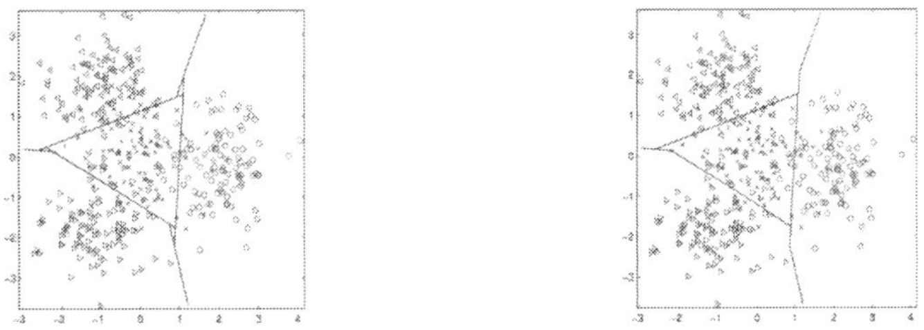

Figure 3. Decision boundaries. left: combining with ECOC code matrix, right: combining with proposed method.

The original code matrix and $\boldsymbol{\theta}_{kr}$ are shown as images in figure 4, where red and blue correspond to positive and negative values, respectively. Scales are normalized since boundaries are scale invariant. It shows that the estimated parameters are similar to the original code matrix, but not the same. It shows how each binary classifier contributes to each class. As the result of estimation, f_r of classes $(1, 2)$ does not contribute to class 1 and f_r of classes $(2, 3)$ does not contribute to class 4. It shows that the estimated sparse parameter shares its structure with code matrix, but the values are different.

4. BT model approach

4.1. BT model approach

Hastie and Tibshirani proposed another approach for combining binary 1-vs.-1 predictors [3].

[1] If each f_r follows the assumption of ECOC (independent and BSC), $\theta_{k0} = 0$, $\theta_{kr} = \beta m_{kr}$ gives an MLE (it is not unique because of the redundancy).

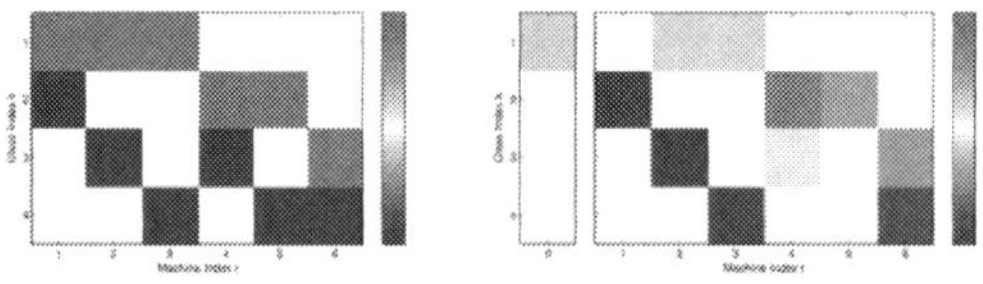

m_{kr} θ_{kr}

Figure 4. Left: code matrix, right: estimated parameters.

We first explain the BT model [6]. Suppose there are K "players," and each of them plays games against another. The BT model assumes the probability that a player k beats another player l is parameterized as follows

$$P(k \text{ beats } l) = \frac{p_k}{p_k + p_l}, \quad p_{k'} > 0 \quad (k' = 1, \cdots, K).$$

This is scale invariant and $\sum_k p_k = 1$ is imposed. When independent results of games are observed as,

$$n_{kl} = \#(\text{games between } k \text{ and } l), \; r_{kl} = \frac{\#(k \text{ beats } l)}{n_{kl}},$$

the log-likelihood function is

$$F(\boldsymbol{p}) = \sum_{k<l} n_{kl}\Big(r_{kl} \ln \frac{p_k}{p_k + p_l} + (1 - r_{kl}) \ln \frac{p_l}{p_k + p_l}\Big),$$

where we assumed there is no tie, that is, $r_{kl} + r_{lk} = 1$. The parameters $\boldsymbol{p} = (p_1, \cdots, p_K)$ can be estimated as the maximizer of the log-likelihood function (MLE)

$$\hat{\boldsymbol{p}} = \operatorname*{argmax}_{\boldsymbol{p}} F(\boldsymbol{p}).$$

In order to combine binary predictors, r_{kl} is replaced with a 1-vs.-1 predictor $q_r(\boldsymbol{x})$ and n_{kl} is set to 1,

$$F_{BT}(\boldsymbol{p}, \boldsymbol{x}) = \sum_{r=1}^{R}\Big(q_r(\boldsymbol{x}) \ln \frac{p_{k_r}}{p_{k_r}+p_{l_r}} + (1 - q_r(\boldsymbol{x})) \ln \frac{p_{l_r}}{p_{k_r}+p_{l_r}}\Big),$$

where, $\{k_r\} = C_r^+$ and $\{l_r\} = C_r^-$. The maximizer of $F_{BT}(\boldsymbol{p}, \boldsymbol{x})$ is regarded as the estimate of $p(y|\boldsymbol{x})$,

$$p(y{=}k|\boldsymbol{x}){=}\hat{p}_k, \; \hat{\boldsymbol{p}}{=}(\hat{p}_1, \cdots, \hat{p}_K){=} \operatorname*{argmax}_{\boldsymbol{p}} F_{BT}(\boldsymbol{p}, \boldsymbol{x}).$$

This idea is generalized to arbitrary combination of predictors [4, 5] by extending the definition of $F_{BT}(\boldsymbol{p}, \boldsymbol{x})$ to $F_{GBT}(\boldsymbol{p}, \boldsymbol{x})$,

$$F_{GBT}(\boldsymbol{p}, \boldsymbol{x}) = \sum_{r=1}^{R}\Big(q_r(\boldsymbol{x}) \ln \frac{\sum_{k \in C_r^+} p_k}{\sum_{l \in C_r} p_l} + (1 - q_r(\boldsymbol{x})) \ln \frac{\sum_{k \in C_r^-} p_k}{\sum_{l \in C_r} p_l}\Big). \tag{8}$$

$$p(y{=}k|\boldsymbol{x}){=}\hat{p}_k, \; \hat{\boldsymbol{p}}{=}(\hat{p}_1, \cdots, \hat{p}_K){=} \operatorname*{argmax}_{\boldsymbol{p}} F_{GBT}(\boldsymbol{p}, \boldsymbol{x}).$$

$F_{GBT}(\boldsymbol{p}, \boldsymbol{x})$ may have multiple maxima, and the condition of $F_{GBT}(\boldsymbol{p}, \boldsymbol{x})$ to have a unique global maximum has been studied [5]. Two important cases are **combining 1-vs.-1 predictors**, and **combining 1-vs.-the rest predictors**. In our experiments, we consider combining 1-vs.-1 predictors. In each case of $F_{BT}(\boldsymbol{p}, \boldsymbol{x})$ and $F_{GBT}(\boldsymbol{p}, \boldsymbol{x})$, we compute $p(y|\boldsymbol{x})$ for all $y \in \{1, \cdots, K\}$ and the label which maximizes the probability is the predicted class.

An important problem of this approach is how to compute the maximizer of $F_{GBT}(\boldsymbol{p}, \boldsymbol{x})$. Some iterative algorithms have been proposed [5, 11]. In our experiments, we used the MM algorithm [11].

4.2. Statistical model

Suppose $F_{GBT}(\boldsymbol{p}, \boldsymbol{x})$ in (8) has a unique global maximum and we have an algorithm to compute it, for example, combining 1-vs.-1 predictors with the MM algorithm. Let us define the following function

$$\boldsymbol{p}(\boldsymbol{x}) = (p_1(\boldsymbol{x}), \cdots, p_K(\boldsymbol{x})) = \underset{\boldsymbol{p}}{\text{argmax}}\, F_{GBT}(\boldsymbol{p}, \boldsymbol{x}).$$

The solution of this optimization problem defines a statistical model as,

$$p(y|\boldsymbol{x}) = p_y(\boldsymbol{x}), \quad y = 1, \cdots, K.$$

Although $p(y|\boldsymbol{x})$ cannot be expressed in closed form, the "argmax" operation has a unique solution and there exists a function. This is the statistical model of the BT model approach.

In the following, we assume each binary predictor $q_r(\boldsymbol{x})$ has a set of parameters $\boldsymbol{\eta}_r = (\eta_{r1}, \cdots, \eta_{rp})^T$, and is differentiable with respect to $\boldsymbol{\eta}_r$. A simple example of $q_r(\boldsymbol{x}; \boldsymbol{\eta}_r)$ is a binary (kernel) logistic regression. The statistical model is redefined as

$$p(y|\boldsymbol{x}; \{\boldsymbol{\eta}_r\}) = \hat{p}_y(\boldsymbol{x}; \{\boldsymbol{\eta}_r\}) \quad \hat{\boldsymbol{p}}(\boldsymbol{x}; \{\boldsymbol{\eta}_r\}) = \underset{\boldsymbol{p}}{\text{argmax}}\, F_{GBT}(\boldsymbol{p}, \boldsymbol{x}, \{\boldsymbol{\eta}_r\})$$
$$F_{GBT}(\boldsymbol{p}, \boldsymbol{x}, \{\boldsymbol{\eta}_r\}) = \sum_{r=1}^{R} \left(q_r(\boldsymbol{x}; \boldsymbol{\eta}_r) \ln \frac{\sum_{k \in C_r^+} p_k}{\sum_{l \in C_r} p_l} + (1 - q_r(\boldsymbol{x}; \boldsymbol{\eta}_r)) \ln \frac{\sum_{k \in C_r^-} p_k}{\sum_{l \in C_r} p_l} \right). \tag{9}$$

This function is not an exponential family, but gives a multinomial distribution on y when $\boldsymbol{x}$ is given.

4.3. Parameter estimation

The BT model approach gives the correct prediction if each binary predictor satisfies the following equation,

$$q_r(\boldsymbol{x}; \boldsymbol{\eta}_r) = P(y \in C_r^+ | y \in C_r, \boldsymbol{x}),$$

where P denotes the true probability. Practically, the parameter $\boldsymbol{\eta}_r$ of $q_r(\boldsymbol{x}; \boldsymbol{\eta}_r)$ is estimated from samples which satisfy $y_t \in C_r$. Therefore, the training data distribution for $q_r(\boldsymbol{x}; \boldsymbol{\eta}_r)$ follows $P(\boldsymbol{x}|y \in C_r)$. Since it is different from $P(\boldsymbol{x})$, the outputs of $q_r(\boldsymbol{x}; \boldsymbol{\eta}_r)$ for the data $P(\boldsymbol{x}|y \notin C_r)$ might not be accurate enough.

Our goal is to estimate $P(y|\boldsymbol{x})$ and the model is $p(y|\boldsymbol{x}; \{\boldsymbol{\eta}_r\})$. A straightforward idea is to estimate the parameters $\{\boldsymbol{\eta}_r\}$ by maximum likelihood estimation of $p(y|\boldsymbol{x}; \{\boldsymbol{\eta}_r\})$. The log-likelihood function is

$$L(\{\boldsymbol{\eta}_r\}) = \sum_t \ln p(y_t|\boldsymbol{x}_t; \{\boldsymbol{\eta}_r\}).$$

The optimization seems difficult since $p(y_t|\boldsymbol{x}_t;\{\boldsymbol{\eta}_r\})$ is not expressed in closed form. In the following, we show how to compute the derivative with respect to $\boldsymbol{\eta}_r$, and the gradient ascent method is used to compute the MLE.

We consider combining 1-vs.-1 binary predictors. Generalization to other combinations is not difficult. Let us fix $\boldsymbol{x}$ and define the following notations.

$$p_k = p(y{=}k|\boldsymbol{x};\{\boldsymbol{\eta}_r\}),\quad \{k\}=C_r^+,\quad \{l\}=C_r^-,\quad q_{kl}=q_r(\boldsymbol{x};\boldsymbol{\eta}_r),\quad q_{lk}=1-q_{kl}$$

Since $\{p_k\}$ maximizes F_{GBT} in (9) under $\sum_k p_k{=}1$,

$$\partial_{p_k}\Big(F_{GBT}+\lambda(1-\sum_k p_k)\Big)=0,\quad \sum_{l\neq k}\Big(\frac{q_{kl}}{p_k}-\frac{1}{p_k+p_l}\Big)=\lambda,\ k=1,\cdots,K, \tag{10}$$

where λ is a Lagrange multiplier. By multiplying p_k to both sides and summing up with k,

$$\sum_k\sum_{l\neq k}\Big(q_{kl}-\frac{p_k}{p_k+p_l}\Big)=\lambda=0.$$

From $\lambda=0$, K equations are defined by (10) as

$$\sum_{l\neq k}\frac{p_k}{p_k+p_l}=\sum_{l\neq k}q_{kl},\quad k=1,\cdots,K. \tag{11}$$

Let us take the derivative of (11) with respect to a parameter η_r. If $\{k,l\}\neq C_r$, the right-hand side is $\partial_{\eta_r}q_{kl}=0$. Let $\delta_r(k,l)$ be the indicator function which takes 1 iff $\{k,l\}=C_r$ and 0 otherwise, and (11) becomes

$$\sum_{l\neq k}\frac{p_kp_l}{(p_k+p_l)^2}(\partial_{\eta_r}l_k-\partial_{\eta_r}l_l)=\delta_r(k,l)\partial_{\eta_r}q_{kl},\quad k=1,\cdots,K, \tag{12}$$

where $l_k=\ln p_k$. Since our objective is to compute the derivative of the log-likelihood function, that is, $\partial_{\eta_r}l_k$, it can be obtained by solving this linear system of equations.

A $K\times K$ matrix A and a K-dimensional vector $\partial\boldsymbol{q}$ are defined as follows,

$$A=\{a_{kl}\},\ a_{kl}=\begin{cases}\displaystyle\sum_{l'\neq k}\frac{p_kp_{l'}}{(p_k+p_{l'})^2}, & k=l\\ \displaystyle-\frac{p_kp_l}{(p_k+p_l)^2}, & k\neq l\end{cases},\quad \partial_{\eta_r}\boldsymbol{q}=(\cdots,0,\underset{\vee}{\overset{k}{(\partial_{\eta_r}q_{kl})}},0,\cdots,0,\underset{\vee}{\overset{l}{(\partial_{\eta_r}q_{lk})}},0,\cdots)^T$$

and K equations in (12) are summarized as follows

$$A\begin{pmatrix}\partial_{\eta_r}l_1\\ \vdots\\ \partial_{\eta_r}l_K\end{pmatrix}=\partial_{\eta_r}\boldsymbol{q}.$$

Unfortunately this linear system of equations cannot be solved since A is singular. It is easily checked by multiplying a K-dimensional vector $\mathbf{1}=(1,\cdots,1)^T$. This comes from the fact F_{GBT}

is scale invariant and it can be avoided by replacing one row of A with the following equation, which is obtained from $\sum_k p_k = 1$,

$$\sum_k p_k \partial_{\boldsymbol{\eta}_r} l_k = 0.$$

Let us define A' and $\partial_{\eta_r} \boldsymbol{q}'$ as follows,

$$A' = \{a'_{kl}\},\ a'_{kl} = \begin{cases} a_{kl} & 1 \le k \le K-1 \\ p_l, & k = K \end{cases}, \tag{13}$$

$$(\partial_{\eta_r} \boldsymbol{q}')_k = \begin{cases} (\partial_{\eta_r} \boldsymbol{q})_k & 1 \le k \le K-1 \\ 0 & k = K \end{cases}, \tag{14}$$

and the following set of equations is derived.

$$A' \begin{pmatrix} \partial_{\eta_r} l_1 \\ \vdots \\ \partial_{\eta_r} l_K \end{pmatrix} = \partial_{\eta_r} \boldsymbol{q}' \tag{15}$$

This linear system of equations can be solved by computing A'^{-1} and $\partial_{\eta_r} l_k {=} \partial_{\eta_r} p(y{=}k|\boldsymbol{x};\{\boldsymbol{\eta}_r\})$ is the derivative of the log-likelihood function.

Let us summarize the procedure to compute the MLE.

(i) Initialize $\{\boldsymbol{\eta}_r\}$ *of* $q_r(\boldsymbol{x};\boldsymbol{\eta}_r)$.

(ii) For $t = 1, \cdots, N$,

(a) Solve (9) *to have*

$$\hat{p}_k = p(y = k|\boldsymbol{x}_t;\{\boldsymbol{\eta}_r\}),\ k = 1, \cdots, K.$$

(b) Compute A' *in* (13) *with* $\{\hat{p}_k\}$ *and its inverse* A'^{-1}. *Let* $\boldsymbol{b}^T$ *be the* y_t*-th row of* A'^{-1}.

(c) For each parameter, compute $\partial_{\eta_r} \boldsymbol{q}'$ *in* (14)*, and take the inner product with* $\boldsymbol{b}$

$$\partial_{\eta_r} \ln p(y_t|\boldsymbol{x}_t;\{\boldsymbol{\eta}_r\}) = \boldsymbol{b} \cdot (\partial_{\eta_r} \boldsymbol{q}'). \tag{16}$$

(iii) Compute (16) *for every* t*, and take the sum. It becomes the derivative of the log-likelihood function with respect to* η_r.

$$\partial_{\eta_r} L(\{\boldsymbol{\eta}_r\}) = \sum_t \partial_{\eta_r} \ln p(y_t|\boldsymbol{x}_t;\{\boldsymbol{\eta}_r\})$$

(iv) A simple update rule for η_r *is*

$$\eta_r^{\text{new}} = \eta_r^{\text{old}} + \mu \partial_{\eta_r} L(\{\boldsymbol{\eta}_r^{\text{old}}\}), \tag{17}$$

where μ *is a small positive real number.*

The regularized ML can also be considered by replacing (17) with

$$\eta_r^{\text{new}} = \eta_r^{\text{old}} + \mu \Big[\partial_{\eta_r} L(\{\boldsymbol{\eta}_r^{\text{old}}\}) - \partial_{\eta_r} R_\lambda(\{\boldsymbol{\eta}_r^{\text{old}}\})\Big],$$

where $R_\lambda(\{\boldsymbol{\eta}_r^{\text{old}}\})$ is the regularizer, one example of which is $\lambda \sum_r |\boldsymbol{\eta}_r|^2$.

4.4. Numerical simulation

We show how the parameter estimation method works with a numerical simulation. We prepared a set of 3-class data (figure 5). One of the classes is separated in two disjoint regions. We consider combining 1-vs.-1 classifiers with the BT model approach.

First, we estimate the binary logistic regressions separately (figure 5.B, right), and combined them. The decision boundaries are shown in figure 5.B, left. As a result of the parameter estimation, we obtain the binary logistic regressions (figure 5.C, right), and combine them. The decision boundaries are shown in figure 5.C, left. The decision boundaries are not a combination

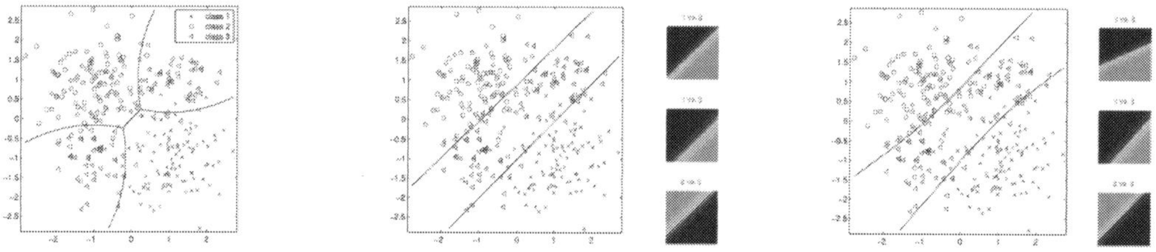

A B C

Figure 5. 3 class data: A. data and the true boundaries computed from the true distribution, B. Combining separately estimated binary logistic regressions, and C. Combining separately estimated binary logistic regressions

of the line segments, which shows this model is different from MLR. We see that as the result of the parameter estimation of BT, the boundaries become curved and closer to the true boundaries.

5. Experiments and discussion

5.1. UCI machine learning repository

The algorithms for the ECOC and BT model approaches are applied to the data of UCI machine learning repository. We used data set "iris" (3 classes, 4 features, 150 data), "wine" (3 classes, 13 features, 178 data), and "glass" (7 classes, 9 features, 214 data).

First, we created 1-vs.-1 binary machines with kernel logistic regressions (KLRs) [9, 10] with l_2 regularizer. The hard decisions of KLRs are used as the binary classifiers while the soft decisions are used as the binary predictors.

The final results are evaluated with the 10-fold cross validation (CV). Even if we do not apply the proposed estimation methods, the CV results are quite good (# of the errors is 6/150, 5/178, and 64/214 for iris, wine, and glass, respectively) for both of the ECOC and BT model approaches, and we cannot observe any significant improvements by our method.

Although the error rates were not improved, we show the estimated code matrices of the ECOC approach. The matrices are averaged over each cross validation trials, and they are shown in figures 6 and 7.

We see that the contribution of each binary machine is quite different from that of the code. We even see that some binary machines are hardly used (16, 17, and 18 of "glass"). Since the results of CV show that the outputs of ECOC do not change by modifying $\{m_{kr}\}$ to $\{\theta_{kr}\}$, the model could be reduced.

5.2. Combining with concave function

We have studied two statistical models, one is $p(y|\boldsymbol{x};\boldsymbol{\theta})$ and the other is $p(y|\boldsymbol{x};\{\boldsymbol{\eta}_r\})$ derived from the ECOC approach and the BT model approach, respectively. In order to see their relation, we reconsider F_{GBT} in (9).

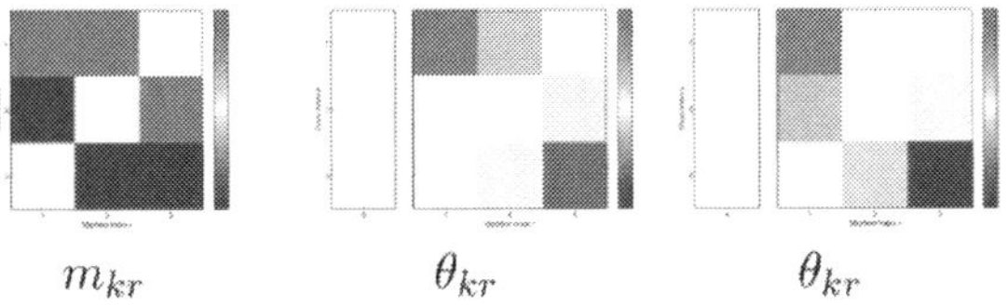

Figure 6. Left: code matrix for "iris" and "wine," center: the estimated parameters for "iris," right: the estimated parameters for "wine."

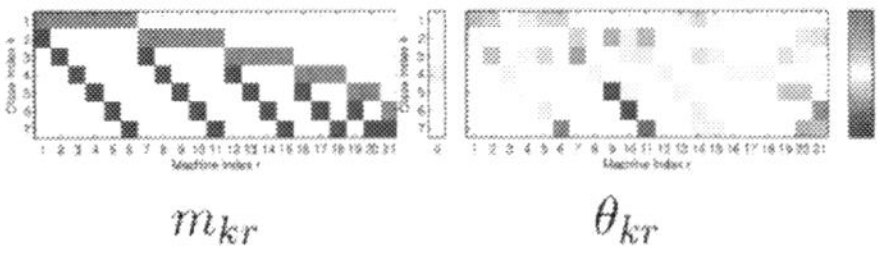

Figure 7. Left: code matrix, right: estimated parameter for "glass."

This function is derived as the log-likelihood function of the GBT model. However, in order to combine the outputs of binary predictors, it is not necessary to utilize the GBT model nor $F_{GBT}(\boldsymbol{p})$. Instead, consider a strictly concave function $F(\boldsymbol{p})$, and let us assume there is a unique maximizer $\hat{\boldsymbol{p}}$. The new distribution of $p(y|\boldsymbol{x})$ is defined by setting $p(y = k|\boldsymbol{x}) = \hat{p}_k$.

Thus, any concave function $F(\boldsymbol{p})$ works for combining binary classifiers. Many concave functions, such as Bregman divergence and f–divergence can be used for this purpose, which is a direction of possible extensions. Let us consider the following concave function

$$F_{ECOC}(\boldsymbol{p}, \boldsymbol{x}, \boldsymbol{\theta}) = \sum_k \exp\Big(\sum_{r=0}^{R} \theta_{kr} f_r(\boldsymbol{x})\Big) \ln p_k.$$

Consider maximizing it under $\sum_k p_k = 1$

$$\hat{\boldsymbol{p}} = \operatorname*{argmax}_{\boldsymbol{p}} F_{ECOC}(\boldsymbol{p}, \boldsymbol{x}, \boldsymbol{\theta}), \quad \text{subj. to } \sum_k p_k = 1.$$

It can be easily solved in closed form

$$\hat{p}_k = \frac{\exp\big(\sum_{r=0}^{R} \theta_{kr} f_r(\boldsymbol{x})\big)}{\sum_l \exp\big(\sum_{r=0}^{R} \theta_{lr} f_r(\boldsymbol{x})\big)}.$$

A distribution $p(y = k|\boldsymbol{x}; \boldsymbol{\theta}) = p_k$ is defined as

$$p(y|\boldsymbol{x}; \boldsymbol{\theta}) = \frac{\exp\big(\sum_k \delta_k(y) \sum_{r=0}^{R} \theta_{kr} f_r(\boldsymbol{x})\big)}{\sum_l \exp\big(\sum_{r=0}^{R} \theta_{lr} f_r(\boldsymbol{x})\big)}.$$

This is the MLR and equivalent to (5). The relation between $p(y|\boldsymbol{x}; \boldsymbol{\theta})$ and $p(y|\boldsymbol{x}; \{\boldsymbol{\eta}_r\})$ is schematically summarized in figure 8.

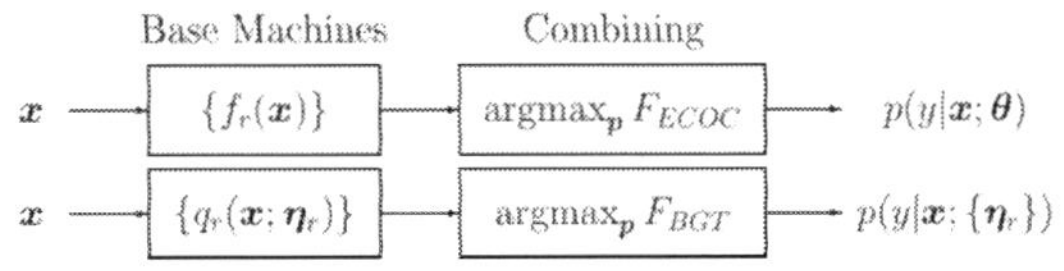

Figure 8. Statistical models of ECOC and BT approaches.

5.3. Neural networks and statistical model

Figure 8 shows that each stochastic model consists of the "base machine" part and the "combining" part. This reminds us of the structure of three-layered neural networks. More precisely, each corresponds to hidden and output layers, respectively, while $\boldsymbol{x}$ is the input.

In section 3, a parameter estimation method of $\boldsymbol{\theta}$ is derived for ECOC. It is the estimation of output layer under the analogy. On the other hand, the estimation of $\{\boldsymbol{\eta}_r\}$ in section 4 corresponds to the estimation of hidden layer, and the algorithm derived in subsection 4.3 is the back propagation algorithm.

We did not assume hidden layer is differentiable in the ECOC approach. On the other hand, we did not assume any parameter in the output layer of the BT model approach. These assumptions are restrictive. We can define f_r with a differentiable function with some parameters. This will bring us back to the neural networks model. It is also possible to add some parameters to the BT model. A naive idea is to generalize F_{GBT} in (9) as

$$F_{GBT}(\boldsymbol{p}, \{q_r\}, \boldsymbol{\alpha}) = \sum_{r=1}^{R} \alpha_r \Big(q_r \ln \frac{\sum_{k \in C_r^+} p_k}{\sum_{l \in C_r} p_l} + (1 - q_r) \ln \frac{\sum_{k \in C_r^-} p_k}{\sum_{l \in C_r} p_l} \Big),$$

under $\sum_r \alpha_r = 1$. Studying these extensions is remained for future works.

For MLR (or ECOC approach), the outputs of hidden layer correspond to the set of sufficient statistics. The key for building a good statistical model is to have a good sufficient statistics. The idea of the ECOC approach is to use a rather small number of "good" sufficient statistics which are binary machines. Another important approach is to prepare a very large number of functions. The kernel based approach [9, 10] is one of such approaches and is another promising method. It is difficult to compare which method is better in general. But preparing the statistical viewpoint will help us to compare and understand these methods.

6. Conclusion

Combining binary machines for multi-class classification problems is a practical and promising idea. Many efficient algorithms have been proposed, and their empirical results and the aspects of computational costs have been extensively studied. However, it is difficult to find sufficient theoretical comparisons.

The main proposal of this paper is to consider the approaches as statistical models. This viewpoint will enable us to compare them theoretically. As a preliminary result, the parameter estimation methods for the ECOC and BT model approaches are naturally derived. Although the work is preliminary, we hope it opens a further understanding of such methods in the future.

Acknowledgement

We are grateful for helpful discussion with Dr. Yuichi Shiraishi. We also thank the anonymous reviewers for valuable feedback. The work was supported by Grant-in-Aid for Scientific Research No. 18079013, MEXT, Japan.

References

[1] Dietterich T G and Bakiri G 1995 *Journal of Artificial Intelligence Research* **2** 263–86
[2] Allwein E L, Schapire R E and Singer Y 2000 *Journal of Machine Learning Research* **1** 113–41
[3] Hastie T and Tibshirani R 1998 *The Annals of Statistics* **26** 451–71
[4] Zadrozny B 2002 *Advances in Neural Information Processing Systems* **14** 1041–48
[5] Huang T K, Weng R C and Lin C J 2006 *Journal of Machine Learning Research* **7** 85–115
[6] Bradley R A and Terry M E 1952 *Biometrika* **39** 324–45

[7] Crammer K and Singer Y 2002 *Machine Learning* **47** 201–33
[8] Zou H and Hastie T 2005 *J. R. Statistical Society, Series B* **67** 301–20
[9] Zhu J and Hastie T 2005 *Journal of Computational and Graphical Statistics* **14** 185–205
[10] Krishnapuram B, Carin L, Figueiredo M A and Hartemink A J 2005 *IEEE Trans. on Pattern Analysis and Machine Intelligence* **27** 957–68
[11] Hunter D R 2004 *The Annals of Statistics* **32** 384–406

Optimal aggregation of noisy observations: A large deviations approach

Tatsuto Murayama and Peter Davis

NTT Communication Science Laboratories, NTT Corporation, 2-4, Hikaridai, Seika-cho, Keihanna, Kyoto 619-0237, Japan

E-mail: `murayama@cslab.kecl.ntt.co.jp`, `davis@cslab.kecl.ntt.co.jp`

Abstract. Sensing and data aggregation tasks in distributed systems should not be considered as separate issues. The quality of collective estimation involves a fundamental tradeoff between sensing quality, which can be increased by increasing the number of sensors, and aggregation quality under a given capacity of the network, which decreases if the number of sensors is too large. In this paper, we examine a system level strategy for optimal aggregation of data from an ensemble of independent sensors. In particular, we consider large scale aggregation from very many sensors, in which case the network capacity diverges to infinity. Then, by applying the large deviations techniques, we conclude the following significant result: larger scale aggregation always outperforms smaller scale aggregation at higher noise levels, while below a critical value of noise, there exist moderate scale aggregation levels at which optimal estimation is realized. At a critical value of noise, there is an abrupt change in the behavior of a parameter characterizing the aggregation strategy, similar to a phase transition in statistical physics.

1. Introduction

The purpose of this paper is to present results which give a new perspective on the growing field of networked sensors. We focus on the issue of reducing sensing error by collecting sensing data from many sensors. In particular, we work on the simple idea that the collective sensing error can be reduced by collecting data from more sensors. However, in distributed sensor networks, collecting data from many sensors usually involves some cost in terms of network resources; total bandwidth, power consumptions, computational capabilities, and so forth [1]. When the network resources and the amount of data from each sensor are fixed, then there will be a limit on the number of sensors, and thus a limit on the collective sensing quality. If it is possible to reduce the amount of sensor data by data compression then the number of sensors, and thus the sensing quality, will depend on the amount of data compression which can be achieved.

Now, there is a fundamental limit on the amount of data reduction which is possible. The limit is called the entropy rate and is determined by the statistical property of the source [2]. However, if it is possible to tolerate some error introduced by the reduction process, then data can be coded and compressed beyond this fundamental limit [3]. This is a key idea in information theory. When we extend our consideration of the collective sensing problem to include the possibility of arbitrary data reduction, then there arises a new type of tradeoff, i.e., the tradeoff between the effect of reducing collective sensing error with increasing number of sensors, and the effect of increasing aggregation error with increasing number of sensors due to the need to reduce the amount of data by data compression below the entropy rate. This tradeoff determines the optimal number of sensors for minimal collective sensing error. This new type of tradeoff and its implications for optimal size of sensor networks is the new problem that we focus on in this paper. We see this as a fundamental and universal problem in sensor networks. Moreover with the growing feasibility of deploying huge numbers of smart sensors with advanced coding and communication abilities, this is a problem of rapidly increasing relevance and significance [4].

In this paper, we consider a fundamental formulation of the problem with only one information source and suppose that all sensors are symmetrical, i.e., exchangeable with respect to their contributions to the final result of aggregation. This allows us to treat the problem in terms of the theory of large deviations. The paper is divided into 5 sections. Section 2 presents our system model which is first introduced in the context of multi terminal information theory. Section 3 briefly summarizes our main result. The detailed proof for the proposition, however, is postponed until the following section 4. Conclusions are given in section 5.

2. System model

2.1. Ensemble of independent sensors

We consider that an observer is interested in observing the Bernoulli(1/2) source X, the state of which can be represented by a series of Ising variables X_μ and their realizations are explicitly denoted by the lower case letters $x_\mu = \pm 1$. We assume that this observer can not directly observe the source. Instead, he deploys a collection of L sensors, labeled by an index a, to independently observe the source and report the results of their observations over a communication network. Assuming a certain level of environmental noise, the individual observations $Y_\mu(a)$ could be different for different sensors. We define a common level of noise $p \in [0, 1/2)$ for our observations

$$\langle \delta(X_\mu, -Y_\mu(a)) \rangle = p \tag{1}$$

with Kronecker's delta δ, where the braket $\langle \cdot \rangle$ denotes the expectation of an argument. Here we also assume that $y_\mu(a) = \pm 1$.

Then we suppose that each sensor can compress (i.e., lossy encode), if necessary, its sensor readings $\boldsymbol{y}(a) = (y_1(a), \cdots, y_M(a))$ into a codeword $\boldsymbol{z}(a) = (z_1(a), \cdots, z_N(a))$ independently. In this paper, we assume that the codeword is represented by a series of Ising variables $Z_\nu(a)$ and thus their realizations are restricted to $z_\nu(a) = \pm 1$ as well. We further assume that the sensors themselves can not share any information about their observations. That is, they are not permitted to communicate with each other to decide what to send beforehand. As a result, the observer must collect the L codewords from all the sensors, each of which separately encodes its own observations $y_\mu(a)$, and use them to estimate the original x_μ for $\mu = 1, \cdots, M$. We assumed here that the lengths of the codewords are the same N, so that all the sensors are identical with respect to the theoretical ability of encoding their observations. That is, regardless of the sensor label, the rate for the lossy encoding is given by $R = N/M$. Therefore, the load level of our network can be measured by the sum rate LR, which should not be greater than the network capacity given by, say, C. We assume that C is a given integer, not a real, in which case our argument will be greatly simplified.

If the sensors were able to share information about their observations before reporting to the observer, then they would be able to smooth out their independent environmental noises entirely as the number of sensors L diverges. Then the observer can figure out all the realizations of X_μ if the network capacity C exceeds 1, which is the entropy rate of the source X. However, if the mutual communications are prohibited, there does not exist any finite value of C for which even infinitely many sensors can transmit all the information [5]. Therefore, our goal should be the semifaithful reconstruction of the original x_μ given the codewords $z_\nu(a)$ under a certain fidelity criterion.

2.2. Exchangeable sensor ansatz

Suppose that $\hat{\boldsymbol{y}}(a) = (\hat{y}_1(a), \cdots, \hat{y}_M(a))$ be best reproductions for the observations obtained by using the codewords, respectively. Assume that the distortion between two sequences are always measured by the Hamming distance per symbol. Then it is easy to see that the distortion is given by, in this case,

$$d(\boldsymbol{y}(a), \hat{\boldsymbol{y}}(a)) = (1/M) \sum_{\mu=1}^{M} \delta(y_\mu(a), \hat{y}_\mu(a))$$

for $a = 1, \cdots, L$. Since we have exchangeable sensors as stated, we can impose that

$$\langle d(Y(a), \hat{Y}(a)) \rangle = D$$

for any given pairs. With this Hamming distortion constraint, the lower bound on the rate $R(D)$ required to describe a variable $Y_\mu(a)$ is given by

$$R(D) = 1 - H_2(D) \ ,$$

where $H_2(D)$ denotes the binary entropy function [6]. This is called the rate distortion function for the Bernoulli(1/2) source.

The observer then collects all the transmitted information $z_\nu(a)$ to calculate the estimate $\hat{x}_\mu$ for the μth symbol of the unknown $\boldsymbol{x} = (x_1, \cdots, x_M)$. To go further, we now restrict ourselves to the case where

$$\langle \delta(Y_\mu(a), -\hat{Y}_\mu(a)) \rangle = D \ . \tag{2}$$

That is, every variable $\hat{Y}_\mu(a)$ in the reproductions is expected to have the same error probability D. Since the observation and communication processes are independent, the above three variables X_μ, $Y_\mu(a)$, and $\hat{Y}_\mu(a)$ form a Markov chain. Here the best estimator for X_μ is $\hat{Y}_\mu(a)$ if $0 \le p, D < 1/2$ holds. Then it is straightforward to get, independently,

$$\langle \delta(X_\mu, -\hat{Y}_\mu(a)) \rangle = \rho \ ,$$

where

$$\rho = p(1-D) + (1-p)D$$

represents the combined error probability for replacing the original x_μ by the available symbols $\hat{y}_\mu(a)$. In other words, the error indicator function $\delta(X_\mu, -\hat{Y}_\mu(a))$ reduces to the Bernoulli random variable that takes the value 1 with probability ρ for $a = 1, \cdots, L$.

2.3. Bayes optimal estimator

Now let us consider the most probable realization of X_μ given a set of evidences $\hat{\boldsymbol{y}}_\mu = (\hat{y}_\mu(1), \cdots, \hat{y}_\mu(L))$. Since $\delta(X_\mu, -\hat{Y}_\mu(a))$ obeys the Bernoulli statistics, it is easy to see that the majority vote procedure gives the best strategy [7]. That is, the optimal estimator should be a mapping

$$\hat{X}_\mu = \mathrm{sgn}\left\{ \sum_{a=1}^{L} \hat{Y}_\mu(a) \right\} \ .$$

Then overall error probability for the estimate $\hat{x}_\mu$ is minimized. The probability of getting more errors than $L/2$ out of L Bernoulli trials is given by

$$P(X_\mu \neq \hat{X}_\mu) = \langle \delta(X_\mu, -\hat{X}_\mu(a)) \rangle \tag{3}$$

$$= \begin{cases} \sum_{l=\frac{L+1}{2}}^{L} Q_\rho(l|L), & (L \text{ is odd}) \\ \sum_{l=\frac{L}{2}+1}^{L} Q_\rho(l|L) + \frac{1}{2} Q_\rho(L/2|L) & (L \text{ is even}) \end{cases}, \tag{4}$$

where

$$Q_\rho(l|L) = \binom{L}{l} \rho^l (1-\rho)^{L-l}$$

denotes the binomial distribution. In principle, we may choose whatever value of L which is compatible with the sum rate constraint of $LR \leq C$. To minimize the error probability for the estimator $\hat{X}_\mu$, however, we should use the largest possible value. Hereafter we assume that L denotes the largest possible value. In particular, suppose that the sensors do not encode their observations. Instead, each sensor simply sends the whole information of the noisy $Y_\mu(a)$. Then, the error probability for the estimator $\hat{X}_\mu$ reduces to

$$P(X_\mu \neq \hat{X}_\mu) = \begin{cases} \sum_{l=\frac{C+1}{2}}^{C} Q_p(l|C), & (C \text{ is odd}) \\ \sum_{l=\frac{C}{2}+1}^{C} Q_p(l|C) + \frac{1}{2} Q_p(C/2|C) & (C \text{ is even}) \end{cases}.$$

3. Statement of results

Assume that a network capacity C is given. Consider that the common data rate R is first allocated to all the sensors. The number of sensors L is thus determined as the maximum value of L satisfying the sum rate constraint $RL < C$. In our system model, it is obvious to say that $P(X_\mu \neq \hat{X}_\mu) \to 0$ as $C \to \infty$. As is shown in section 4, it is not hard to refine the above statement of convergence and to prove that $P(X_\mu \neq \hat{X}_\mu)$ decays to 0 exponentially fast as $C \to \infty$. By analogy with large deviation theory [8], we define the exponential rate of decay by

$$I_p(R) = -\lim_{C\to\infty} \frac{1}{C} \ln P(X_\mu \neq \hat{X}_\mu) \quad (0 < R \leq 1) .$$

The decay rate $I_p(R)$ describes the limiting behavior of the system from a macroscopic level, on which the rate R could be used as a control parameter. The case of $R = 1$ reduces to a naive aggregation scheme in which the sensors just send their noisy observations to the observer. For this smallest aggregation, we aggregate data from only $L = C$ sensors. Hereafter, we call this scheme the level-1 aggregation. For a given $R > 0$, the level-R aggregation is defined in which every sensor encodes its observations at the rate of R independently. As an extension of the definition of $I_p(R)$ for $R > 0$, we could naturally define the level-0 decay rate as

$$I_p(0) = -\lim_{C\to\infty} \frac{1}{C} \lim_{R\to 0} \ln P(X_\mu \neq \hat{X}_\mu) .$$

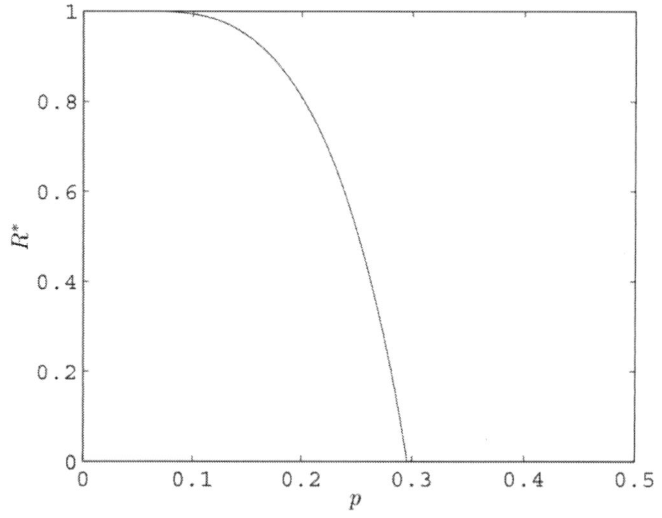

Figure 1. Optimal aggregation levels for ensemble of independent sensors in noisy environment. The p is a given noise level, while R^* is the optimal value of R, the data rate per sensor, which maximizes the decay rate $I_p(R)$ of error probability with increase of network capacity C.

Assume that $D(R)$ denotes the distortion rate function, which is the inverse function of $R(D)$. Suppose that $\alpha(p, R) = (1 - 2p)(1 - 2D(R))$ for $0 \leq p < 1/2$ and $0 < R \leq 1$. Then, the main result of this paper is given below.

Proposition 3.1 (Decay rate) *We have*

$$I_p(R) = \begin{cases} (1-2p)^2 \ln 2 & \text{(R=0)} \\ \dfrac{\alpha(p,R)^2}{2R(1-\alpha(p,R))(1+\alpha(p,R))} & (0 < R \leq 1) \end{cases}. \tag{5}$$

The maximum of $I_p(R)$ is of great interest from an engineering point of view. That is, we prefer larger values of $I_p(R)$. Therefore, we examine the optimal levels defined by $R^* = \text{argmax}_{0 \leq R \leq 1} I_p(R)$. The optimal aggregation, for a given p, is called the level-R^* aggregation.

We now examine the behavior of formula (5) which gives the optimal levels R^* for the noise p. As is seen in figure 1, the optimal aggregation scale diverges, i.e., the optimal data rate R^* per sensor vanishes for noise levels larger than the critical point $p_0 = 0.295$. In this noisy region, we want the system to be as large as possible. The larger the system we have, the smaller the error probability. By definition, the optimal aggregation is said to be level-0. In contrast, we can always find the non-zero optimal levels below p_0. In particular, if the noise level is below $p_1 = 0.066$, our investigations indicate that the level-1 aggregation is optimal. Moderate aggregation levels could be optimal in the intermediate noise levels between the two critical points. It is also worth noticing that the behavior of R^* of p is reminiscent of that of order parameters at a continuous phase transition in statistical mechanics [9]. The analytical results presented here are also consistent with numerical simulations for the system size $C = 10^3$.

Since the optimal levels R^* are unique values for each noise p, we can plot the optimal decay rate $I_p(R^*)$ as is given in figure 2. The optimal rate $I_p(R^*)$ describes the limiting behavior of

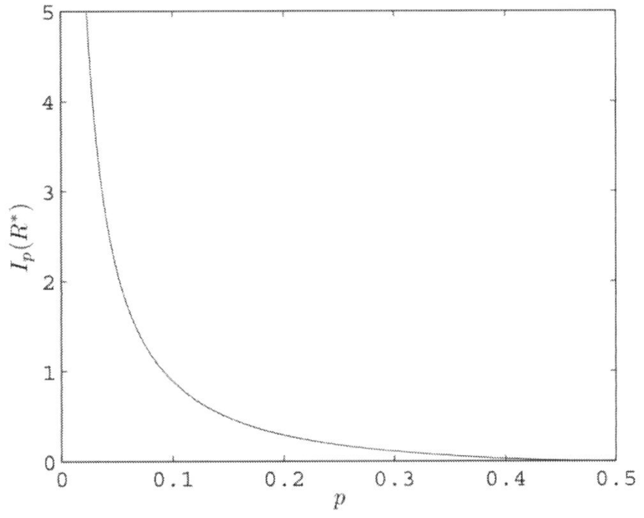

Figure 2. Maximum decay rates for error probability of final decision. The $I_p(R^*)$ is the largest decay rate at the noise level p, which is given by the optimal data rate R^* per sensor.

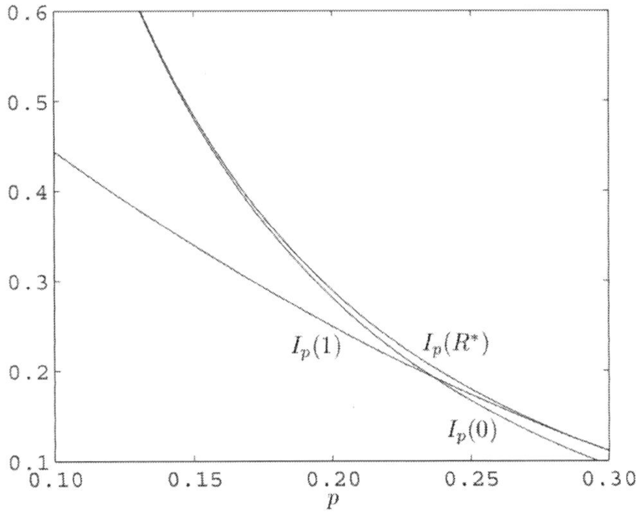

Figure 3. The asymptotics of $I_p(0)$ and $I_p(1)$. The analysis indicates that the combination of these two options can provide a suboptimal aggregation strategy.

the smallest error probability $P(X_\mu \neq \hat{X}_\mu)$ in terms of macroscopic variables. Clearly, it is a strongly decreasing function of the noise p. An approximation of $I_p(R^*)$ by the combination of $I_p(0)$ and $I_p(1)$ is also presented in figure 3. However the crossover point of $I_p(0)$ and $I_p(1)$ as observed in figure 3 is quite different from the value obtained by numerics [10]. We think that this inconsistency should be due to the use of Gaussian approximation and that more rigorous approach should be taken toward the quantitative argument.

4. Large deviations analysis

This section is devoted to presenting the large deviations analysis which gives proposition 3.1. For sufficiently large L, the binomial distribution $Q_\rho(l|L)$ is well approximated by the Gaussian distribution $\mathrm{N}(L\rho, L\rho(1-\rho))$ with mean $L\rho$ and variance $L\rho(1-\rho)$ [11]. Changing the variable

$$s = \frac{l - L\rho}{\sqrt{L\rho(1-\rho)}}$$

enables us to use the central limit theorem to get

$$P(X_\mu \neq \hat{X}_\mu) \sim \int_{\lambda_1}^{\lambda_2} \frac{\mathrm{d}s}{\sqrt{2\pi}} e^{-s^2/2} , \tag{6}$$

where we denote, respectively,

$$\lambda_1 = \frac{1/2 - \rho}{\sqrt{\rho(1-\rho)}}\sqrt{L} , \quad \lambda_2 = \frac{1-\rho}{\sqrt{\rho(1-\rho)}}\sqrt{L} .$$

Since every sensor can achieve the optimal rate $R(D)$, we may evaluate the number of sensors L as $C/R(D)$. Write $\beta(p,D) = (1-2p)(1-2D)$ for simplicity. Together with an identity

$$\frac{1}{2} - \rho = (1-2p)\left(\frac{1}{2} - D\right) ,$$

we have

$$\lambda_1 = \frac{\beta(p,D)\sqrt{C}}{\sqrt{R(D)(1-\beta(p,D))(1+\beta(p,D))}} , \quad \lambda_2 = \frac{(1+\beta(p,D))\sqrt{C}}{\sqrt{R(D)(1-\beta(p,D))(1+\beta(p,D))}} .$$

In particular, the asymptotics of $R(D)$ yields, respectively,

$$\lambda_1 \sim \frac{(1-2p)\sqrt{2C\ln 2}}{\sqrt{(1-\beta(p,D))(1+\beta(p,D))}} , \quad \lambda_2 \sim \frac{(1+\beta(p,D))\sqrt{C\ln 2}}{\sqrt{(1-\beta(p,D)(1+\beta(p,D))}} \left(\frac{1}{2} - D\right)^{-1} .$$

Therefore λ_1 takes a finite value for every C given, while λ_2 could diverge if we consider the level-0 aggregation scheme. These formulas assure us that $I_p(0)$ is well defined.

Now consider

$$\Phi(\lambda) = \int_\lambda^\infty \frac{\mathrm{d}s}{\sqrt{2\pi}} e^{-s^2/2}$$

and write

$$\int_{\lambda_1}^{\lambda_2} \frac{\mathrm{d}s}{\sqrt{2\pi}} e^{-s^2/2} = \Phi(\lambda_1) - \Phi(\lambda_2) .$$

The following lemma tells us that the asymptotic behavior of the definite integral is governed by the asymptotics of $\Phi(\lambda_1)$. The proof is given in the appendix.

Lemma 4.1 (Asymptotic expansion) *We have*

$$\Phi(\lambda) = \tfrac{1}{\sqrt{2\pi}} e^{-\lambda^2/2} \left\{ \tfrac{1}{\lambda} - F_2(\lambda) \right\} , \tag{7}$$

where the second term is given by

$$F_2(\lambda) = e^{\lambda^2/2} \int_\lambda^\infty \frac{\mathrm{d}s}{s^2} e^{-s^2/2} .$$

Here we have $1/\lambda \gg F_2(\lambda)$ as $\lambda \to \infty$ [12]. Then simple algebra gives

$$\Phi(\lambda_1) - \Phi(\lambda_2) \sim \frac{e^{-\lambda_1^2/2}}{\sqrt{2\pi}\lambda_1}\left\{1 - \frac{\lambda_1}{\lambda_2}e^{-(\lambda_2^2-\lambda_1^2)/2}\right\} .$$

Taking the logarithm results in

$$\ln P(X_\mu \neq \hat{X}_\mu) \sim -\frac{\lambda_1^2}{2}$$

which proves the proposition 3.1.

5. Conclusions

It has been shown that the optimal aggregation strategy for a system of sensors exhibits a kind of phase transition with respect to the noise level. We expect this result to be useful when engineering a large sensor network. We described the critical behavior of the ratio $R = C/L$, the data rate per sensor which is the ratio of the network communication capacity C and the number of sensors L.

The analysis shows that in the high noise region beyond a critical value of noise p_0, the ratio R should converge to zero in order to reduce collective estimation error. This means that we should deploy very many sensors $L \gg C$ in the large C limit. In contrast, if the noise level is lower than the critical point, the ratio R should take a positive value. In this case, the number of sensors scales as $L = \mathcal{O}(C)$. The authors hope that this brief paper will trigger interest in this kind of asymptotic property of the sensory aggregation and applications of the large deviations approach in the field of systems science.

References

[1] Akyildiz I F, Su W, Sankarasubramaniam Y and Cayirci E 2002 *IEEE Communications Magazine* **40** 102–14
[2] Shannon C E 1948 *Bell Syst. Tech. J* **27** 379–423
[3] Shannon C E 1959 *IRE National Convention Record* **4** 142–63
[4] Culler D E and Mulder H 2004 *Scientific American* **290** 84–91
[5] Berger T, Zhang Z and Viswanathan H 1996 *IEEE Transactions on Information Theory* **42** 887–902
[6] Cover T M and Thomas J A 1991 *Elements of Information Theory* (New York: Wiley)
[7] MacKay D J C 2003 *Information Theory, Inference, and Learning Algorithms* (Cambridge: Cambridge University Press)
[8] Ellis R S 1985 *Entropy, Large Deviations, and Statistical Mechanics* (Berlin: Springer)
[9] Monasson R, Zecchina R, Kirkpatrick S, Selman B and Troyansky L 1999 *Nature* **400** 133–7
[10] Murayama T and Davis P 2008 *Journal of Physics: Conference Series* **95** 012010
[11] Hays W L 1981 *Statistics* (New York: Holt, Rinehart and Winston)
[12] Copson E T 2004 *Asymptotic Expansions* (Cambridge: Cambridge University Press)

Proof of lemma 4.1

Define a series of integrals,

$$F_n(\lambda) = e^{\lambda^2/2}\int_\lambda^\infty \frac{ds}{s^n}e^{-s^2/2} ,$$

the notations of which is also consistent with the preceding $F_2(\lambda)$. By definitions, it is an easy matter to check that

$$\Phi(\lambda) = \frac{F_0(\lambda)}{\sqrt{2\pi}}e^{-\lambda^2/2} .$$

Journal of Physics: Conference Series **233** (2010) 012007 doi:10.1088/1742-6596/233/1/012007

Since elementary calculus gives

$$F_n(\lambda) = -e^{\lambda^2/2} \int_\lambda^\infty \frac{\mathrm{d}s}{s^{n+1}} \frac{\mathrm{d}}{\mathrm{d}s} e^{-s^2/2} ,$$

the integration by parts results in the recurrence formula

$$F_n(\lambda) = \frac{1}{\lambda^{n+1}} - (n+1) F_{n+2}(\lambda) .$$

Therefore, we find the asymptotic expansion

$$\Phi(\lambda) = \frac{1}{\sqrt{2\pi}} e^{-\lambda^2/2} \left\{ \frac{1}{\lambda} - \frac{1}{\lambda^3} + \frac{3!}{\lambda^5} + \cdots + (-1)^n \frac{(2n-1)!!}{\lambda^{2n+1}} + (-1)^{n+1} (2n+1)!! F_{2n+2}(\lambda) \right\}$$

which is also exact for n equals 0. The conclusion follows.

International Workshop on Statistical-Mechanical Informatics 2010 IOP Publishing
Journal of Physics: Conference Series **233** (2010) 012008 doi:10.1088/1742-6596/233/1/012008

Energy-gap analysis of quantum spin-glass transitions at zero temperature

Kazutaka Takahashi and Yoshiki Matsuda

Department of Physics, Tokyo Institute of Technology, Tokyo 152–8551, Japan

Abstract. We study for random quantum spin systems the energy gap between the ground and first excited states to clarify a relation to the spin-glass–paramagnetic phase transition. We find that for the transverse Sherrington-Kirkpatrick model the vanishing of the averaged gap does not identify the transition. A power-law form of the gap distribution function leads to power-law distributions of the linear, spin-glass, and nonlinear susceptibilities. They diverge at different points, which we attribute to a quantum "Griffiths" mechanism. On the other hand, the classical treatment is justified for the transverse random energy model and the phase transition can be found by a sudden change of the ground state.

1. Introduction

Spin glasses are known as nontrivial states of random spin systems and a lot of studies have revealed their interesting properties [1, 2, 3]. Now, the mean field theory of the classical spin-glass has been established and is recognized as a useful tool for attacking various random systems. As one of such applications, quantum systems have been studied for many years since the work by Bray and Moore [4]. Recently, much attention has been paid to quantum systems for their applications to information processing such as quantum annealing [5, 6, 7]. In contrast to classical systems, mean field analysis has a serious limitation and a reliable analytical method is needed.

At finite temperature, state of the system is determined by the free energy. A competition between energy and entropy induces a phase transition between different phases. At zero temperature, the system is determined by the energy only and the entropy does not play any role. The quantum fluctuations become important at low temperature and the system falls into its ground state. Then, the phase transition is realized by a change of the ground state, which can be found by looking at the energy gap between the ground and first excited states. Although the mechanism of a quantum phase transition is simpler than that of a thermodynamic phase transition, it is rather difficult to treat quantum fluctuation effects even for clean systems [8]. The effect of disorder further makes the analysis even more difficult. As in the classical system, the mean field theory of quantum spin-glass systems has been developed. However, a lot of studies use the static approximation [4] which neglects the quantum fluctuation effects. Although this approximation is justified for not-too-low temperatures, no reliable result can be obtained at zero temperature.

In the present paper, we give a detailed study of quantum phase transitions in terms of the energy gap [9]. As typical random quantum systems which exhibit a spin-glass phase transition, we treat the Sherrington-Kirkpatrick (SK) model [10] and random energy model (REM) [11]

with a transverse field. The transverse REM can be solved exactly and shows a discontinuous transition [12]. The transverse SK model has not been solved exactly, but many studies indicate a continuous transition. Since these models without transverse field can be treated exactly by a mean field theory, most of previous studies took a similar approach. We study these systems by using the energy gap analysis. In section 2, we define the models and consider the energy gap by using perturbation theory. The energy gap is also examined numerically and we see that the gap strongly fluctuates from sample to sample. In order to see a phase transition, several kinds of susceptibilities are studied in terms of the energy gap in section 3. We find there a strong quantum fluctuation effect. After discussing a sufficient condition for this effect to occur in section 4, we discuss its mechanism in section 5. The final section 6 is devoted to give conclusions, perspectives, and possible applications.

2. Energy level distribution

In this section, we define quantum spin-glass models and examine the energy gap between the ground and first excited states.

2.1. Model

We treat the p-body interacting spin-glass model with transverse field

$$\hat{H} = -\sum_{i_1<i_2<\cdots<i_p}^{N} J_{i_1 i_2 \cdots i_p}\sigma_{i_1}^z\sigma_{i_2}^z\cdots\sigma_{i_p}^z - \Gamma\sum_{i=1}^{N}\sigma_i^x, \tag{1}$$

where $\sigma_i^{z,x}$ are Pauli matrices on site i, N is the number of the site, and Γ is the transverse field. We take an ensemble average over random interaction $J_{i_1 i_2\cdots i_p}$ with a probability distribution

$$\mathrm{P}(J_{i_1\cdots i_p}) = \sqrt{\frac{N^{p-1}}{\pi p! J^2}}\exp\left\{-\frac{N^{p-1}}{p!J^2}\left(J_{i_1 i_2\cdots i_p}\right)^2\right\}, \tag{2}$$

where J is the strength of the random fluctuations. Here we study the cases of $p=2$ (SK model) and $p\to\infty$ (REM), and the case of other-p will be reported elsewhere. The quantum REM can be solved exactly and at zero temperature the system shows a discontinuous phase transition between the spin-glass and paramagnetic phases at $\Gamma/J=\sqrt{\ln 2}\sim 0.83$ [12]. The SK model has not been solved exactly, but a lot of studies reported a continuous transition around the point $\Gamma/J = 1 \sim 2$ when the temperature is equal to zero. A naive perturbative calculation gives $\Gamma/J=1$ and the static approximation gives $\Gamma/J=2$ [13, 14], while more advanced methods indicate a point around $\Gamma/J\sim 1.5$ [15, 16, 17, 18, 19, 20].

2.2. Perturbative calculation of the energy gap

Here we estimate for the SK model the energy gap between the ground and first excited states by using a perturbative expansion with respect to J/Γ. When the randomness J is absent, the Hamiltonian is exactly solved to give energy levels $E_n^{(0)} = -(N-2n)\Gamma$ with nonnegative integer n. Each of the levels has a degeneracy $N!/n!(N-n)!$ which is split into many nondegenerate ones for finite-J. The unperturbed ground state is denoted by $n=0$ and the first excited state belongs to the sector with $n=1$. The Hamiltonian of the $n=1$ sector is given by $H_{ij}^{(1)} = -(N-2)\Gamma - J_{ij}$. Since J_{ij} are Gaussian random variables, the energy level distribution is given by that of the Gaussian orthogonal ensemble [21]. The energy levels distribute in a semicircle form and the lowest energy level is given by $E_1 = -(N-2)\Gamma - 2J$. Neglecting the correlations between blocks with different n's, we obtain the energy gap $\Delta\sim 2(\Gamma - J)$ and the degenerate point $\Gamma = J$.

The subleading corrections are calculated as follows. The ground state $n = 0$ is coupled to $N(N-1)/2$-levels in the sector $n = 2$ by random interactions J_{ij}. Up to second order in perturbation theory, we obtain in average,

$$E_0^{(2)} \sim -N\Gamma - \frac{\sum_{i<j}^N J_{ij}^2}{E_2^{(0)} - E_0^{(0)}} \sim -N\Gamma - \frac{J^2}{8\Gamma}(N-1). \tag{3}$$

In the same way, the first excited state is directly coupled to the sector $n = 3$. The correction is roughly estimated as

$$E_1^{(2)} \sim -(N-2)\Gamma - 2J - \frac{J^2}{8\Gamma}(N-3). \tag{4}$$

As a result we have the energy gap

$$\Delta \sim 2(\Gamma - J) + \frac{J^2}{4\Gamma}, \tag{5}$$

and the degenerate point $\Gamma/J = (2+\sqrt{2})/4 \sim 0.85$. Although the present analysis is based on a perturbative expansion and the degenerate point cannot be estimated exactly, it is probable to say that the averaged degenerate point does not coincide with the phase transition point around 1.5. On the other hand, a similar perturbative calculation was carried out for the transverse REM [22] and a good estimate of the phase transition point was obtained.

2.3. Numerical result of the energy gap

We estimate the energy gap by using the numerical calculation. The Hamiltonian of the size 2^N is diagonalized by using the Lanczos method and the average is taken over more than 20000 samples. The number of the site is taken to be $N = 8 - 22$. We note that for the SK model, the Hamiltonian commutes with $P = \prod_{i=1}^N \sigma_i^x$ and the matrix splits into two blocks. This allows us to diagonalize their two matrices separately. For the REM, we take the Derrida's original Hamiltonian for the random part and do not use the original Hamiltonian (1) with $p \to \infty$. The Gaussian random variables with the average $[E_i] = 0$ and the variance $[E_i^2] = NJ^2/2$ are produced 2^N times and are used for the diagonal part of the Hamiltonian in σ_z-basis.

In figure 1, we plot the averaged energy gap. For the SK model, the gap is an increasing function of Γ and becomes negligible at values smaller than $\Gamma/J \sim 0.6$. This result is consistent with the perturbative calculation. From this figure, we see that it is hard to identify a phase transition point around $\Gamma/J \sim 1.5$ even if the fluctuations are taken into account. On the other hand, the gap shows a complicated behavior for the REM. The averaged gap becomes minimum around the phase transition point and the fluctuations behave differently on both sides of the point. These results imply that it is possible to find a phase transition from the averaged energy gap for the REM, while it is hard for the SK model.

In order to see the difference more in detail, we show the energy gap distribution function in figure 2. For the SK model, we find long tails of the distribution function. It is hard to imagine that these fluctuations vanish at the thermodynamic limit and we expect a significant role of fluctuations. A more complicated behavior can be observed for the REM. A single peak is replaced by another one, which implies a discontinuous transition between two different states.

These analytical and numerical results show that for the SK model the fluctuations play a significant role and the averaged gap is not a useful quantity to describe the phase transition. Concerning the REM, the averaged gap is expected to be useful although the fluctuations are large as well. Therefore, it is important to understand where this difference comes from.

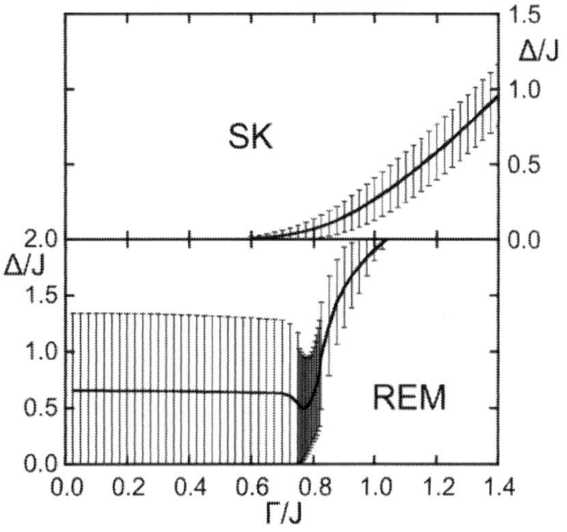

Figure 1. Averaged energy gap $\Delta = [E_1 - E_0]$ between the ground and first excited states for the SK model and REM with $N = 22$. The variance of the energy gap $\pm\sqrt{[(E_1 - E_0)^2] - \Delta^2}$ is denoted by the error bar.

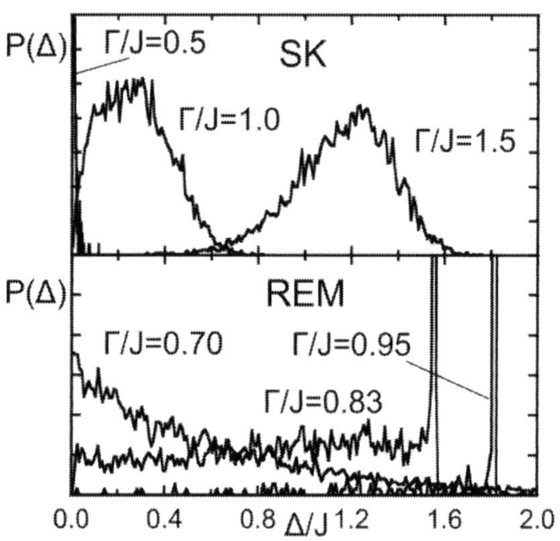

Figure 2. Gap distribution function for the SK model and REM with $N = 22$. For the REM, $\Gamma/J \sim 0.83$ corresponds to the phase transition point.

3. Susceptibility

In the standard theory of critical phenomena, phase transitions are identified by a singularity of response functions such as the magnetic susceptibility. In this section, we discuss the linear, spin-glass, and nonlinear susceptibilities. For our purpose to find a transition in terms of the energy gap, the spectral representation is useful and is discussed in detail.

3.1. Spectral representation

The magnetization of the system is given by the response of the longitudinal magnetic field h whose contribution to the Hamiltonian is given as $\hat{H} \to \hat{H} - h\sum_i \sigma_i^z$. We have

$$m = \frac{1}{N\beta}\frac{\partial}{\partial h}[\ln Z], \tag{6}$$

where $Z = \mathrm{Tr}\, e^{-\beta\hat{H}}$ is the partition function and $\beta = 1/T$ is the inverse temperature. The magnetization can be expanded with respect to h as $m = \chi_l h - \chi_{nl}h^3 + \cdots$, which defines the linear (χ_l) and nonlinear (χ_{nl}) susceptibilities [23, 24].

Our analysis is based on the imaginary time formalism [8, 25]. The linear susceptibility is expressed as

$$\chi_l = \frac{2}{N\beta}\sum_{i,j=1}^{N}\int_0^\beta d\tau \int_0^\tau d\tau' \left[\frac{1}{Z}\mathrm{Tr}\, e^{-\beta\hat{H}}\sigma_i^z(\tau)\sigma_j^z(\tau')\right] - \beta q, \tag{7}$$

where $\sigma_i^z(\tau) = e^{\hat{H}\tau}\sigma_i^z e^{-\hat{H}\tau}$, τ is the imaginary time running from 0 to β, and q is the spin-glass order parameter. Inserting the completeness relation of eigenstates of the Hamiltonian, we obtain that the first term of the right hand side of equation (7) is equal to

$$\frac{\beta}{N}\left[\frac{1}{Z}\sum_{mn(E_m=E_n)} e^{-\beta E_n}|\langle n|\sum_i \sigma_i^z|m\rangle|^2\right] + \frac{2}{N}\left[\frac{1}{Z}\sum_{mn(E_m\neq E_n)} \frac{e^{-\beta E_n}}{E_m - E_n}|\langle n|\sum_i \sigma_i^z|m\rangle|^2\right], \tag{8}$$

where $|n\rangle$ is an eigenstate with the energy E_n. The first term comes from degenerate states and the second from nondegenerate ones. The first term is equal to the static part $\omega = 0$ of the spectral function

$$\chi(\omega) = \frac{1}{N}\left[\frac{1}{Z}\sum_{nm}\delta(\omega - E_m + E_n)e^{-\beta E_n}|\langle n|\sum_i \sigma_i^z|m\rangle|^2\right]. \tag{9}$$

For the static part $\chi = \chi(\omega = 0)$, the energy is conserved through a virtual process in imaginary time and a classical treatment is possible. Then, the linear susceptibility is expressed by a sum of classical and quantum parts. The classical part $\chi_{\rm l}^{\rm (c)}$ is equal to $\beta(\chi - q)$ and can be treated by the static approximation. It corresponds to the linear susceptibility for the classical system given by $\chi_{\rm l} = \beta(1-q)$ [2]. On the other hand, the quantum part is expressed as the second term of equation (8). At the zero temperature limit, we assume that the ground state $|0\rangle$ with the energy E_0 is nondegenerate and obtain for the quantum part

$$\chi_{\rm l}^{\rm (q)} = \frac{2}{N}\left[\sum_{n\neq 0}\frac{|\langle 0|\sum_i \sigma_i^z|n\rangle|^2}{E_n - E_0}\right]. \tag{10}$$

In the same way, the quantum part of the nonlinear susceptibility is given by

$$\begin{aligned}\chi_{\rm nl}^{\rm (q)} &= \frac{4}{N}\left[\sum_{n,m\neq 0}\frac{|\langle 0|\sum_i \sigma_i^z|n\rangle|^2}{E_n - E_0}\frac{|\langle 0|\sum_j \sigma_j^z|m\rangle|^2}{(E_m - E_0)^2}\right] \\ &-\frac{4}{N}\left[\sum_{n,n',n''\neq 0}\frac{\langle 0|\sum_i \sigma_i^z|n\rangle\langle n|\sum_j \sigma_j^z|n'\rangle\langle n'|\sum_k \sigma_k^z|n''\rangle\langle n''|\sum_l \sigma_l^z|0\rangle}{(E_n - E_0)(E_{n'} - E_0)(E_{n''} - E_0)}\right]. \end{aligned}\tag{11}$$

In classical spin-glass systems, the nonlinear susceptibility is an important quantity since it is related directly to the spin-glass susceptibility $\chi_{\rm sg}$ as $\chi_{\rm nl} = \beta(\chi_{\rm sg} - 2\beta^2/3)$. A spin-glass transition which is characterized by the divergence of $\chi_{\rm sg}$ can be observed directly by $\chi_{\rm nl}$. However, in the present case, the quantum part of the spin-glass susceptibility is given by

$$\chi_{\rm sg}^{\rm (q)} = \frac{4}{N}\left[\left(\sum_{n\neq 0}\frac{|\langle 0|\sum_i \sigma_i^z|n\rangle|^2}{E_n - E_0}\right)^2\right]. \tag{12}$$

$\chi_{\rm nl}^{\rm (q)}$ and $\chi_{\rm sg}^{\rm (q)}$ behave differently and there is no direct relation between them. The spin-glass susceptibility is not equivalent to the nonlinear susceptibility, which is a crucial difference from the classical system.

If the transition is of quantum nature, the singularity comes from the quantum part. We see from above expressions that the three susceptibilities are divergent when the gap between the ground and first excited states collapses into zero. This is understood as the standard mechanism of a quantum phase transition as we mention in section 1. However, in random systems, the gap vanishing point fluctuates from sample to sample. If the fluctuation is small, the transition point can be given by the averaged gap vanishing point. But, as we see in section 2, the fluctuation is not negligible and the vanishing point does not coincide with the transition point reported in previous studies. If the gap distribution function behaves in a power-law form $\mathrm{P}(\Delta) \sim \Delta^k$, we can find that three susceptibilities diverge at different points. Since $\chi_{\rm l}$, $\chi_{\rm sg}$, and $\chi_{\rm nl}$ are roughly proportional to the gap as $1/\Delta^{1,2,3}$ respectively, they diverge at $k = 0, 1, 2$ respectively. We note that this behavior is specific to random and quantum systems. The quantum nature is necessary to obtain the spectral representations and a power-law distribution of the gap can be obtained only for random systems.

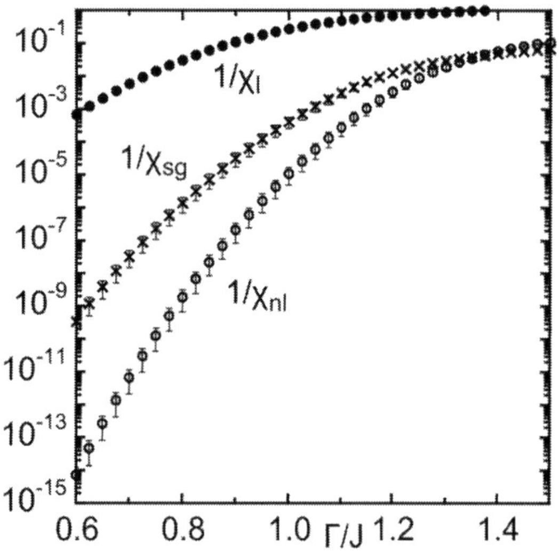

Figure 3. Inverse susceptibilities (quantum part) for the SK model with $N = 18$.

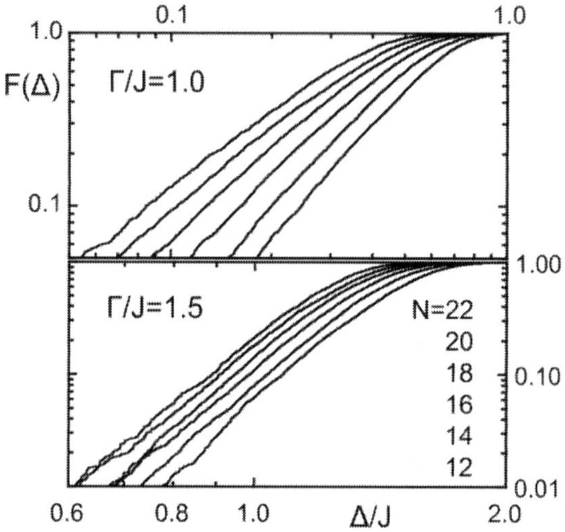

Figure 4. The cumulative distribution function of the energy gap $F(\Delta) = \int_0^{\Delta} d\Delta' P(\Delta')$ (N = 12,14,16,18,20, and 22 from below to above) for the SK model.

3.2. Numerical result

In order to see whether the above scenario holds, we analyze numerically the spectral representation of the susceptibilities. In figure 3, we plot the quantum part of $\chi_{\rm l}$, $\chi_{\rm sg}$, and $\chi_{\rm nl}$ for the SK model. They become very large at small transverse fields and there is a relevant difference between $\chi_{\rm sg}^{(q)}$ and $\chi_{\rm nl}^{(q)}$, which support our expectation. We also plot the cumulative distribution function of the energy gap in figure 4. The size dependence of the function implies that the power-law persists in the thermodynamic limit.

We also plot the distribution function of the susceptibilities in figure 5. The distribution function P(χ) shows a power-law dependence at large-χ as P(χ) $\sim \chi^{-s-1}$. In this form of the distribution function, the averaged susceptibility diverges when $s < 1$. We analyze the distribution functions for each of the susceptibilities and find that the divergent point can be different with each other, which is consistent with the analysis of the gap distribution. Although our system size is not so large enough to give a reliable extrapolation value, our result of the spin-glass transition, identified by the divergence of $\chi_{\rm sg}$, supports the results in [15, 16, 17, 18, 19, 20]. Previous studies did not pay any attention to the difference between the spin-glass and nonlinear susceptibilities and it may be worth reconsidering the results more carefully.

We note that the present numerical analysis is only for the quantum part. To find an observable physical value of susceptibilities, we must include the classical part. The classical part of the susceptibilities can be analyzed by the static approximation. For the linear susceptibility, $\chi_{\rm l}^{(c)} = \beta(\chi - q)$ only gives a cusp and does not show any strong singularity. The classical nonlinear and spin-glass susceptibilities can be divergent and we must study closely which part between the classical and quantum parts is important. It can be considered that the classical part is important at the spin-glass phase and the quantum part is dominant at the paramagnetic phase. We treat a related problem in the next section and a detailed study is presented elsewhere.

3.3. Effect of higher excited states

Up to here, our attention has been paid to the gap between the ground and first excited states. The spin-glass system has a complicated ground state which is described by a replica-symmetry-

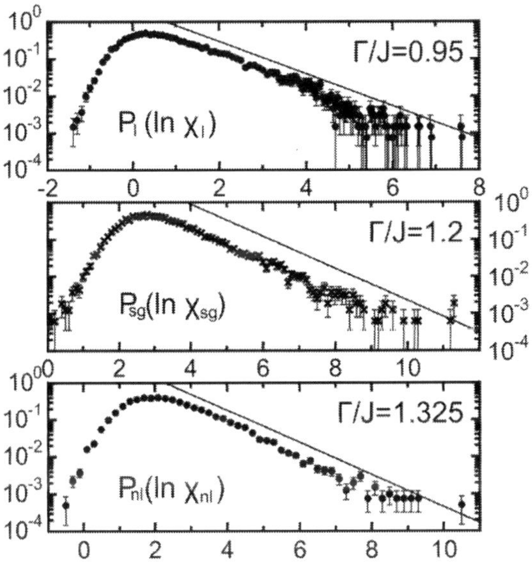

Figure 5. Susceptibility distribution functions for the SK model with $N = 18$. The straight lines are $P(\chi) \sim 1/\chi^2$.

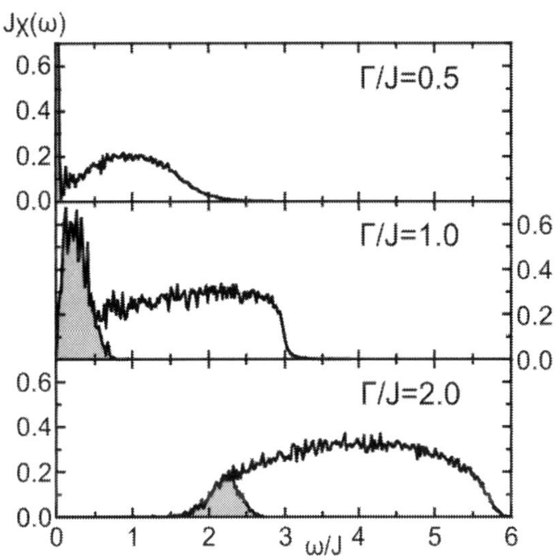

Figure 6. Spectral function $\chi(\omega)$ for the SK model ($N = 20$). The shaded parts are contributions from the first excited state.

broken order parameter, and it might be natural to expect that at the transition point a vast amount of degeneracy occurs at the same time. In order to see effects of higher order excited states, we plot the spectral function (9) for the SK model in figure 6. For large-Γ, the function was analytically calculated as [26]

$$\chi(\omega) = \frac{\sqrt{4J^2 - (\omega - 2\Gamma)^2}}{2\pi J^2}\theta(2J - |\omega - 2\Gamma|). \tag{13}$$

In this case, the contribution from the first excited state cannot be distinguished from those of other states. This property drastically changes when Γ is small. We see that a single peak from the first excited states is separated from other contributions and plays a dominant role for the phase transition. Actually, when the transverse field is absent, we have a symmetry that the Hamiltonian is unchanged under the gauge transformation $\sigma_i^z \to -\sigma_i^z$. As a result, the energy level is doubly degenerate in the classical limit. Thus, for the present model, the first excited state plays a dominant role for the phase transition.

This result implies that the second term of the right hand side of equation (11) gives a subleading correction since the expression includes matrix elements of higher order excited states ($|n'\rangle$). Numerically we confirm that this term is not important to find a divergent point, which makes numerical calculations easier.

4. Localization of state

In the previous section, we have treated the SK model only. The spectral representation is a general relation and is independent of the explicit form of the Hamiltonian. However, it is evident from the known exact solution that the above scenario cannot be applied to the REM. Actually the classical treatment is justified for the REM and the quantum part of the susceptibility does not give any significant contributions. This can be understood by examining the inverse participation ratio (IPR) defined as

$$I = \sum_n |\langle n|\sigma_i^z|0\rangle|^4 . \tag{14}$$

This function describe how many states are included in the state $\sigma_i^z|0\rangle = \sum_{n=1}^{M} c_n|n\rangle$. If it includes M-states, $c_n \sim 1/\sqrt{M}$ and we obtain $I \sim 1/M$. Thus this function is a useful quantity to see whether the state is localized in eigenstate space [27].

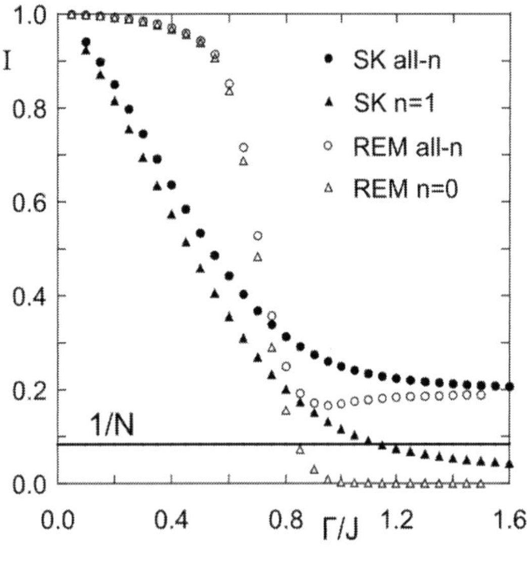

Figure 7. Inverse participation ratio (N=12). The summation over n in equation (14) is restricted to $n = 0(1)$ for a plot with caption $n = 0(1)$.

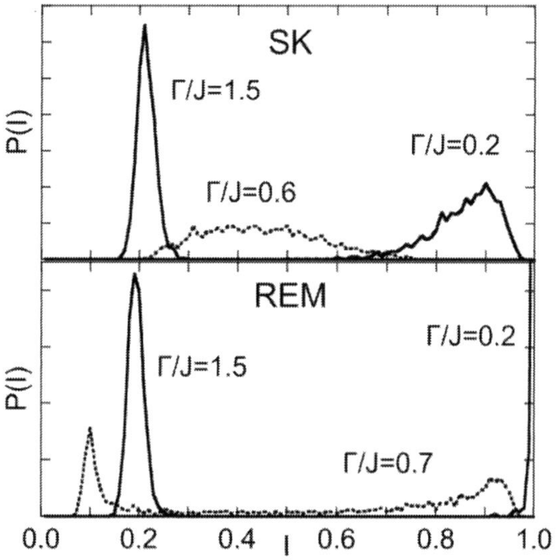

Figure 8. Distribution function of the inverse participation ratio (N=12).

In figure 7, we plot the averaged IPR for both models and find a significant difference between them. For the SK model, the IPR decreases uniformly from 1 to $1/N$ as a function of Γ. This behavior reflects the fact that the gauge symmetry is important at small-Γ and, at large-Γ, the one spin flip changes the ground state to the sector $n = 1$ with N-levels. $\sigma_i^z|0\rangle$ mainly consists of the first excited state. There is no such symmetry for the REM and the IPR becomes minimum around the phase transition point. At large-Γ the IPR gradually approaches to unity. $\sigma_i^z|0\rangle$ almost stays in the ground state $|0\rangle$ and the excitations to the higher energy levels are small. This result shows that the quantum fluctuation effect can be negligible for the REM. This is the reason why the discussion in the previous section cannot be applied to the REM. The state of the system is determined by the ground state.

We also find for the IPR distribution function in figure 8 that the IPR is always larger than $1/N$. This shows that all states 2^N do not contribute to determine the state of the system, which may allow us to construct a reduced effective model. As in the energy gap distribution, a double peak structure is observed for the REM, while it is not for the SK model.

5. Quantum Griffiths singularity

The power-law distribution of the susceptibility was also found in finite-dimensional systems [28, 29, 30, 31, 32, 33]. In two and three dimensional systems, the nonlinear susceptibility diverges at a point of Γ which is larger than the phase transition point. This was understood the Griffiths singularity effect [34, 35]. Rare configurations of ordered clusters in a disordered state give a singular contribution to the susceptibility. A similar effect is known in disordered electron systems as anomalously localized states [36].

Since the origin of the singularity comes from ordered clusters in spatial dimensions, this effect should be absent in the present infinite-ranged models. The authors in [29, 30, 32, 33] called this mechanism as the quantum Griffiths singularity. However, the essential mechanism is the same as the classical one. The quantum nature is relevant for how this mechanism is related to physical quantities.

Our finding is that, even for infinite-ranged systems, a similar power-law behavior can be observed for the susceptibility distribution. We cannot attribute this behavior to the ordered clusters as in the finite dimensional case. The effect must be a purely quantum mechanical one which can be expressed by fluctuations in imaginary time. Therefore, it is natural to expect that ordered clusters are formed in the imaginary time direction. Compared to the spatial dimension, the crucial difference is that there is no disorder in this direction. Theoretically, the imaginary time correlation arises only when the ensemble average is carried out for the free energy, which implies that the present Griffiths-like effect can be obtained for random quantum systems. The randomness induces quantum fluctuations, which allows a formation of ordered clusters. If we accept this scenario, it is not difficult to obtain a power-law form of the gap distribution function. Referred to the classical Griffiths mechanism [8], we expect an exponentially small gap $\Delta \sim \Delta_0 e^{-c\tau/(k+1)}$ in τ is formed with a very small probability $e^{-c\tau}$, where Δ_0, k, and c are some constants. Summing over τ, we obtain

$$\mathrm{P}(\Delta) \sim \int d\tau e^{-c\tau} \delta\left(\Delta - \Delta_0 e^{-\frac{c}{k+1}\tau}\right) \sim \Delta^k. \tag{15}$$

Since $0 \leq \tau \leq \beta$, the power law can be obtained only for large-β. This feature is consistent with our expectation that the present effect is a purely quantum one and becomes important at low temperatures.

This mechanism is not specific to the present infinite-ranged models and the finite dimensional systems can share the same property. For the finite dimensional systems, the classical and quantum Griffiths singularities appear at the same time. They must be treated in a different way and their contributions to physical quantities are expected to be different. It is an important issue to be clarified in future studies.

6. Conclusions

In conclusion, we have discussed quantum spin-glass phase transitions in terms of the energy gap. We found for the transverse SK model that the gap distribution shows a power-law behavior near the origin. As a result, depending on its exponent, the linear, spin-glass, and nonlinear susceptibilities diverge at different points. On the other hand, the transverse REM does not show such behavior. Although the gap distribution shows a broad form, the state of the system can be described by the ground state and the classical treatment is justified for the REM. The difference can be understood by the symmetry of the system. The role of symmetry can be more closely studied by generalizing the model (1) to arbitrary-p. Details are presented elsewhere.

The present results are considered a universal feature that can be seen in broad random quantum systems since the mechanism of the quantum fluctuation effects induced by disorder is independent of the details of the Hamiltonian. Our result for the SK model shows that the self-averaging property does not hold at zero temperature and the quantum phase transition is determined by rare configurations of disorder. Random quantum systems can be totally different from classical thermodynamic systems. In order to reveal such properties, we consider that the present approach is useful and can be a supplement to the dynamical mean field theory as discussed in [17, 20]. Work in this direction is under progress and we hope that such universal properties are clarified in near future.

Acknowledgments
We are grateful to K. Takeda for useful discussions and comments. Y.M. acknowledges support from the Japan Society for the Promotion of Science.

References

[1] Mézard M, Parisi G, and Virasoro M A 1987 *Spin Glass Theory and Beyond* (Singapore: World Scientific)
[2] Fischer K H and Hertz J A 1991 *Spin Glasses* (Cambridge: Cambridge University Press)
[3] Nishimori H 2001 *Statistical Physics of Spin Glasses and Information Processing: An Introduction* (Oxford: Oxford University Press)
[4] Bray A J and Moore M A 1980 *J. Phys. C* **13** L655
[5] Finnila A B, Gomez M A, Sebenik C, Stenson C, and Doll J D 1994 *Chem. Phys. Lett.* **219** 343
[6] Kadowaki T and Nishimori H 1998 *Phys. Rev.* E **58** 5355
[7] Farhi E, Goldstone J, Gutmann S, Lapan J, Lundgren A, and Preda D 2001 *Science* **292** 472
[8] Sachdev S 1999 *Quantum Phase Transitions* (Cambridge: Cambridge University Press)
[9] Takahashi K and Matsuda Y 2010 *J. Phys. Soc. Jpn.* **79** 043712
[10] Sherrington D and Kirkpatrick S 1975 *Phys. Rev. Lett.* **35** 1792
[11] Derrida B 1980 *Phys. Rev. Lett.* **45** 79
[12] Goldschmidt Y Y 1990 *Phys. Rev.* B **41** 4858
[13] Usadel K D 1986 *Solid State Commun.* **58** 629
[14] Thirumalai D, Li Q, and Kirkpatrick T R 1989 *J. Phys. A* **22** 3339
[15] Yamamoto T and Ishii H 1987 *J. Phys. C* **20** 6053
[16] Goldschmidt Y Y and Lai P Y 1990 *Phys. Rev. Lett.* **64** 2467
[17] Miller J and Huse D A 1993 *Phys. Rev. Lett.* **70** 3147
[18] Arrachea L and Rozenberg M J 2001 *Phys. Rev. Lett.* **86** 5172
[19] Arrachea L, Dalidovich D, Dobrosavljević V, and Rozenberg M J 2004 *Phys. Rev.* B **69** 064419
[20] Takahashi K 2007 *Phys. Rev.* B **76** 184422
[21] Mehta M L 2004 *Random Matrices* 3rd ed. (Boston: Academic Press)
[22] Jörg T, Krzakala F, Kurchan J, and Maggs A C 2008 *Phys. Rev. Lett.* **101** 147204
[23] Chalupa J 1977 *Solid State Commun.* **22** 315
[24] Suzuki M 1977 Prog. *Theor. Phys.* **58** 1151
[25] Negele J W and Orland H 1988 *Quantum Many-Particle Systems* (Reading, MA: Addison-Wesley)
[26] Takahashi K and Takeda K 2008 *Phys. Rev.* B **78** 174415
[27] Wegner F 1980 *Z. Phys.* B **36** 209
[28] Fisher D S 1992 *Phys. Rev. Lett.* **69** 534
[29] Guo M, Bhatt R N, and Huse D A 1994 *Phys. Rev. Lett.* **72** 4137
[30] Rieger H and Young A P 1994 *Phys. Rev. Lett.* **72** 4141
[31] Fisher D S 1995 *Phys. Rev.* B **51** 6411
[32] Rieger H and Young A P 1996 *Phys. Rev.* B **54** 3328
[33] Guo M, Bhatt R N, and Huse D A 1996 *Phys. Rev.* B **54** 3336
[34] Griffiths R B 1969 *Phys. Rev. Lett.* **23** 17
[35] McCoy B M 1969 *Phys. Rev. Lett.* **23** 383
[36] Altshuler B L, Kravtsov V E, and Lerner I V 1987 *Pisma Zh. Eskp. Teor. Fiz.* **45** 160 [1987 *JETP Lett.* **45** 199]

Clusters of resource consuming nodes in transportation networks

C. H. Yeung and K. Y. Michael Wong

Department of Physics, The Hong Kong University of Science and Technology, Hong Kong, China

E-mail: `phbill@ust.hk`, `phkywong@ust.hk`

Abstract. We study the problem of optimal source location in transportation networks. When the installation cost of source nodes is comparable to the transportation cost, regimes characterized by discrete fractions of resource consuming nodes emerge and resemble the Devil's staircase, corresponding to different clusters of consumer nodes. When the fraction of deficient nodes increases, frustration arises in choices of locating extra source nodes and numerous sub-optimal configurations emerge, resembling typical glassy behavior.

1. Introduction

The physics of spin glasses is highly relevant to the study of optimization problems. Formally, objective functions in optimization problems can be mapped to spin glass Hamiltonians, enabling the analysis of optimization problems using statistical physical techniques. Moreover, many optimization problems are associated with algorithmic hardness, resembling the thermodynamics of spin glasses in the glassy phase. These similarities further enhance the application of techniques including the replica method [1, 2] and Bethe approximation on optimization problems. Successful examples are found in the MAX-SAT or satisfiability problem [3] and the minimal vertex cover problem [4, 5, 6].

Recently, resource allocation problems in transportation networks have been considered from these approaches [9, 10, 11]. These problems are important in load balancing in computer networks [7] and network flow of commodities [8]. In this paper, we address the problem of optimal source location in transportation networks. In this context, transportation networks consist of either *surplus* or *deficient* nodes, and the optimization task is to locate those deficient nodes that need to be converted to *source* nodes, so that networkwide satisfaction with minimal costs is achieved by flow of resources along the links. This problem has wide applications. In wireless sensor networks, for example, signals collected by the population of sensors are transmitted to base stations through the network formed by the sensors. However, due to the limited power of the sensors, it is advantageous to reduce the signal traffic by installing extra access points so that the service lifetimes of the sensors can be prolonged. These access points can be considered as source nodes which provide service to the sensors. Since the installation of access points requires various overheads in terms of harware, software, management, finance, and so on, the optimal locations of access points involves balancing the installation cost of the extra access points and the transportation cost of the signal traffic.

In general, the installation cost required to convert a deficient node into a source node is step-like. This high nonlinearity gives rise to unique behaviors and a physical picture absent in the previous models [12]. When the installation cost is comparable to the transportation cost, the total cost is optimized either by saving the transportation cost (while increasing the installation cost), or decreasing the installation cost (while increasing the transportation cost). Frustration arises and numerous sub-optimal configurations of source nodes emerge, resembling typical glassy behavior. When the installation cost changes, regimes with different clustering patterns of resource consuming nodes are observed, resembling the Devil's staircase observed in the circle map and other dynamical systems [14].

We analyze the problem through the cavity approach [1, 2], which is a tool developed in the context of spin glasses. In contrast to problems with discrete variables, the cavity fields in the present problem are in the form of functions. Such functional fields result in recursions of functions, complicating the analyses. We report here how these functional recursions are converted to simple recursions of probabilities, which allow us to derive the emergence condition for typical glassy behavior. The underlying physical picture of this model has some resemblance with the random field Ising model [15, 16, 17], the vertex cover problems [4, 5], and the Bethe lattice glass model [18, 19, 20], but it also has its own distinctive features. We refer interested readers to [12] for more details about these connections. More details of our study can be found in [13].

2. Model Formulation

We consider a network of N nodes, labelled $i = 1 \ldots N$. Each node i is connected randomly to a set $\mathcal{L}_i$ of K neighbors. Each node i is either a *surplus* and *deficient* node of resources, as specified by its *capacity* Λ_i; positive and negative values of Λ_i correspond to excess and shortage of resources respectively. The capacities Λ_i are randomly drawn from a distribution $\rho(\Lambda_i)$. With network applications in mind, we consider a bimodal distribution in which $\Lambda_i = A(\gg 1)$ with probability ϕ_s, and $\Lambda_i = -1$ with probability $\phi_d \equiv 1 - \phi_s$. Surplus nodes are inherently *source nodes*, providing resources to other nodes. By default, the deficient nodes are *consumer nodes* receiving resources. However, to minimize cost functions that include transportation costs, it is often desirable to convert some deficient nodes into source nodes as well. Hence in general, the task is to optimally locate these *extra* source nodes.

Denoting as $y_{ij} \equiv -y_{ji}$ the current of resources from node j to node i, we consider the minimization of the cost function

$$E = \frac{u^2}{2} \sum_i \Theta(-\xi_i) + \sum_{(ij)} \frac{y_{ij}^2}{2}. \tag{1}$$

$\xi_i \equiv \Lambda_i + \sum_{j \in \mathcal{L}_i} y_{ij}$ is the final resource of node i, and $\Theta(x) = 1$ when $x > 0$, and 0 otherwise. The link connecting nodes i and j is denoted as (ij). The first term corresponds to the installation cost of an extra source node at deficient node i, since the condition $\Theta(-\xi_i) = 1$ indicates that the deficient node needs to be converted to a source node to satisfy the demand on resources from its neighbors. The most unique feature of this cost function is the discontinuity in its magnitude when the final resource changes from zero to negative, representing various installation overheads. In a more general context, one may envisage installation costs whose magnitudes depend on the shortages of resource $-\xi_i$, which is beyond the scope of the present study. As we shall see, the step-like installation cost already generates an extremely rich picture of network behavior.

Note that the cost function in Eq. (1) is identical to that in [12], but the present setting is far more relevant to network applications. The previous results in [12] have direct correspondence to the present case if the unsatisfied nodes are considered to be the extra source nodes. For example, the single-sat regime studied in [12] corresponds to the case that each consumer is surrounded by source nodes, since the installation cost is low compared with the transportation cost. When the installation cost is gradually raised, resource provision is achieved with less source nodes, but optimization requires the consumers to be located in clusters surrounded by source nodes, forming the clusters observed in [12].

3. Major Simulation Results and the Consumer Clusters

We first look for phase transitions in the model by numerical simulations. To formulate an algorithm, we introduce a variable $s_i = \pm 1$ for each node i and restrict its resources by $s_i(\Lambda_i + \sum_{j\in\mathcal{L}_i} y_{ij}) \geq 0$. Introducing Lagrange multipliers μ_i for the resource constraint, we minimize the Lagrangian

$$L = \frac{u^2}{2}\sum_i \frac{s_i+1}{2} + \sum_{(ij)}\frac{y_{ij}^2}{2} + \sum_i \mu_i s_i \xi_i \qquad (2)$$

with the Kühn-Tucker conditions $\mu_i s_i \xi_i = 0$ and $\mu_i \leq 0$. Optimizing L with respect to y_{ij}, one obtains

$$y_{ij} = \mu_j s_j - \mu_i s_i, \qquad (3)$$

$$\mu_i = \min\left[0, \frac{1}{s_i K}\left(\Lambda_i + \sum_{j\in\mathcal{L}_i} \mu_j s_j\right)\right]. \qquad (4)$$

Given a particular set of $\{s_i\}$, we iterate these equations to find the corresponding set of $\{\mu_i\}$. The set of optimal $\{s_i\}$ is found by an approach similar to the the *GSAT* algorithm [21], by comparing the Lagrangian in Eq. (2) for each choice of $\{s_i\}$. In each step of this algorithm, a cluster of N_{flip} nodes is randomly selected. The network energies involving the different configurations of this cluster are compared, and the cluster configuration is updated to the one that yields the lowest network energy.

As shown in Fig. 1 for $\phi_d = 0.9$ and $K = 3$, two phases can be identified: (1) *all-source* phase for $u^{-1} \geq \sqrt{3}$, in which all nodes are assigned to be source nodes due to the very high transportation cost; (2) *partial-source* phase for $0 < u^{-1} < \sqrt{3}$, in which only some nodes are assigned to be source nodes. (In [12] we also identified a phase transition $u^{-1} = 0$ to an all-consumer phase.)

The fraction of source nodes is a discontinuous function of u^{-1}, showing abrupt jumps at threshold values of u^{-1}. The step size of the curve decreases as u^{-1} increases, and gradually becomes unresolvable by the numerical experiments. This resembles the *Devil's staircase* observed in the circle map and other dynamical systems [14]. These threshold values of u^{-1} mark the positions at which certain configurations of the source and consumer nodes become energetically stable. Similar features are observed in RFIM due to the formation of ferromagnetic clusters resultant from the competition between the strengths of couplings and random fields [16, 17].

Measuring the average maximum cluster size of the consumers in the samples, we observe abrupt jumps of the cluster size at the same threshold values. This indicates that new types of clusters are formed at each jump, as sketched in the top of Fig. 1. The observed threshold values can be calculated by considering the energies of consumer clusters surrounded by source nodes as shown in Fig. 2, obtained by the minimization of Eq. (1). By comparing the energy of different configurations, we have

$$\mathcal{E}_0 \geq \mathcal{E}_1 \quad \text{(singlet)},$$

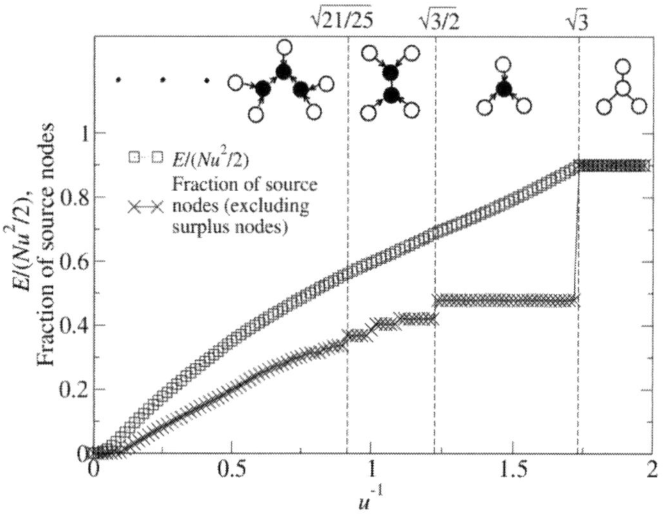

Figure 1. Simulation results of average energy per node and the fraction of consumer nodes. Parameters: $K = 3$, $\phi_d = 0.9$, $N = 100$, 100 samples, $1000N$ steps and $N_{flip} = 4$. New clusters formed on increasing u^{-1} are sketched at the top, with filled and unfilled circles representing consumer and source nodes respectively.

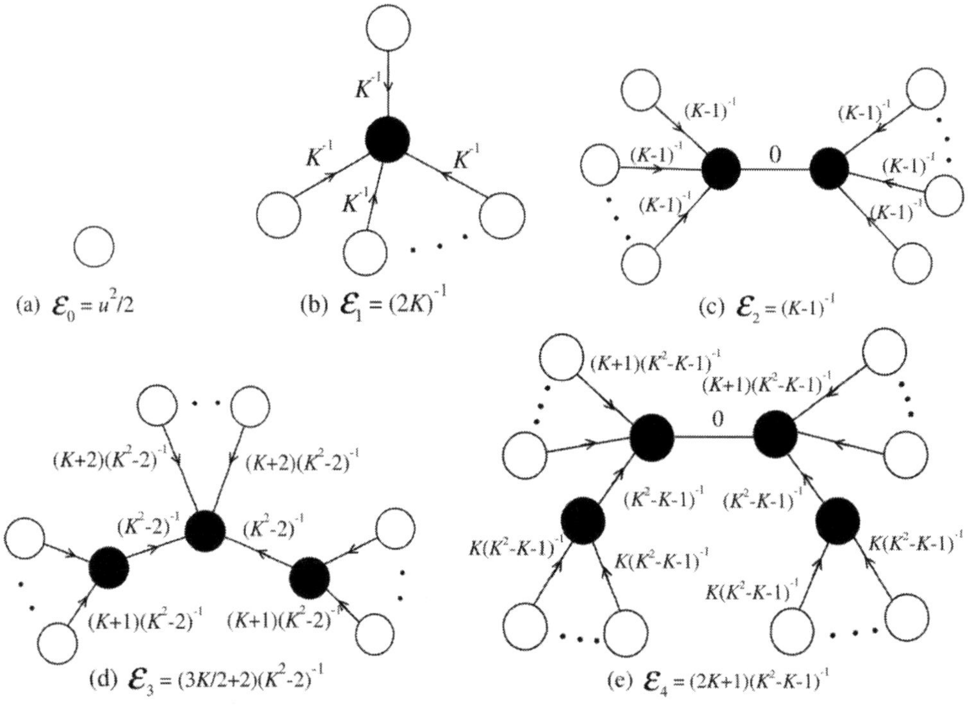

Figure 2. Cluster of (a) source node, (b) singlet consumer, (c) doublet consumers, (d) triplet consumers and (e) quadruplet consumers.

$$\begin{aligned}\mathcal{E}_0 + \mathcal{E}_1 &\geq \mathcal{E}_2 \quad \text{(doublet)},\\ \mathcal{E}_0 + 2\mathcal{E}_1 &\geq \mathcal{E}_3 \quad \text{(triplet)},\\ \mathcal{E}_0 + \mathcal{E}_1 + \mathcal{E}_2 &\geq \mathcal{E}_4 \quad \text{(quadruplet)},\end{aligned}$$

resulting in the threshold values

$$\begin{aligned} u^{-1} &\leq \sqrt{K} \quad \text{(singlet)},\\ u^{-1} &\leq \sqrt{\frac{K(K-1)}{K+1}} \quad \text{(doublet)},\\ u^{-1} &\leq \sqrt{\frac{K(K^2-2)}{(K+2)^2}} \quad \text{(triplet)},\\ u^{-1} &\leq \sqrt{\frac{K(K-1)(K^2-K-1)}{(K+1)(K^2+K-1)}} \quad \text{(quadruplet)}.\end{aligned}$$

These results agree with those obtained through the cavity approach in Section 4. Further analyses of larger consumer clusters lead to additional regimes as u^{-1} increases. These threshold values are closer to one another when u^{-1} increases, leading to a decrease of step size in the fraction of source nodes in Fig. 1, resembling the *Devil's staircase* [14].

The threshold values are shown in Fig. 1 for $K = 3$. We call the regime $\sqrt{3/2} < u^{-1} < \sqrt{3}$ with isolated consumer the *singlet* regime, and $\sqrt{21/25} < u^{-1} < \sqrt{3/2}$ the *doublet* regime.

4. The Cavity Approach

We start the analysis in the simple case with only deficient nodes in the networks (i.e. $\Lambda_i = -1$ for all i). Since the network has a tree-like structure locally, we carry out the analysis using the Bethe approximation. As shown in Fig. 3, we identify node i to be the ancestor of node j, and nodes labelled by k to be the descendents of j. We call this a sub-tree terminated at vertex node j in the absence of i. A new sub-tree is constructed by connecting $K-1$ sub-trees to a newly added vertex. We describe the physical properties of a node in the absence of its ancestor, leading to a *cavity* on the node. Energy functions of the sub-trees are passed from the vertices of the sub-trees to the next layer, resulting in their recursion relations. We note the equivalence of this cavity approach with message passing algorithms [22], in which a message passed from a vertex describes the conditional probability of the vertex states in the absence of the receiving node. In the cavity approach for discrete variables [3, 23], these messages are usually the local fields in the absence of the ancestors of the vertices, known as the cavity fields. Here, since the variables in the cost function of Eq. (1) are continuous, the messages become energy functions [9, 10]. Let $E_j(y_{ij})$ be the energy of the tree terminated at a link from vertex j to its ancestor i, when a current y_{ij} is drawn from j to i. One can express $E_j(y_{ij})$ in terms of the energies of its descendents $k = 1, \ldots, K-1$,

$$E_j(y_{ij}) = \min_{\{y_{jk}\}} \left[\sum_{k=1}^{K-1} E_k(y_{jk}) + \frac{u^2}{2}\Theta(-\xi_j(y_{ij})) + \frac{y_{ij}^2}{2} \right], \tag{5}$$

where $\xi_j(y_{ij}) = \Lambda_j + \sum_k y_{jk} - y_{ij}$. We call $E_k(y_{jk})$ the *cavity energy functions* which is sent from the vertices to the next layer as shown in Fig. 3.

Due to the quadratic form of transportation cost in Eq. (1), we have demonstrated that the energy $E_k(y_{jk})$ is a continuous piecewise-quadratic function [12]. Each segment of the function has the quadratic form

$$f_{n_k}^k(y_{jk}) = a_{n_k}^k (y_{jk} - \tilde{y}_{n_k}^k)^2 + d_{n_k}^k, \tag{6}$$

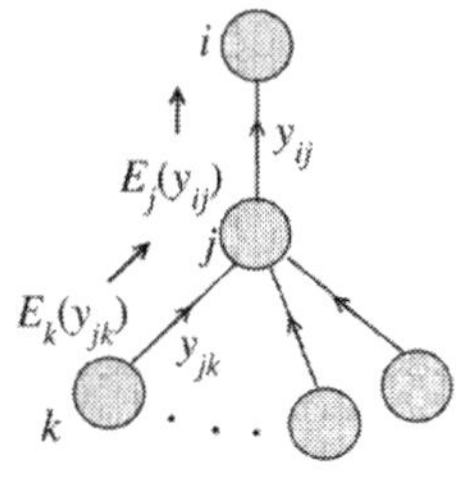

Figure 3. The tree-like assumption in cavity approaches. Node i and nodes k are respectively the ancestor and descendents of node j. y_{ij} denotes the flow of resources from j to i. Negative y_{ij} and y_{jk} correspond to flows in the opposite direction.

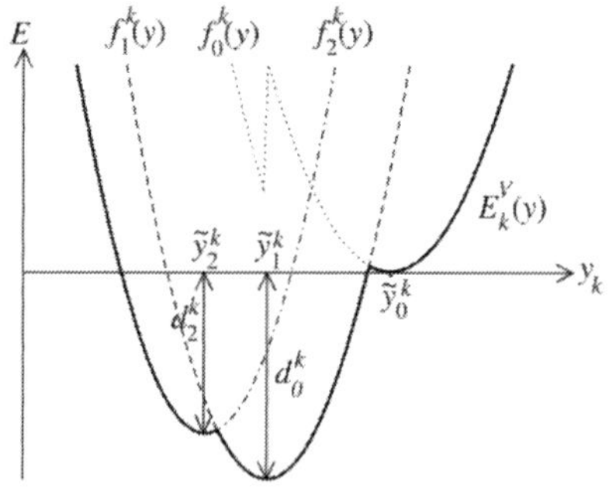

Figure 4. An example of $E_k^V(y)$ composed of three quadratic functions $f_{n_k}^k(y)$ labelled by $n_k = 0, 1, 2$. Each composite function is characterized by its minimum position $\tilde{y}_n^k$, minimum value d_n^k, and curvature a_n^k)

Descendents	$\tilde{y}^j_{n_j}$	$a^j_{n_j}$	$d^j_{n_j}$	n_j
$n_k = 0,\ n_l = 0$	$-\frac{1}{3}$	$\frac{3}{4}$	$\frac{1}{6} + d_0^k + d_0^l$	1
$n_k = 1,\ n_l = 0$	$-\frac{1}{2}$	$\frac{4}{5}$	$\frac{1}{3} + d_1^k + d_0^l$	2

Table 1. Generation of the composite functions labelled by $n_j = 1$ and 2 from descendents k and l for $K = 3$. $\Lambda_i = -1$ for deficient nodes.

where $n_k = 0, 1, 2, \ldots$ is the label of the n_k-th *composite function* of $E_k(y_{jk})$, and is characterized by the coefficients $a_{n_k}^k$, $\tilde{y}_{n_k}^k$ and $d_{n_k}^k$. An example of the cavity energy function $E_k(y_{jk})$ is shown in Fig. 4, composing of the three quadratic functions labelled by $n_k = 0, 1, 2$. The composite function with $n_k = 0$ is given by $\tilde{y}_0^k = 0$, $a_0^k = 1/2$, and corresponds to the case of resource provision. Other composite functions can be generated by plugging the composite functions into Eq. (5) recursively. The first two resultant composite functions, for the case $K = 3$ and obtained with the constraint $y_j = \Lambda_j + y_k + y_l$ active, are given in Table 1.

To determine the composite functions selected by the minimization in Eq. (5), we also need to compare the functions in Table 1 with the composite function in case of resource provision, labelled by $n_j = 0$ and obtained from Eq. (5) with $\tilde{y}_0^j = 0$, $a_0^j = 1/2$ and $d_0^j = u^2/2 + d_{\min}^k + d_{\min}^l$ where $d_{\min}^k = \min_{\{n_k\}} d_{n_k}^k$. (For negative Λ, the composite function with $y_j < \Lambda_j + y_k + y_l$ does not contribute.) Since the network behavior only depends on the energy differences among the composite functions, we consider recursion relations for the energy differences $\epsilon_{n_j}^j \equiv d_{n_j}^j - d_0^j$. In particular, in the singlet regime where the composite functions with $n_j = 0$ and 1 have the

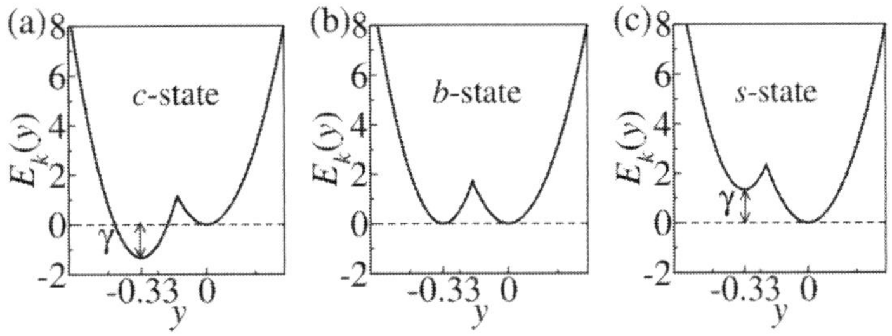

Figure 5. The closed set of $E_k(y)$ at $K = 3$, $u^{-1} = -5/3$ with $\gamma = u^2/2 - 1/2c$, corresponding to (a) the c-state E_c^V, (b) the b-state E_b^V and (c) the s-state E_s^V.

lowest energies, we have $\epsilon_0^j = 0$,

$$\epsilon_1^j = -\gamma - \min(0, \epsilon_1^k) - \min(0, \epsilon_1^l), \tag{7}$$

where $\gamma \equiv u^2/2 - 1/6$.

In general, the iteration of Eq. (5) results in a distribution of the cavity energy functions $E_k(y)$. However, in the singlet regime $\sqrt{3/2} \le u^{-1} < \sqrt{3}$ for $K = 3$, there are 3 functional forms of $E_k(y)$ forming a closed set of solutions to the recursion relation in Eq. (5), each composing of the composite functions with $n_k = 0$ and 1 only. Indeed, since $\gamma \ge 0$ in this regime, the values $\epsilon_1^k = -\gamma, 0, \gamma$ form a closed set under the iteration of Eq. (7). Furthermore, higher composite functions do not contribute in this regime.

The three functions with $\epsilon_1^k = \gamma, -\gamma$ and 0 are referred to as the s-, c- and b-states, as shown in Fig. 5(a-c). Respectively, they correspond to states favoring resource provision (as a source), consumption, and bistability, and have absolute minima at $y = 0$, $y = -1/3$, and both $y = 0$ and $-1/3$. Furthermore, numerical iterations of Eq. (5) starting with random $E_k(y)$ show that this set of solutions is stable. For general values of K, the closed set consists of more states having absolute minima at $y = 0$, and local minima at different energies at $y = -K^{-1}$, similar to the c-state in Fig. 5.

Considering the recursion relations among the K states, we find that all states with absolute minima at $y_k = 0$ (i.e., $\epsilon_1^k \ge 0$) behave in the same way. Thus, we denote the s- and b-states by the S-state, and the c-state by the C-state. Their recursion relations are given by

$$\begin{aligned} &S + \cdots + S \to C, \\ &\text{all other combinations} \to S. \end{aligned} \tag{8}$$

The closure property of the K states greatly simplifies the calculation. The physical states of a node and its links can be obtained by feeding K vertices to a central node. The recursion rules imply that optimization is achieved by the close packing of consumers in the network, with the constraint that they do not form clusters. This reduces the problem to the vertex cover problem [4, 5, 6]. Since there is at most one consumer at the end of each link, the maximization of the number of consumers is equivalent to the minimization of the covered set size in the vertex cover problem. Alternatively, the model can be considered as the close packing limit of a lattice glass model [18]. Both the vertex cover problem and the lattice glass model exhibit glassy behavior, and both phases with single ground state and multiple metastable states are found therein, corresponding to the computationally easy and hard phases respectively.

5. Networks with Both Surplus and Deficient Nodes

Next, we consider the cases where both surplus and deficient nodes are present in the networks. The recursion relations for the deficient nodes follow the close-packing rule of Eq. (8), whereas

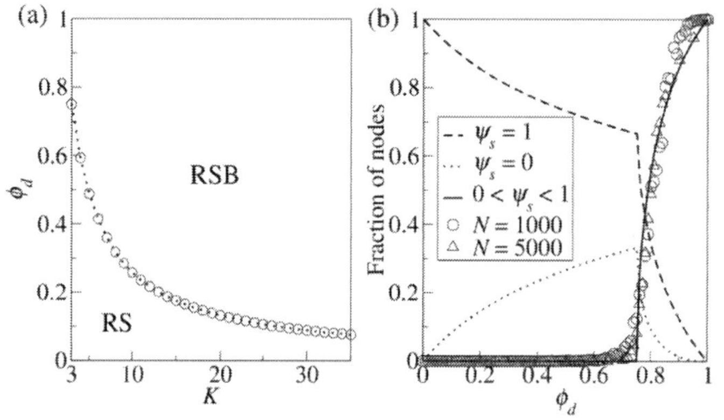

Figure 6. (a) The phase diagram. (b) The fraction of nodes with $\psi_s^{k\to j} = 0$, 1, and $0 < \psi_s^{k\to j} < 1$ for $K = 3$. The symbols represent the simulated fraction of non-converging BP messages.

the surplus nodes are always in the S-state. We let $\psi_s^{k\to j}$ be the probability that node k is in the S-state with the absence of node j. Its recursion relation can be written as

$$\psi_s^{j\to i} \quad = \delta_{\Lambda_j,-1}\Big(1 - \prod_{k=1}^{K-1} \psi_s^{k\to j}\Big), + \delta_{\Lambda_j,A}. \tag{9}$$

This implies that the probability η_s of finding a cavity source node can be obtained from the equation $\eta_s = \phi_s + \phi_d(1 - \eta_s^{K-1})$. At low ϕ_d, a stable fixed point solution with all $\psi^{k\to j} = 0$ or 1 exists. From the perspectives of the replica analysis [1], this corresponds to the so-called replica symmetric (RS) ansatz, in which the network behavior is dominated by a single ground state. In terms of algorithms, the optimal network states in this regime can be obtained by the BP algorithm, initializing the messages to 0 or 1.

The stability of the RS solution can be studied by considering the propagation of fluctuations $\langle(\delta\psi_s^{k\to j})^2\rangle$ under the recursion relation Eq. (9) [25]. This leads to the Almeida-Thouless (AT) stability condition, which reads

$$(K-1)\phi_d\langle\psi_s^2\rangle^{K-2} \le 1. \tag{10}$$

In the RS regime, $\langle\psi_s^2\rangle = \langle\psi_s\rangle = \eta_s$ for $\psi_s^{k\to j} = 0$ or 1. Simple algebra leads us to the AT line

$$\phi_d^{\mathrm{AT}} = \frac{K^{K-2}}{(K-1)^{K-1}} \tag{11}$$

which separates the RS and the RSB phases in the space of K and ϕ_d as shown in Fig. 6(a).

When $\phi_d > \phi_d^{\mathrm{AT}}$, nodes with $0 < \psi_s^{k\to j} < 1$ exist, as shown in Fig. 6(b). The fraction of $\psi_s^{j\to i}$ not converging to 0 and 1 can be deduced from Eq. (9), and simulation results of non-converging BP messages have an excellent agreement with it, as shown in Fig. 6(b). This implies that nodes with free states start to exist, analogous to the free states in the non-trivial solution of vertex cover [4, 5] and graph coloring [23, 26]. This characterizes the region with multiple solutions of optimal source locations, and the network behavior is described by the so-called full replica symmetry-breaking (RSB) ansatz, which assumes an infinite hierarchy of states. We have considered the one-step RSB approximation, assuming that the number $\Gamma(E)$ of states with energy E is distributed according to $\Gamma(E) \propto \exp[N\Sigma(\mathsf{e})]$, where $\mathsf{e} = E/N$ [12]. $\Sigma(\mathsf{e})$ is referred

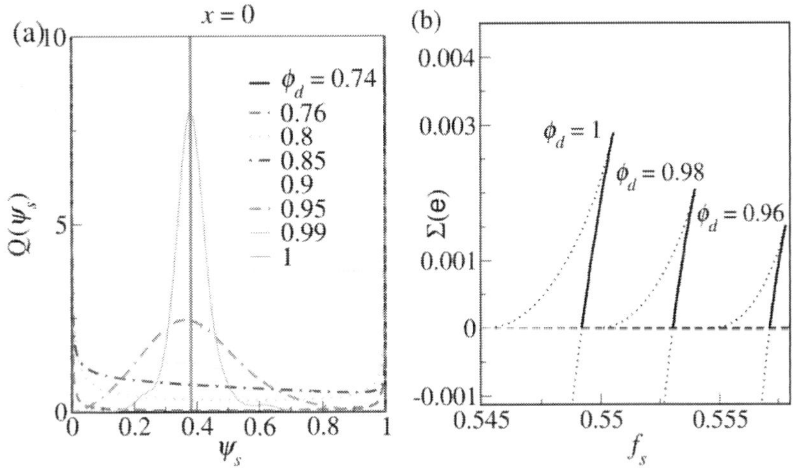

Figure 7. (a) The distribution $Q(\psi_s)$ in the singlet regime for $K = 3$ and $x = 0$. (b) The 1RSB complexity $\Sigma(\mathsf{e})$ for $K = 3$ and $\phi_d = 0.96, 0.98, 1$.

to as the *complexity* or the *configurational entropy* [24]. For small changes in the average energy, we can write $\Sigma(\mathsf{e}) = x(\mathsf{e} - \mathsf{e}^R)$, where e^R is the reference energy. The recursion relation Eq. (9) is then replaced by

$$\begin{aligned} \psi_s^{j\to i} &= \delta_{\Lambda_j,-1}\left(1 - \frac{1}{\mathcal{Z}^{j\to i}}\prod_{k\in\mathcal{L}_j\setminus\{i\}}\psi_s^{k\to j}\right) + \delta_{\Lambda_j,A}, \\ \mathcal{Z}^{j\to i} &= \prod_{k\in\mathcal{L}_j\setminus\{i\}}\left[e^{x\gamma} - (e^{x\gamma}-1)\psi_s^{k\to j}\right]. \end{aligned} \quad (12)$$

The stable distribution $Q(\psi_s)$ of ψ_s for $K = 3$ and $x = 0$ is shown in Fig. 7(a). When $\phi_d \le \phi_d^{\rm AT}$, there are no fractional components of ψ_s, and $Q(\psi_s) = \langle\psi_s\rangle\delta(\psi_s - 1) + (1 - \langle\psi_s\rangle)\delta(\psi_s)$. When $\phi_d > \phi_d^{\rm AT}$, non-zero components of $0 < \psi_s < 1$ exist. When $\phi_d = 1$, there is no disorder in the capacities. All vertices are identical and $Q(\psi_s) = \delta(\psi_s - \eta_s)$. This means that among the different states of the system, all vertices are equally probable to be in the S state with probability η_s.

The complexity $\Sigma(\mathsf{e})$ is shown in Fig. 7(b). The abscissa f_s is the fraction of source nodes related to the average energy e through $f_s = (\mathsf{e} - 1/2K)/\gamma$. The physical segments of Σ are indicated by the solid segments with $\Sigma \ge 0$ in Fig. 7(b). The energy e_s with $\Sigma(\mathsf{e}_s) = 0$ corresponds to the lowest energy among the states with non-vanishing complexity, which is considered as the typical ground state in the picture of 1RSB. The typical behaviors of the system is thus described by Eq. (12) with the value of x satisfying $\Sigma(\mathsf{e}_s) = 0$. The energy e_d with the highest complexity is believed to be the point where search algorithms get trapped, giving rise to dynamical transitions. (However, recent work on the coloring problem shows that the efficacy of search algorithms is not affected by the dynamical transition [27]. We leave this issue for future studies.)

We have verified that the 1RSB ansatz yields better energy estimates than RS. We have done simulations at $\phi_d = 1$ using a decimated BP algorithm [12]. The fractions f_s of source nodes are found to be 0.551 (simulation), 0.551 (1RSB at $\mathsf{e} = \mathsf{e}_d$), 0.549 (1RSB at $\mathsf{e} = \mathsf{e}_s$) and 0.545 (RS) at $K = 3$ and $N = 5000$.

6. Conclusion

We have studied the problem of optimal source location on transportation networks with step-like source node installation cost. As the installation cost increases, the number of source nodes decreases. Regimes characterized by discrete fractions of source nodes emerge and resemble the Devil's staircase, revealing the optimal flow pattern of resources supplying different consumer clusters. In the singlet regime, which has a correspondence with the vertex cover problem and the lattice glass models, a closed set of only a few cavity energy functions is able to describe the physics of the system. In the case with surplus and deficient nodes, an increase in the fraction of deficient nodes induces a glassy phase transition incorporating the picture of numerous sub-optimal solutions. The one-step replica symmetry-breaking ansatz yields better agreement with simulations than the replica symmetric ansatz.

Acknowledgments

We acknowledge D. Saad for discussions. This work is supported by the Research Grants Council of Hong Kong (grant numbers HKUST 603607 and HKUST 604008).

References

[1] Mézard M, Parisi G and Virasoro M A 1987 *Spin Glass Theory and Beyond* (World Scientific, Singapore)
[2] Nishimori H 2001 *Statistical Physics of Spin Glasses and Information Processing* (Oxford University Press, Oxford, UK)
[3] Mézard M and Zecchina R 2002 *Phys. Rev.* E **66** 056126
[4] Weigt M and Hartmann A K 2000 *Phys. Rev. Lett.* **84** 6118
[5] Weigt M and Hartmann A K 2001 *Phys. Rev.* E **63** 056127
[6] Zhou H 2005 *Phys. Rev. Lett.* **94** 217203
[7] Kopparapu C 2002 *Load Balancing Servers, Firewalls and Caches* (Wiley, New York, NY)
[8] Rardin R L 1998 *Optimization in Operations Research* (Prentice Hall, Englewood Cliffs, NJ)
[9] Wong K Y M and Saad D 2006 *Phys. Rev.* E **74** 010104
[10] Wong K Y M and Saad D 2007 *Phys. Rev.* E **76** 011115
[11] Yeung C H and Wong K Y M 2009 *J. Stat. Mech* P03029
[12] Yeung C H and Wong K Y M 2009 *Phys. Rev.* E **80** 021102
[13] Yeung C H and Wong K Y M 2010 arXiv:1001.4060, submitted to *J. Stat. Mech.*
[14] Devaney R L 1989 *An Introduction to Chaotic Dynamical Systems* (Addison-Wesley, Redwood City, CA)
[15] Imry Y and Ma S K 1975 *Phys. Rev. Lett.* **50** 1399
[16] Bruinsma R and Aeppli G 1983 *Phys. Rev. Lett.* **50** 1494
[17] Bruinsma R 1984 *Phys. Rev.* B **30** 289
[18] Biroli G and Mézard M 2002 *Phys. Rev. Lett.* **88** 025501
[19] Rivoire O, Biroli G, Martin O C and Mézard M 2004 *Eur. Phys. J.* B **37** 55
[20] Krzakala F, Tarzia M and Zdeborová L 2008 *Phys. Rev. Lett.* **101** 165702
[21] Selman B, Kautz H and Cohen B 1996 *DIMACS Series in Discrete Mathematics and Theoretical Computer Science* **26** 521
[22] Mackay D J C 2003 *Information Theory, Inference, and Learning Algorithms* (Cambridge University Press, Cambridge, UK)
[23] Braunstein A, Mulet R, Pagnani A, Weigt M and Zecchina R 2003 *Phys. Rev.* E **68** 036702
[24] Mézard M and Parisi G 2001 *Eur. Phys. J.* B **20** 217
[25] Thouless D J 1986 *Phys. Rev. Lett.* **56** 1082
[26] Mulet R, Pagnani A, Weigt M and Zecchina R 2002 *Phys. Rev. Lett.* **89** 268701
[27] Zdeborová L and Krzakala F 2007 *Phys. Rev.* E **76** 031131

International Workshop on Statistical-Mechanical Informatics 2010 IOP Publishing
Journal of Physics: Conference Series **233** (2010) 012010 doi:10.1088/1742-6596/233/1/012010

Deterministic flows of order-parameters in stochastic processes of quantum Monte Carlo method

Jun-ichi Inoue

Complex Systems Engineering, Graduate School of Information Science and Technology, Hokkaido University, N14-W9, Kita-ku, Sapporo 060-0814, Japan

E-mail: `j_inoue@complex.eng.hokudai.ac.jp`

Abstract. In terms of the stochastic process of quantum-mechanical version of Markov chain Monte Carlo method (the MCMC), we analytically derive macroscopically deterministic flow equations of order parameters such as spontaneous magnetization in infinite-range ($d(=\infty)$-dimensional) quantum spin systems. By means of the Trotter decomposition, we consider the transition probability of Glauber-type dynamics of microscopic states for the corresponding $(d+1)$-dimensional classical system. Under the static approximation, differential equations with respect to macroscopic order parameters are explicitly obtained from the master equation that describes the microscopic-law. In the steady state, we show that the equations are identical to the saddle point equations for the equilibrium state of the same system. The equation for the dynamical Ising model is recovered in the classical limit. We also check the validity of the static approximation by making use of computer simulations for finite size systems and discuss several possible extensions of our approach to disordered spin systems for statistical-mechanical informatics. Especially, we shall use our procedure to evaluate the decoding process of Bayesian image restoration. With the assistance of the concept of dynamical replica theory (the DRT), we derive the zero-temperature flow equation of image restoration measure showing some 'non-monotonic' behaviour in its time evolution.

1. Introduction

In Bayesian statistics and statistical inference within the framework, to evaluate the expectation of microscopic labels (such as 'pixels' or 'bits') over the posterior distribution of these microscopic states or to calculate the marginal probability is very important procedure to solve various problems in the research field of 'massive' probabilistic information processing [1]. For most of the cases, it is very hard for us to carry out the task except for several limited cases of solvable probabilistic models. Recently, to overcome the difficulty, deterministic approaches based on mean-field approximation including the belief-propagation succeeded in dealing with the problems efficiently [2, 3].

On the other hand, in a lot of research fields, the Markov chain Monte Carlo method (the MCMC) has been widely used to calculate these expectations or to make marginal distributions by sampling the important states that contribute effectively to the macroscopic quantities [4]. Generally speaking, the MCMC takes a long time to wait until the Markovian stochastic process starts to generate the microscopic states from well-approximated posterior distribution. Especially, for some classes of probabilistic models which are categorized in the so-called *spin glasses* [5, 6, 7], the time consuming is sometimes very serious to make attempt to proceed the desired information processing within a realistic time. However, even for such cases, various improvements based on several important concepts have been proposed and succeeded in carrying out the numerical calculations [8, 9]. From the view point of statistical physics, transitions between microscopic states are controlled by a specific hyper-parameter, namely, 'temperature' of the system and by cooling the temperature slowly enough during the Markovian stochastic process, one can get the lowest energy states efficiently. This type of optimization tool based on 'thermal fluctuation' is referred to as *simulated annealing* [10, 11].

Recently, the simulated annealing has been extended to the quantum-mechanical version. This new type of simulated annealing called as *quantum annealing* [12, 13, 14, 15, 16] is based on the adiabatic theorem of the quantum system that evolves according to Schrödinger equations. To use the quantum annealing, or more generally, to utilize the quantum fluctuation for combinatorial optimization problems or massive information processing dealing with a huge number of particles such as image restoration or error-correcting codes, the approach by solving the Schrödinger equations is apparently limited and we should look for different ways to simulate the quantum systems.

As the most effective and efficient way, the quantum Monte Carlo method [17] was established. The method is based on the following Suzuki-Trotter decomposition for non-commutative two operators $\mathcal{A}$ and $\mathcal{B}$:

$$\mathrm{tr}\exp\left(\mathcal{A}+\mathcal{B}\right) = \lim_{M\to\infty} \mathrm{tr}\left(\exp\left(\frac{\mathcal{A}}{M}\right)\exp\left(\frac{\mathcal{B}}{M}\right)\right)^{M} \tag{1}$$

Namely, the d-dimensional quantum system is mapped to the corresponding $(d+1)$-dimensional classical spin systems. This approach is very powerful and a lot of researches have succeeded in exploring the quantum phases in strongly correlated quantum systems. However, when we simulate the quantum system at zero temperature in which quantum effect is essential, we encounter some technical difficulties although several sophisticated algorithms were proposed [18]. From the view point of information science, it is very informative for us to evaluate the process of information processing and if one seeks to utilize the quantum fluctuation to solve the problems, we should use the 'zero-temperature dynamics'. For this purpose, it seems that we need some tractable 'bench mark tests' to investigate the 'dynamical process' of information processing such as quantum annealing by using quantum fluctuation at zero temperature.

In classical system, Coolen and Ruijgrok [19] proposed a way to derive the differential equations with respect to order-parameters of the system from the microscopic master equations for the so-called Hopfield model [20] as an associative memory in which a finite number of patterns are embedded. The procedure was extended by Coolen and Sherrington [21], Coolen, Laughton and Sherrington [22] to more complicated spin systems categorized in the infinite-range (or mean-field) models including the Sherrington-Kirkpatrick spin glasses [23]. The so-called dynamical replica theory (DRT) was now established as a strong approach to investigate the dynamics in the classical disordered spin systems . On the considering the matter, it seems to be very important for us to extend their approach to quantum systems evolving stochastically according to the quantum Monte Carlo dynamics.

In this paper, in terms of the stochastic process of quantum-mechanical version of Markov chain Monte Carlo method, we analytically derive macroscopically deterministic flow equations of order parameters such as spontaneous magnetization in infinite-range ($d(=\infty)$-dimensional) quantum spin systems. By means of the Trotter decomposition, we consider the transition probability of Glauber-type dynamics of microscopic states for the corresponding $(d+1)$-dimensional classical system. Under the static approximation, differential equations with respect to macroscopic order parameters are explicitly obtained from the master equation that describes the microscopic-law. In the steady state, we show that the equations are identical to the saddle point equations (equations of states) for the equilibrium state of the same system. We easily find that the equation for the dynamical Ising model is recovered in the classical limit. We also check the validity of the static approximation by computer simulations for finite size systems and discuss several possible extensions of our approach to disordered spin systems for statistical-mechanical informatics [24]. Especially, we shall use our procedure to evaluate the decoding process of Bayesian image restoration [25, 26, 27, 28, 29]. With the assistance of the concept of dynamical replica theory (the DRT) [21, 22], we derive the zero-temperature flow equation of image restoration measure, namely, overlap function between a given original image and the degraded one, showing some 'non-monotonic' behaviour in its time evolution.

This paper is organized as follows. In Section 2, we explain our formulation to describe the deterministic flow equation from master equation for a simplest quantum spin system. In Section 3, we apply our approach to evaluate the decoding process of Bayesian image restoration. Some 'non-monotonic' behaviour in time evolution of image restoration measure is observed by the analysis of flow equations. The last section contains some remarks.

2. The model system and formulation

In this section, we derive the differential equations with respect to several order parameters for a simplest quantum spin system, namely, a class of the infinite range transverse Ising model [30, 31] described by the following Hamiltonian:

$$H = -\frac{1}{N}\sum_{i,j=1}^{N} J_{ij}\sigma_i^z\sigma_j^z - h\sum_{i=1}^{N}\tau_i\sigma_i^z - \Gamma\sum_{i=1}^{N}\sigma_i^x. \quad (2)$$

where σ_i^z and σ_i^x denote the Pauli matrices given by

$$\sigma_i^z = \begin{pmatrix} 1 & 0 \\ 0 & -1 \end{pmatrix},\ \sigma_i^x = \begin{pmatrix} 0 & 1 \\ 1 & 0 \end{pmatrix}.$$

It should be noticed that from the view point of Bayesian statistics, the above Hamiltonian corresponds to the logarithm of the posterior distribution. For various choices of parameters $\{J_{ij}\}, h$ and $\{\tau\}$, one can model the problem of information processing appropriately. For instance, for the choice of $J_{ij} = J, h \neq 0$, the above Hamiltonian describes image restoration [26, 28, 29] from a given set of degraded images $\boldsymbol{\tau} = (\tau_1, \cdots, \tau_N)$ by means of Markov random fields. For $J_{ij} \neq 0, h = 0$, it corresponds to the logarithm of the posterior of the Sourlas codes [32, 33] sending the parity check of two-body interactions $\xi_i\xi_j\ \forall(i,j)$ through the binary symmetric channel (BSC). For the Sherrington-Kirkpatrick model [23], we may choose J_{ij} obeying the Gaussian with J_0 mean and $\tilde{J}^2$ variance. Especially, for the choice of the so-called Hebb rule $J_{ij} = \sum_{\mu=1}^{\alpha N}\xi_i^\mu\xi_j^\mu, h = 0$, it becomes energy function of the Hopfield model [20] in which extensive number of patterns (loading rate α) are embedded.

In this section, we focus on the simplest case of $J_{ij} = J > 0, h = 0$ ferromagnetic transverse Ising model [34, 35]. Let us start our argument from the effective Hamiltonian which is decomposed by the Suzuki-Trotter formula (1).

$$\beta H = -\sum_{k=1}^{M}\sum_{i=1}^{N}\beta\phi_i(\boldsymbol{\sigma}_k : \sigma_i(k+1))\sigma_i(k) = -\frac{\beta J}{MN}\sum_{k,ij}\sigma_i(k)\sigma_j(k) - B\sum_{k,i}\sigma_i(k)\sigma_i(k+1) \quad (3)$$

where k means the Trotter index and M denotes the number of the Trotter slices. We also defined the parameter B as $B \equiv (1/2)\log\coth(\beta\Gamma/M)$ and a microscopic spin state on the k-th Trotter slice by $\boldsymbol{\sigma}_k \equiv (\sigma_1(k), \sigma_2(k), \cdots, \sigma_N(k))$, $\sigma_i(k) \in \{+1, -1\}$.

2.1. The Glauber dynamics and its transition probability

In the expression of the effective Hamiltonian (3), $\beta\phi_i(\boldsymbol{\sigma}_k, \sigma_i(k+1))$ is a local field on the cite i in the k-th Trotter slice, which is explicitly given by

$$\beta\phi_i(\boldsymbol{\sigma}_k : \sigma_i(k+1)) = \frac{\beta J}{MN}\sum_j \sigma_j(k) + B\sigma_i(k+1).$$

It should be noticed that the effective Hamiltonian (3) is a classical system under the Trotter decomposition. Therefore, the quantum Monte Carlo dynamics should be described by the Glauber-type stochastic update rule whose transition probability is explicitly written by

$$w_i(\boldsymbol{\sigma}_k) = \frac{1}{2}\left[1 - \sigma_i(k)\tanh(\beta\phi_i(\boldsymbol{\sigma}_k : \sigma(k+1)))\right].$$

Namely, this means that

$$\sigma_i(k) = \begin{cases} +1 & \text{with prob.} \ \frac{\exp[\beta\phi_i(\boldsymbol{\sigma}_k:\sigma_i(k+1))]}{\exp[\beta\phi_i(\boldsymbol{\sigma}_k:\sigma_i(k+1))]+\exp[-\beta\phi_i(\boldsymbol{\sigma}_k:\sigma_i(k+1))]} \\ -1 & \text{with prob.} \ \frac{\exp[-\beta\phi_i(\boldsymbol{\sigma}_k:\sigma_i(k+1))]}{\exp[\beta\phi_i(\boldsymbol{\sigma}_k:\sigma_i(k+1))]+\exp[-\beta\phi_i(\boldsymbol{\sigma}_k:\sigma_i(k+1))]} \end{cases} \tag{4}$$

Obviously, in the classical limit, namely, when B goes to infinity as $\Gamma \to 0$ and $\phi_i(\boldsymbol{\sigma}_k : \sigma_i(k+1)) \sim B\sigma_i(k+1)$, $\sigma_i(k)$ is identical to $\sigma_i(k+1)$ with probability 1. Thus, for small Γ, we have $\boldsymbol{\sigma}_k = \boldsymbol{\sigma}_{k+1}$ with a relatively high probability. In this sense, we can regard the term $B\sigma_i(k+1)$ as an 'external field' from the nearest Trotter slice $k+1$. Here we use the periodic boundary condition for the Trotter direction, that is, $\boldsymbol{\sigma}_1 = \boldsymbol{\sigma}_{M+1}$.

2.2. The master equation

Then, the master equation for the probability of the microscopic states on the k-th Trotter slice $p_t(\boldsymbol{\sigma}_k)$ is written by

$$\frac{dp_t(\boldsymbol{\sigma}_k)}{dt} = \sum_{i=1}^{N}\left[p_t(F_i^{(k)}(\boldsymbol{\sigma}_k))w_i(F_i^{(k)}(\boldsymbol{\sigma})) - p_t(\boldsymbol{\sigma}_k)w_i(\boldsymbol{\sigma}_k)\right] \tag{5}$$

where $p_t(\boldsymbol{\sigma}_k)$ denotes a probability that the system in the k-th Trotter slice is in a microscopic state $\boldsymbol{\sigma}_k$ at time t. We also defined an operator $F_i^{(k)}$ to flip a single spin on the cite i in the k-th Trotter slice as $\sigma_i(k) \to -\sigma_i(k)$.

2.3. From master equation to deterministic flow of order parameter

We next introduce the spontaneous magnetization in the k-th Trotter slice $m_k = N^{-1}\sum_i \sigma_i(k)$ as a relevant order parameter. The probability that the system is described by the magnetization m_k at time t is given in terms of the probability $p_t(\boldsymbol{\sigma}_k)$ for a given realization of the microscopic state as $P_t(m_k) = \sum_{\boldsymbol{\sigma}_k} p_t(\boldsymbol{\sigma}_k)\delta(m_k - m_k(\boldsymbol{\sigma}_k))$. Taking the derivative of this equation with respect to t and substituting (5) into the result, we obtain

$$\begin{aligned}
\frac{dP_t(m_k)}{dt} &= \frac{\partial}{\partial m_k}\sum_{\boldsymbol{\sigma}_k} p_t(\boldsymbol{\sigma}_k)\sum_i \frac{\sigma_i(k)}{N}\left[1 - \sigma_i(k)\tanh(\beta\phi_i(\boldsymbol{\sigma}_k : \sigma_i(k+1)))\right]\delta(m_k - m_k(\boldsymbol{\sigma}_k)) \\
&= \frac{\partial}{\partial m_k}\sum_{\boldsymbol{\sigma}_k} p_t(\boldsymbol{\sigma}_k)\left\{\frac{1}{N}\sum_i \sigma_i(k)\right\}\delta(m_k - m_k(\boldsymbol{\sigma}_k)) \\
&- \frac{\partial}{\partial m_k}\sum_{\boldsymbol{\sigma}_k} p_t(\boldsymbol{\sigma}_k)\frac{1}{N}\sum_i \tanh[\beta\phi_i(\boldsymbol{\sigma}_k : \sigma_i(k+1))]\delta(m_k - m_k(\boldsymbol{\sigma}_k)) \\
&= \frac{\partial}{\partial m_k}\{m_k P_t(m_k)\} - \frac{\partial}{\partial m_k}\left\{P_t(m_k)\frac{1}{N}\sum_i \tanh[\beta\phi_i(\boldsymbol{\sigma}_k : \sigma_i(k+1))]\right\}
\end{aligned} \tag{6}$$

by neglecting $\mathcal{O}(N^{-1})$ term. To continue the derivation of deterministic flow equations of order parameters, we use the following assumption:

$$\lim_{N\to\infty} \frac{1}{N}\sum_i \tanh[\beta\phi_i(\boldsymbol{\sigma}_k : \sigma_i(k+1))] = \langle \tanh[\beta\phi(k)]\rangle_{\backslash\sigma(k)} \quad (7)$$

where we defined effective single site local field $\beta\phi(k) \equiv (\beta J/M)m_k + B\sigma(k+1)$ and the average

$$\begin{aligned}\langle\cdots\rangle_{\backslash\sigma(k)} &\equiv \lim_{M\to\infty}\sum_{\sigma(1)}\cdots\sum_{\sigma(k-1)}\sum_{\sigma(k+1)}\cdots\sum_{\sigma(M)}(\cdots)p(\sigma(1),\cdots,\sigma(k-1),\sigma(k+1),\cdots,\sigma(M)) \\ &= \lim_{M\to\infty}\frac{\sum_{\sigma(1)}\cdots\sum_{\sigma(k-1)}\sum_{\sigma(k+1)}\cdots\sum_{\sigma(M)}(\cdots)\exp[\beta\sum_{l\neq k}^{M}\phi(l)\sigma(l)]}{\sum_{\sigma(1)}\cdots\sum_{\sigma(k-1)}\sum_{\sigma(k+1)}\cdots\sum_{\sigma(M)}\exp[\beta\sum_{l\neq k}^{M}\phi(l)\sigma(l)]}.\end{aligned}$$

Namely, here we shall assume that in the thermodynamical limit $N\to\infty$, the physical quantity in a particular choice (realization) of the Trotter slice, say, the k-th Trotter slice $N^{-1}\sum_i \tanh[\beta\phi_i(\boldsymbol{\sigma}_k : \sigma_i(k+1))]$ is identical to its own average over all possible paths in the imaginary-time axis for both 'past' $\{\sigma(1),\sigma(2),\cdots,\sigma(k-1)\}$ and 'future' $\{\sigma(k+1),\cdots,\sigma(\infty)\}$. The weight of each path is proportional to

$$\exp\left[\beta\sum_{l\neq k}^{M}\phi(l)\sigma(l)\right] = \exp\left[\frac{\beta J}{M}\sum_{l\neq k}m_l\sigma(l) + B\sum_{l\neq k}\sigma(l)\sigma(l+1)\right].$$

In order to grasp the physical meaning of the quantity (7), it might be helpful for us to notice that $\tanh[\beta\phi(k)]$ can be rewritten as

$$\tanh[\beta\phi(k)] = \frac{\sum_{\sigma(k)=\pm1}\sigma(k)\exp[\beta\phi(k)\sigma(k)]}{\sum_{\sigma(k)=\pm1}\exp[\beta\phi(k)\sigma(k)]} = \sum_{\sigma(k)=\pm1}\sigma(k)p(\sigma(k)|\sigma(1),\cdots,\sigma(k-1)).$$

Thus, the quantity $\tanh[\beta\phi(k)]$ means a conditional expectation over the effective single spin $\sigma(k)$ (in the thermodynamic limit $N\to\infty$) for a given 'past' $\{\sigma(1),\cdots,\sigma(k-1)\}$ and it becomes a function of $\sigma(k+1)$. After averaging over the 'past' with the weight $P(\sigma(1),\cdots,\sigma(k))$ and 'future' with the weight $P(\sigma(k+1),\cdots,\sigma(\infty))$, we obtain

$$\begin{aligned}\langle\tanh[\beta\phi(k)]\rangle_{\backslash\sigma(k)} &= \lim_{M\to\infty}\frac{\mathrm{tr}_{\{\sigma\}}\sigma(k)\exp[\beta\sum_{l=1}^{M}\phi(l)\sigma(l)]}{\mathrm{tr}_{\{\sigma\}}\exp[\beta\sum_{l=1}^{M}\phi(l)\sigma(l)]} \\ &= \lim_{M\to\infty}\frac{\mathrm{tr}_{\{\sigma\}}\sigma(k)\exp[\frac{\beta J}{M}\sum_{l=1}^{M}m_l\sigma(l) + B\sum_{l=1}^{M}\sigma(l)\sigma(l+1)]}{\mathrm{tr}_{\{\sigma\}}\exp[\frac{\beta J}{M}\sum_{l=1}^{M}m_l\sigma(l) + B\sum_{l=1}^{M}\sigma(l)\sigma(l+1)]} \equiv \langle\sigma(k)\rangle_{path}\end{aligned}$$

where we defined the summation of all possible paths by $\mathrm{tr}_{\{\sigma\}}(\cdots) \equiv \sum_{\sigma(1)=\pm1}\cdots\sum_{\sigma(M)=\pm1}(\cdots)$. We should notice that in the classical limit $B\to\infty$ $(\Gamma\to 0)$, the weight of the path in the imaginary-time axis is dominated by $\sigma(1)=\cdots=\sigma(M), m_1=\cdots m_k=m$ as we always see it in the path integral approach of quantum mechanics in the limit of the zero Planck constant as $\hbar\to 0$ [36, 37] , namely, $P(\sigma(1),\cdots,\sigma(M)) = \prod_{l=1}^{M}\delta(\sigma(l)-\sigma)$ and we have

$$\lim_{B\to 0\,(\Gamma\to 0)}\langle\sigma(k)\rangle_{path} = \langle\sigma(k)\rangle_{classical\,path} = \frac{\sum_\sigma \sigma\exp(\beta Jm\sigma)}{\sum_\sigma \exp(\beta Jm\sigma)} = \tanh(\beta Jm). \quad (8)$$

Substituting the result into (6) and carrying out the integral with respect to m_k after multiplying the m_k, we immediately obtain the spontaneous magnetization flow for the dynamical Ising model; $dm/dt = -m + \tanh(\beta Jm)$. Near the critical point, it behaves as $dm/dt = -(1-\beta J)m - (\beta Jm)^3/3 + \mathcal{O}(m^5)$. From this equation, we easily find that spontaneous magnetization shows well-known time-dependent behaviour $m(t) \simeq m(0)\,\mathrm{e}^{-t/(1-\beta J)^{-1}}$ around the critical point $\beta \simeq \beta_c = J^{-1}$, and at the critical point, the relaxation time diverges as $(1-\beta_c J)^{-1}$ resulting in $m(t) \simeq t^{-1/2}$ (critical slowing down with dynamical exponent $\nu' = 1/2$).

Therefore quantum fluctuation comes from the finite B ($\Gamma > 0$). For the quantum case, the equation (6) leads to

$$\frac{dP_t(m_k)}{dt} = \frac{\partial}{\partial m_k}\{m_k P_t(m_k)\} - \frac{\partial}{\partial m_k}\{P_t(m_k)\langle\sigma(k)\rangle_{path}\} \tag{9}$$

In order to obtain the deterministic equation of order parameter, we should use the static approximation $m_k = m \ \forall(k)$. Under this assumption and using the inverse procedure of the Suzuki-Trotter decomposition (1) :

$$\lim_{M\to\infty} Z_M \equiv \lim_{M\to\infty} \mathrm{tr}_{\{\sigma\}}\exp\left[\frac{\beta Jm}{M}\sum_k \sigma(k) + B\sum_k \sigma(k)\sigma(k+1)\right] = \mathrm{tr}\exp[\beta Jm\sigma_z + \beta\Gamma\sigma_x]$$

we immediately have $\langle\sigma(k)\rangle_{path} = \lim_{M\to\infty}\langle M^{-1}\sum_k \sigma(k)\rangle_{path}$ as

$$\begin{aligned}\langle\sigma(k)\rangle_{path} &= \lim_{M\to\infty}\frac{\mathrm{tr}_{\{\sigma\}}\frac{1}{M}\sum_k \sigma(k)\exp[\frac{\beta Jm}{M}\sum_{l=1}^M \sigma(l) + B\sum_{l=1}^M \sigma(l)\sigma(l+1)]}{\mathrm{tr}_{\{\sigma\}}\exp[\frac{\beta Jm}{M}\sum_{l=1}^M \sigma(l) + B\sum_{l=1}^M \sigma(l)\sigma(l+1)]} \\ &= \lim_{M\to\infty}\frac{\partial \log Z_M}{\partial(\beta Jm)} = \frac{Jm}{\sqrt{(Jm)^2+\Gamma^2}}\tanh\beta\sqrt{(Jm)^2+\Gamma^2}\end{aligned}$$

and equation (9) leads to

$$\frac{P_t(m)}{dt} = \frac{\partial}{\partial m}\{mP_t(m)\} - \frac{\partial}{\partial m}\left\{P_t(m)\frac{Jm}{\sqrt{(Jm)^2+\Gamma^2}}\tanh\sqrt{(Jm)^2+\Gamma^2}\right\}.$$

Finally, substituting the form $P_t(m) = \delta(m - m(t))$ and making the integral by part with respect to m after multiplying itself m, we obtain the following deterministic equation.

$$\frac{dm}{dt} = -m + \frac{Jm}{\sqrt{(Jm)^2+\Gamma^2}}\tanh\beta\sqrt{(Jm)^2+\Gamma^2} \tag{10}$$

It is easy to see that the steady state $dm/dt = 0$ is nothing but the equilibrium state described by the equation of state $m = Jm\{\sqrt{(Jm)^2+\Gamma^2}\}^{-1}\tanh\beta\sqrt{(Jm)^2+\Gamma^2}$. In Figure 1(left), we plot the typical behaviour of zero-temperature dynamics (equation (10) with $\beta = \infty$) far from the critical point $\Gamma_c = J = 1$ of quantum phase transition. We easily find that the dynamics exponentially converges to the steady state. The right panel denotes the zero-temperature dynamics at the critical point. The inset shows the log-log plot of $m(t)$ indicating that the dynamical exponent in the critical slowing down is $\nu' = 1/2$. This fact is directly confirmed from equation (10) with $\beta = \infty$ near the critical point $dm/dt \simeq -(1-J\Gamma^{-1})m - (Jm)^3/2\Gamma^3 + \mathcal{O}(m^5)$, namely, $m(t)$ behaves around the critical point $\Gamma \simeq \Gamma_c = J$ as $m(t) = m(0)\,\mathrm{e}^{-t/(1-J\Gamma^{-1})^{-1}}$. At the critical point, the relaxation time diverges as $\tau_\Gamma \equiv (1-J\Gamma_c^{-1})^{-1}$ resulting in the critical slowing down as $m(t) \simeq \mathrm{e}^{-t/\tau_\Gamma} \to m(t) \simeq t^{-\nu'}$, $\nu' = 1/2$. Of course, the exponent is the same as that of the 'mean-field model' universality class.

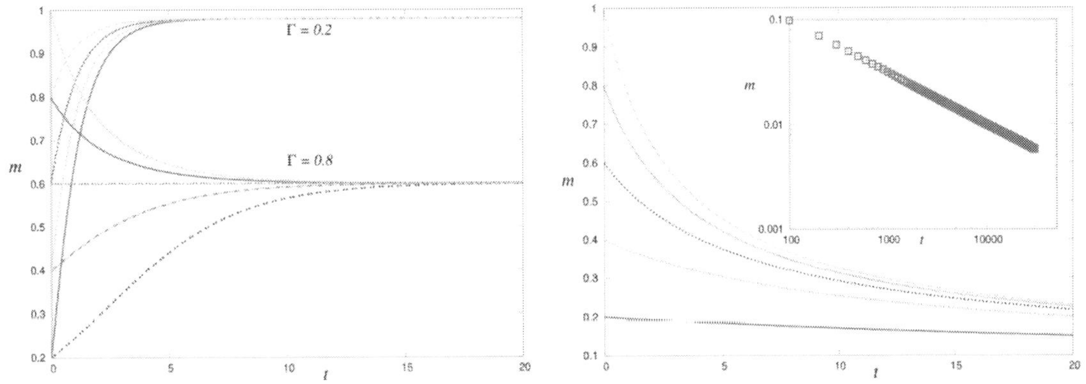

Figure 1. Typical behaviour of zero-temperature dynamics described by (10) with $\beta = \infty$ far from the critical point $\Gamma_c = J = 1$ of quantum phase transition (left). The right panel denotes the zero-temperature dynamics at the critical point. The inset shows the log-log plot of $m(t)$ indicating that the dynamical exponent in the critical slowing down is $\nu' = 1/2$.

2.4. *On the validity of static approximation*

Without the static approximation, the following deterministic flow equations for each Trotter slice is obtained by substituting the form $P_t(m_k) = \delta(m_k - m_k(t))$ and using the same way as deriving (10) as follows.

$$\frac{dm_k}{dt} = -m_k + \lim_{M\to\infty} \frac{\mathrm{tr}_{\{\sigma\}}\sigma(k)\exp[\frac{\beta J}{M}\sum_{l=1}^{M} m_l\sigma(l) + B\sum_{l=1}^{M}\sigma(l)\sigma(l+1)]}{\mathrm{tr}_{\{\sigma\}}\exp[\frac{\beta J}{M}\sum_{l=1}^{M} m_l\sigma(l) + B\sum_{l=1}^{M}\sigma(l)\sigma(l+1)]} \tag{11}$$

Obviously, the equation (11) is symmetric for the choice of k as long as we use the periodic boundary condition $\boldsymbol{\sigma}_1 = \boldsymbol{\sigma}_{M+1}$. This might be a justification to assume that the static approximation is correct at least for the present pure Ising system.

To confirm this argument, we carry out computer simulation for finite size system having $N = 400$ spins. We observe the time evolving process of the histogram $P(m_k)$ which is calculated from the $M = N = 400$ copies of the Trotter slices. We show the result in Figure 2. In this simulation, we chose the initial configuration in each Trotter slice randomly (we set each spin variable $\sigma_i(k)$ to $+1$ with a fixed probability p) and choose the inverse temperature $\beta = 2$ for $\Gamma = 0.5$ and $\Gamma = 0.6$. The time unit (the duration) of the update of $P(m_k)$ is chosen as 1 Monte Carlo step (MCS). From both panels in Figure 2, we find that at the beginning, the $P(m_k)$ is distributed due to the random set-up of the initial configuration, however, the fluctuation rapidly (eventually) shrinks leading up to the delta function around MCS ~ 100. After that, the $P(m_k)$ evolves as a delta function with the peak located at the value of spontaneous magnetization which is explicitly indicated in the inset of each panel. It should be noted that we evaluated the value of order parameter at the time point in the Runge-Kutta method. Thus, the duration between the points to be evaluated is not the MCS but the Runge-Kutta step. Of course, some statistical errors for the finite system should be taken into account, however, the limited result here seems to support the validity of the static approximation even in the dynamical process.

3. Disordered systems: An application for Statistical-Mechanical Informatics

It is easy for us to extend the above formulation to some class of disordered spin systems, that is to say, the infinite-range random field Ising model which is often used to check the performance of image restoration analytically [26, 28, 29] as a bench mark test.

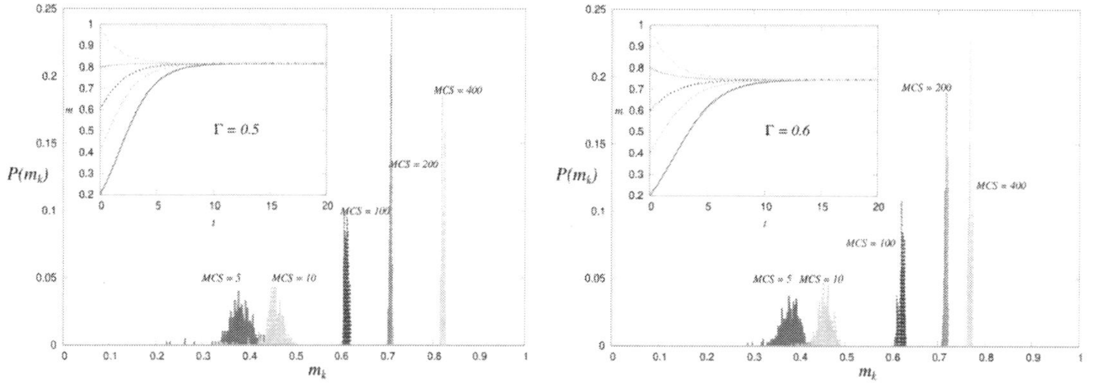

Figure 2. Time evolution of the distribution $P(m_k)$ calculated for finite size system with $N = M = 400$. We choose the inverse temperature $\beta = 2$ for $\Gamma = 0.5$ (left) and $\Gamma = 0.6$ (right). The inset in each panel denotes the deterministic flows of spontaneous magnetization calculated by (10) for corresponding parameter sets.

Here we consider a given original image $\boldsymbol{\xi} \equiv (\xi_1, \cdots, \xi_N)$ which is generated from the infinite-range ferromagnetic Ising model whose Gibbs measure (distribution for effective single spin) is described by $P(\xi_i) = \mathrm{e}^{\beta_s m_0 \xi_i}/\{2\cosh(\beta_s m_0)\}$, where m_0 denotes spontaneous magnetization at temperature β_s^{-1}. A snapshot $\boldsymbol{\xi}$ from the distribution is degraded by additive white Gaussian noise (AWGN) with mean $a_0\xi_i$ and variance a^2, namely, each pixel in the degraded image $\boldsymbol{\tau} = (\tau_1, \cdots, \tau_N)$ is obtained by $\tau_i = a_0\xi_i + ax_i$ with $x_i \sim \mathcal{N}(0,1)$. From the Bayesian inference view point, we assume that the posterior $P(\boldsymbol{\sigma}|\boldsymbol{\tau})$ (here we define $\boldsymbol{\sigma}$ as an estimate of the original image $\boldsymbol{\xi}$) might be proportional to the logarithm of the effective Hamiltonian (2) with $J_{ij} = J\ \forall(i,j)$ and $h \neq 0$. The first and the second terms appearing in the right hand side of (2) correspond to the prior distribution and the likelihood function, respectively. Whereas, the third term is introduced to utilize quantum fluctuation to construct the Bayes estimate for each pixel ('majority-vote decision' on each pixel), namely, $\mathrm{sgn}(\langle\sigma_i^z\rangle)$.

In this section, we attempt to describe the recovering process of original image through the deterministic flows of several relevant order-parameters and image restoration measure, namely, the overlap function $\mathcal{M} \equiv N^{-1}\sum_i \xi_i\, \mathrm{sgn}(\langle\sigma_i^z\rangle)$.

For the above set-up of the problem, the local field on the site i in the k-th Trotter slice now leads to

$$\phi_i(\boldsymbol{\sigma}_k : \sigma_i(k+1), \boldsymbol{\tau}) = \frac{J}{NM}\sum_j \sigma_j(k) + \frac{h}{M}\tau_i\sigma_i(k) + B\sigma_i(k+1).$$

As relevant order parameters, we choose $\mu_k = N^{-1}\sum_i \tau_i\sigma_i(k)$ and magnetization m_k. Then, we derive the differential equation with respect to $P_t(m_k, \mu_k) = \sum_{\boldsymbol{\sigma}_k} p_t(\boldsymbol{\sigma}_k)\delta(m_k - m_k(\boldsymbol{\sigma}_k))\delta(\mu_k - \mu_k(\boldsymbol{\sigma}_k))$ as follows.

$$\begin{aligned}\frac{dP_t(m_k,\mu_k)}{dt} &= \frac{\partial}{\partial m_k}\{m_k P_t(m_k,\mu_k)\} + \frac{\partial}{\partial \mu_k}\{\mu_k P_t(m_k,\mu_k)\} \\ &- \frac{\partial}{\partial m_k}\left\{P_t(m_k,\mu_k)\frac{1}{N}\sum_i \tanh[\beta\phi_i(\boldsymbol{\sigma}_k : \sigma_i(k+1),\boldsymbol{\tau})]\right\} \\ &- \frac{\partial}{\partial \mu_k}\left\{P_t(m_k,\mu_k)\frac{1}{N}\sum_i \tau_i\tanh[\beta\phi_i(\boldsymbol{\sigma}_k : \sigma_i(k+1),\boldsymbol{\tau})]\right\} \qquad (12)\end{aligned}$$

By assuming the self-averaging properties on the following physical quantities over both all possible paths in the imaginary-time axis and input data; original images and degrading processes (a particular realization of the quantity is identical to the average value and its deviation from the average eventually vanishes in the limit $N \to \infty$), we have

$$\begin{aligned} N^{-1}\sum_i \tanh[\beta\phi_i(\boldsymbol{\sigma}_k : \sigma_i(k+1), \boldsymbol{\tau})] &= [\langle\sigma(k)\rangle_{*path}]_{data} \\ N^{-1}\sum_i \tau_i \tanh[\beta\phi_i(\boldsymbol{\sigma}_k : \sigma_i(k+1), \boldsymbol{\tau})] &= [\tau\langle\sigma(k)\rangle_{*path}]_{data} \end{aligned}$$

where we defined the two different kinds of the averages by

$$\begin{aligned} \langle\cdots\rangle_{*path} &\equiv \lim_{M\to\infty} \frac{\mathrm{tr}_{\{\sigma\}}(\cdots)\exp[\frac{\beta}{M}\sum_{l=1}^M (Jm_l + h\tau)\sigma(l) + B\sum_{l=1}^M \sigma(l)\sigma(l+1)]}{\mathrm{tr}_{\{\sigma\}}\exp[\frac{\beta}{M}\sum_{l=1}^M (Jm_l + h\tau)\sigma(l) + B\sum_{l=1}^M \sigma(l)\sigma(l+1)]} \\ [\cdots]_{data} &\equiv \frac{\sum_\xi \mathrm{e}^{\beta_s m_0 \xi}}{2\cosh(\beta_s m_0)}\int_{-\infty}^{\infty}(\cdots)\exp\left[-\frac{(\tau - a_0\xi)^2}{2a^2}\right]d\tau. \end{aligned} \tag{13}$$

Under the static approximation, we obtain

$$\begin{aligned} \frac{dP_t(m,\mu)}{dt} &= \frac{\partial}{\partial m}\{mP_t(m,\mu)\} + \frac{\partial}{\partial\mu}\{\mu P_t(m,\mu)\} \\ &- \frac{\partial}{\partial m}\left\{P_t(m,\mu)\frac{\sum_\xi \mathrm{e}^{\beta_s m_0\xi}}{2\cosh(\beta_s m_0)}\int_{-\infty}^{\infty} Dx \frac{\Xi_m^{(a,a_0)}(\xi,x)\tanh\beta\sqrt{\{\Xi_m^{(a,a_0)}(\xi,x)\}^2 + \Gamma^2}}{\sqrt{\{\Xi_m^{(a,a_0)}(\xi,x)\}^2 + \Gamma^2}}\right\} \\ &- \frac{\partial}{\partial\mu}\left\{P_t(m,\mu)\frac{\sum_\xi \mathrm{e}^{\beta_s m_0\xi}}{2\cosh(\beta_s m_0)}\int_{-\infty}^{\infty} Dx \frac{(a_0\xi + ax)\Xi_m^{(a,a_0)}(\xi,x)\tanh\beta\sqrt{\{\Xi_m^{(a,a_0)}(\xi,x)\}^2 + \Gamma^2}}{\sqrt{\{\Xi_m^{(a,a_0)}(\xi,x)\}^2 + \Gamma^2}}\right\} \end{aligned}$$

with $\Xi_m^{(a,a_0)}(\xi,x) \equiv Jm + ha_0\xi + hax$ and $Dx \equiv dx\,\mathrm{e}^{-x^2/2}/\sqrt{2\pi}$. Using the same way as the pure Ising system discussed in the previous section, we finally obtain the deterministic flow equations of the order-parameters m and μ as follows.

$$\frac{dm}{dt} = -m + \frac{\sum_\xi \mathrm{e}^{\beta_s m_0\xi}}{2\cosh(\beta_s m_0)}\int_{-\infty}^{\infty} Dx \frac{\Xi_m^{(a,a_0)}(\xi,x)\tanh\beta\sqrt{\{\Xi_m^{(a,a_0)}(\xi,x)\}^2 + \Gamma^2}}{\sqrt{\{\Xi_m^{(a,a_0)}(\xi,x)\}^2 + \Gamma^2}} \tag{14}$$

$$\frac{d\mu}{dt} = -\mu + \frac{\sum_\xi \mathrm{e}^{\beta_s m_0\xi}}{2\cosh(\beta_s m_0)}\int_{-\infty}^{\infty} Dx \frac{(a_0\xi + ax)\Xi_m^{(a,a_0)}(\xi,x)\tanh\beta\sqrt{\{\Xi_m^{(a,a_0)}(\xi,x)\}^2 + \Gamma^2}}{\sqrt{\{\Xi_m^{(a,a_0)}(\xi,x)\}^2 + \Gamma^2}} \tag{15}$$

For the solution of the above deterministic flows (m,μ) at time t, the overlap between the original image and degraded image is measured by

$$\mathcal{M}(m,\mu) = \frac{\sum_\xi \mathrm{e}^{\beta_s m_0\xi}}{2\cosh(\beta_s m_0)}\int_{-\infty}^{\infty} Dx\,\mathrm{sgn}[\hat{m} + (a_0\xi + ax)\hat{\mu}] \tag{16}$$

where $(\hat{m},\hat{\mu})$ is a solution of the following coupled equations

$$m = \frac{\sum_\xi \mathrm{e}^{\beta_s m_0\xi}}{2\cosh(\beta_s m_0)}\int_{-\infty}^{\infty} Dx\,\tanh[\hat{m} + (a_0\xi + ax)\hat{\mu}] \tag{17}$$

$$\mu = \frac{\sum_\xi \mathrm{e}^{\beta_s m_0\xi}}{2\cosh(\beta_s m_0)}\int_{-\infty}^{\infty} Dx\,[\hat{m} + (a_0\xi + ax)\hat{\mu}]\tanh[\hat{m} + (a_0\xi + ax)\hat{\mu}] \tag{18}$$

for a given point on the trajectory (m, μ) at time t. To obtain the overlap function (16) and (17)(18), we used the concept of dynamical replica theory (the DRT) [21, 22], namely, 'equipartitioning' and 'self-averaging' of the $\mathcal{M}$ during the evolution in time. As the derivation is a bit complicated, we shall show the detail in Appendix A.

We solve the equations (14)(15) with (16)-(18) numerically and show the results in Figure 3. We choose the set of the parameters for the original image as $\beta_s^{-1} = 0.9$ ($m_0 = 0.523$) and $a_0 = a = 1$ which means that the corresponding optimal hyper-parameters are $h = a_0/a^2 = 1$ and $J^{-1} = \beta^{-1}$. We consider the zero-temperature restoration dynamics in which the fluctuation to make the Bayesian estimate is only quantum-mechanical one (it is controlled by the amplitude of quantum-mechanical tunneling Γ). In the left panel, the deterministic trajectories in the space (m, μ) are plotted for $\Gamma = 0.6$. The state of system in which the image restoration is successfully achieved is in ferromagnetic phase. Thus, order-parameters m and μ converge to the fixed point exponentially (there is no critical slowing down in this model system). In the right panel, we show the time evolution of the image restoration measure $\mathcal{M}$ for several values of Γ. From this panel, we find that some 'non-monotonic' behaviour is observed at the initial stage of the dynamics when we fail to set the amplitude to its optimal value ($\Gamma \sim 0.6$). Similar behaviour was reported in the Bayesian image restoration via thermal (classical) fluctuation [38].

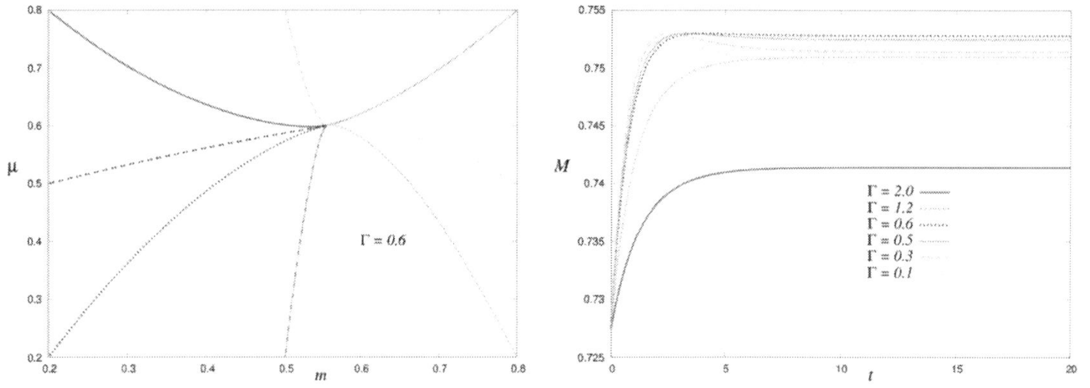

Figure 3. Trajectories in the phase space (m, μ) are plotted for $\Gamma = 0.6$ (left). The parameters are chosen as $\beta_s^{-1} = 0.9$ ($m_0 = 0.523$), $a_0 = a = 1$ and the corresponding hyper-parameters as $h = a_0/a^2 = 1$ and $J^{-1} = \beta^{-1}$. The right panel shows the evolution of image restoration measure, overlap function $\mathcal{M}$ in time for several values of the amplitude of quantum-mechanical tunneling Γ.

In the Bayesian framework, it is desired for us to obtain the estimate of each pixel and hyper-parameters simultaneously. In such case, we use the so-called EM algorithm based on the maximization of marginal likelihood criteria [39]. The quantum-mechanical extension and the formulation presented here is applicable to the simultaneous estimation for both micro and macro parameters.

4. Concluding remarks

In this paper, for a simplest quantum spin systems, we showed a formulation to describe the macroscopically deterministic flows of order parameters from the master equation whose transition probability is given by the Glauber-type. Under the static approximation, differential equations with respect to macroscopic order parameters were explicitly obtained from the master equation describing the microscopic-law. In the steady state, we found that the equations are identical to the saddle point equations for the equilibrium state of the same system. We also checked the validity of the static approximation by computer simulations and found that the result supports the validity of the approximation. Several possible extensions of our approach to disordered spin systems for statistical-mechanical informatics was discussed. Especially, we used our procedure to evaluate the decoding process of Bayesian image restoration. With the assistance of the concept of dynamical replica theory (the DRT), we derived the zero-temperature flow equation of image restoration measure showing some 'non-monotonic' behaviour in its time evolution. Of course, by using the present approach, one can evaluate the 'inhomogeneous' Markovian stochastic process of quantum Monte Carlo method (in which amplitude Γ is time-dependent) such as quantum annealing. In the next step of the present study, we are planning to extend this formulation to the probabilistic information processing described by spin glasses such as quantum Hopfield model [40, 41] including a peculiar type of antiferromagnet [42].

Appendix A. Derivation of overlap function: A dynamical approach

In this appendix, we show the derivation of overlap function for image restoration problem (16)-(18) by using a useful concept of dynamical replica theory (the DRT) [21, 22].

The problem here comes from the fact that the quantity $\mathcal{M} = N^{-1}\sum_i [\xi_i \,\mathrm{sgn}(\langle \sigma_i(k)\rangle_{path})]_{data}$ containing a single site expectation $\langle \sigma_i(k)\rangle_{path}$ is not observable in computer simulations of a single system. Therefore, we must introduce infinite number of 'virtual copies' ('real replicas' or 'ensembles') to evaluate the overlap function by using the law of large number. Namely, we evaluate the 'real-time' dependence of the overlap function as follows.

$$\begin{aligned}\mathcal{M}(t) &\equiv \lim_{L,N\to\infty}\frac{1}{N}\sum_i \xi_i \,\mathrm{sgn}\left(\frac{1}{L}\sum_{\rho=1}^{L}\sigma_{it}^{\rho}(k)\right) = \left[\left\langle \xi\,\mathrm{sgn}\left(\frac{1}{L}\sum_{\rho=1}^{L}\sigma_{1t}^{\rho}(k)\right)\right\rangle_{copy}\right]_{data} \\ &= \mathcal{M}(m_k(t),\mu_k(t))\end{aligned}$$

where $\rho = 1,\cdots,L$ denote the copies of the original system in which $p_t(\boldsymbol{\sigma}^{\rho}(k))$ is the same. From the law of large number, we can expect $\lim_{L\to\infty} L^{-1}\sum_{\rho=1}^{L}\sigma_{it}^{l}(k) = \langle \sigma_i(k)\rangle_{copy}$. To calculate the expectation $\langle\cdots\rangle_{copy}$, we assume the equipartitioning in the (m_k,μ_k)-subsells. Namely, explicit time-dependence of the expectation through $p_t(\boldsymbol{\sigma}^{\rho}(k))$ is now removed within the subsells as

$$\begin{aligned}\langle\cdots\rangle_{copy} &= \frac{\sum_{\{\boldsymbol{\sigma}^{\rho}(k)\}}\prod_{\rho=1}^{L} p_t(\boldsymbol{\sigma}^{\rho}(k))(\cdots)\delta(m_k - m_k(\boldsymbol{\sigma}^{\rho}(k)))\delta(\mu_k - \mu_k(\boldsymbol{\sigma}^{\rho}(k)))}{\sum_{\{\boldsymbol{\sigma}^{\rho}(k)\}}\prod_{\rho=1}^{L} p_t(\boldsymbol{\sigma}^{\rho}(k))\delta(m_k - m_k(\boldsymbol{\sigma}^{\rho}(k)))\delta(\mu_k - \mu_k(\boldsymbol{\sigma}^{\rho}(k)))} \\ &= \frac{\sum_{\{\boldsymbol{\sigma}^{\rho}(k)\}}\prod_{\rho=1}^{L}(\cdots)\delta(m_k - m_k(\boldsymbol{\sigma}^{\rho}(k)))\delta(\mu_k - \mu_k(\boldsymbol{\sigma}^{\rho}(k)))}{\sum_{\{\boldsymbol{\sigma}^{\rho}(k)\}}\prod_{\rho=1}^{L}\delta(m_k - m_k(\boldsymbol{\sigma}^{\rho}(k)))\delta(\mu_k - \mu_k(\boldsymbol{\sigma}^{\rho}(k)))}.\end{aligned}$$

Then, assuming the self-averaging on the $\mathcal{M}$, we have $\mathcal{M}_L(m_k,\mu_k)$ in the limit of $N\to\infty$ as

$$\begin{aligned}&\mathcal{M}_L(m_k,\mu_k) \\ &= \lim_{N\to\infty}\left[\frac{\sum_{\boldsymbol{\sigma}^1(k),\cdots,\boldsymbol{\sigma}^L(k)}\prod_{\rho=1}^{L}\delta(m_k - m_k(\boldsymbol{\sigma}^{\rho}(k)))\delta(\mu_k - \mu_k(\boldsymbol{\sigma}^{\rho}(k)))\xi_1\mathrm{sgn}\left(\frac{1}{L}\sum_{\rho=1}^{L}\sigma_1^{\rho}(k)\right)}{\sum_{\boldsymbol{\sigma}^1(k),\cdots,\boldsymbol{\sigma}^L(k)}\prod_{\rho=1}^{L}\delta(m_k - m_k(\boldsymbol{\sigma}^{\rho}(k)))\delta(\mu_k - \mu_k(\boldsymbol{\sigma}^{\rho}(k)))}\right]_{data} \\ &= \lim_{n\to 0}\lim_{N\to\infty}\left[\sum_{\{\boldsymbol{\sigma}^{\rho\alpha}(k)\}}\xi_1\,\mathrm{sgn}\left(\frac{1}{L}\sum_{\rho=1}^{L}\sigma_1^{\rho 1}(k)\right)\prod_{\alpha=1}^{n}\prod_{\rho=1}^{L}\delta(m_k - m_k(\boldsymbol{\sigma}^{\rho\alpha}(k)))\delta(\mu_k - \mu_k(\boldsymbol{\sigma}^{\rho\alpha}(k)))\right]_{data}\end{aligned}$$

where we used the fact

$$\left[\frac{\mathrm{tr}_{\{s\}}\mathcal{P}\psi}{\mathrm{tr}_{\{s\}}\mathcal{P}}\right]_{data} = \lim_{n\to 0}\left[\frac{(\mathrm{tr}_{\{s\}}\mathcal{P}\psi)(\mathrm{tr}_{\{s\}}\mathcal{P})^n}{\mathrm{tr}_{\{s\}}\mathcal{P}}\right]_{data} = \lim_{n\to 0}\left[\mathrm{tr}_{\{s^{\alpha}\}}\psi(s^1)\prod_{\alpha=1}^{n}\mathcal{P}(\{s^{\alpha}\})\right]_{data}.$$

and introduced the replica index $\alpha = 1, \cdots, n$ to carry out the average $[\cdots]_{data}$ by standard replica trick. It should be noted that $\boldsymbol{\sigma} \equiv (\sigma_1(k), \cdots, \sigma_N(k))$ and $\{\cdots\}$ denotes the set of spin variables for all possible combinations of copies and replicas (ρ, α). By using the integral representation for the delta-function, we obtain

$$\begin{aligned}\mathcal{M}_L(m_k, \mu_k) &= \lim_{n\to 0}\lim_{N\to\infty} \int_{-\infty}^{\infty} \prod_{\rho\alpha} \frac{d\hat{m}_k^{\rho\alpha} d\hat{\mu}_k^{\rho\alpha}}{2\pi N} \mathrm{e}^{-iN\sum_{\rho\alpha}(m_k \hat{m}_k^{\rho\alpha} + \mu_k \hat{\mu}_k^{\rho\alpha})} \\ &\times \sum_{\{\boldsymbol{\sigma}^{\rho\alpha}(k)\}} \xi_1 \,\mathrm{sgn}\left(\frac{1}{L}\sum_{\rho=1}^{L} \sigma_1^{\rho 1}(k)\right) \mathrm{e}^{-i\sum_{\rho\alpha} \hat{m}_k^{\rho\alpha} \sum_i \sigma_i^{\rho\alpha}(k)} \left[\mathrm{e}^{-i\sum_{\rho\alpha}\hat{\mu}_k^{\rho\alpha}\sum_i \tau_i \sigma_i^{\rho\alpha}(k)}\right]_{data}.\end{aligned}$$

Now, under the static approximation, the quantum fluctuation appears through the quantities $m_k = m$ and $\mu_k = \mu$ which obeys (14)(15) at any time t. Therefore, from now on, we cancel the k-dependence. By simple transformation of the variables as $-i\hat{\mu}^{\rho\alpha} \mapsto \hat{\mu}^{\rho\alpha}, -i\hat{m}^{\rho\alpha} \mapsto \hat{m}^{\rho\alpha}$ and carrying out the data average $[\cdots]_{data}$ over the effective single site distribution (13), one obtains

$$\begin{aligned}\mathcal{M}_L(m, \mu) &= \lim_{n\to 0}\lim_{N\to 0} \int_{-\infty}^{\infty} \prod_{\rho\alpha} \frac{d\hat{m}^{\rho\alpha} d\hat{\mu}^{\rho\alpha}}{2\pi N} \mathrm{e}^{-N\sum_{\rho\alpha}(m\hat{m}_k^{\rho\alpha} + \mu\hat{\mu}_k^{\rho\alpha})} \\ &\times \left\{\frac{\sum_\xi \mathrm{e}^{\beta_s m_0 \xi}}{2\cosh(\beta_s m_0)} \int_{-\infty}^{\infty} \prod_{\rho\alpha} \{2\cosh[\hat{m}^{\rho\alpha} + (a_0\xi + ax)\hat{\mu}^{\rho\alpha}]\}\right\}^N \\ &\times \frac{\sum_{\sigma^{\rho\alpha}} \xi_1 \mathrm{sgn}\left(\frac{1}{L}\sum_{\rho=1}^{L}\sigma_1^{\rho 1}\right) \exp[\sum_{\rho\alpha} \hat{m}^{\rho\alpha}\sigma^{\rho\alpha} + (a\xi + ax)\sum_{\rho\alpha}\hat{\mu}^{\rho\alpha}\sigma^{\rho\alpha}]}{\sum_{\sigma^{\rho\alpha}} \exp[\sum_{\rho\alpha}\hat{m}^{\rho\alpha} + (a\xi + ax)\sum_{\rho\alpha}\hat{\mu}^{\rho\alpha}\sigma^{\rho\alpha}]} \\ &\times \lim_{n\to 0}\lim_{N\to 0} \int_{-\infty}^{\infty} d\hat{\boldsymbol{m}} d\hat{\boldsymbol{\mu}} \exp[N\Psi(\hat{\boldsymbol{m}}, \hat{\boldsymbol{\mu}})] \\ &\times \frac{\sum_{\sigma^{\rho\alpha}} \xi_1 \mathrm{sgn}\left(\frac{1}{L}\sum_{\rho=1}^{L}\sigma_1^{\rho 1}\right) \exp[\sum_{\rho\alpha} \hat{m}^{\rho\alpha}\sigma^{\rho\alpha} + (a\xi + ax)\sum_{\rho\alpha}\hat{\mu}^{\rho\alpha}\sigma^{\rho\alpha}]}{\sum_{\sigma^{\rho\alpha}} \exp[\sum_{\rho\alpha}\hat{m}^{\rho\alpha} + (a\xi + ax)\sum_{\rho\alpha}\hat{\mu}^{\rho\alpha}\sigma^{\rho\alpha}]} \quad \text{(A.1)}\end{aligned}$$

with

$$\Psi(\hat{\boldsymbol{m}}, \hat{\boldsymbol{\mu}}) \equiv -\sum_{\rho\alpha}(m\hat{m}^{\rho\alpha} + \mu\hat{\mu}^{\rho\alpha}) + \log\left[\frac{\sum_\xi \mathrm{e}^{\beta_s m_0 \xi}}{2\cosh(\beta_s m_0)} \int_{-\infty}^{\infty} \prod_{\rho\alpha} \{2\cosh[\hat{m}^{\rho\alpha} + (a_0\xi + ax)\hat{\mu}^{\rho\alpha}]\}\right].$$

We should notice that in the above expression of the overlap function $\mathcal{M}_L(m, \mu)$, the condition $\xi_1 = \mathrm{sgn}(L^{-1}\sum_{\rho=1}^{L}\sigma_{\rho 1}(k))$ leading to 'perfect image restoration', namely, $\lim_{L\to\infty}\mathcal{M}_L = 1$ immediately gives $\lim_{n\to 0}\lim_{N\to 0}\int_{-\infty}^{\infty} d\hat{\boldsymbol{m}} d\hat{\boldsymbol{\mu}} \exp[N\Psi(\hat{\boldsymbol{m}}, \hat{\boldsymbol{\mu}})] = 1$. Therefore, it is easy to find that the part $\lim_{n\to 0}\lim_{N\to 0}\int_{-\infty}^{\infty} d\hat{\boldsymbol{m}} d\hat{\boldsymbol{\mu}} \exp[N\Psi(\hat{\boldsymbol{m}}, \hat{\boldsymbol{\mu}})]$ appearing in (A.1) is a normalization factor. Thus, the overlap function derived by the above dynamical approach now leads to

$$\mathcal{M}_L(m, \mu) = \frac{\sum_{\sigma^{\rho\alpha}} \xi_1 \,\mathrm{sgn}\left(\frac{1}{L}\sum_{\rho=1}^{L}\sigma_1^{\rho 1}\right) \exp[\sum_{\rho\alpha} \hat{m}^{\rho\alpha}\sigma^{\rho\alpha} + (a\xi + ax)\sum_{\rho\alpha}\hat{\mu}^{\rho\alpha}\sigma^{\rho\alpha}]}{\sum_{\sigma^{\rho\alpha}} \exp[\sum_{\rho\alpha}\hat{m}^{\rho\alpha}\sigma^{\rho\alpha} + (a\xi + ax)\sum_{\rho\alpha}\hat{\mu}^{\rho\alpha}\sigma^{\rho\alpha}]}$$

where $\hat{\boldsymbol{m}}$ or $\hat{\boldsymbol{\mu}}$ should be chosen as a saddle point of the function $\Psi(\hat{\boldsymbol{m}}, \hat{\boldsymbol{\mu}})$. Assuming the replica symmetric and the copy symmetric solution $\hat{m}^{\rho\alpha} = \hat{m}, \hat{\mu}^{\rho\alpha} = \hat{\mu}\ \forall(\rho, \alpha)$, we obtain the function to be optimized at the saddle point.

$$\Psi_{RS} = \lim_{n\to 0}\lim_{N,L\to\infty} \frac{\Psi(\hat{\boldsymbol{m}}, \hat{\boldsymbol{\mu}})}{nL} = -m\hat{m} - \mu\hat{\mu} + \frac{\sum_\xi \mathrm{e}^{\beta_s m_0 \xi}}{2\cosh(\beta_s m_0)} \int_{-\infty}^{\infty} Dx \cosh[\hat{m} + (a_0\xi + ax)\hat{\mu}]$$

Obviously we find that the saddle point is obtained by the equations (17) and (18). Then, the overlap function is evaluated as

$$\begin{aligned}
\mathcal{M}_L(m,\mu) &= \frac{\sum_\xi \xi\, \mathrm{e}^{\beta_s m_0 \xi}}{2\cosh(\beta_s m_0)} \int_{-\infty}^{\infty} Dx \sum_{\sigma^1,\cdots,\sigma^L} \mathrm{sgn}\left(\frac{1}{L}\sum_{\rho=1}^{L}\sigma_1^{\rho 1}\right) \prod_{\rho=1}^{L} \frac{\mathrm{e}^{\{\hat{m}+(a_0\xi+ax)\hat{\mu}\}\sigma_1^{\rho 1}}}{2\cosh[\hat{m}+(a_0\xi+ax)\hat{\mu}]} \\
&= \int_{-\infty}^{\infty} dz \frac{\sum_\xi \xi\, \mathrm{e}^{\beta_s m_0 \xi}}{2\cosh(\beta_s m_0)} \int_{-\infty}^{\infty} Dx \sum_{\sigma^1,\cdots,\sigma^L} \mathrm{sgn}(z)\delta\left(z - \frac{1}{L}\sum_{\rho=1}^{L}\sigma_1^{\rho 1}\right) \\
&\times \prod_{\rho=1}^{L} \frac{\mathrm{e}^{\hat{m}+(a_0\xi+ax)\hat{\mu}}}{2\cosh[\hat{m}+(a_0\xi+ax)\hat{\mu}]} \equiv \int_{-\infty}^{\infty} dz\, \mathrm{sgn}(z)\mathcal{P}_L(z)
\end{aligned}$$

with

$$\begin{aligned}
\mathcal{P}_L(z) &\equiv \frac{\sum_\xi \xi\, \mathrm{e}^{\beta_s m_0 \xi}}{2\cosh(\beta_s m_0)} \int_{-\infty}^{\infty} Dx \sum_{\sigma^1,\cdots,\sigma^L} \int_{-\infty}^{\infty} \frac{dy}{2\pi} \mathrm{e}^{iy(z-L^{-1}\sum_{\rho=1}^{L}\sigma_1^{\rho})} \prod_{\rho=1}^{L} \frac{\mathrm{e}^{\{\hat{m}+(a_0\xi+ax)\hat{\mu}\}\sigma_1^{\rho 1}}}{2\cosh[\hat{m}+(a_0\xi+ax)\hat{\mu}]} \\
&= \int_{-\infty}^{\infty} \frac{dy}{2\pi}\mathrm{e}^{iyz} \frac{\sum_\xi \xi\, \mathrm{e}^{\beta_s m_0 \xi}}{2\cosh(\beta_s m_0)} \int_{-\infty}^{\infty} Dx \left\{\frac{\cosh[\hat{m}+(a_0\xi+ax)\hat{\mu}-iL^{-1}y]}{\cosh[\hat{m}+(a_0\xi+ax)\hat{\mu}]}\right\}^L \\
&= \int_{-\infty}^{\infty} \frac{dy}{2\pi}\mathrm{e}^{iyz} \frac{\sum_\xi \xi\, \mathrm{e}^{\beta_s m_0 \xi}}{2\cosh(\beta_s m_0)} \int_{-\infty}^{\infty} Dx \left\{\cos(yL^{-1}) - i\sin(yL^{-1})\tanh[\hat{m}+(a_0\xi+ax)\hat{\mu}]\right\}^L .
\end{aligned}$$

Thus, we have

$$\begin{aligned}
\mathcal{P}(z) &\equiv \lim_{L\to\infty} \mathcal{P}_L(z) = \frac{\sum_\xi \xi\, \mathrm{e}^{\beta_s m_0 \xi}}{2\cosh(\beta_s m_0)} \int_{-\infty}^{\infty} Dx \int_{-\infty}^{\infty} \frac{dy}{2\pi} \mathrm{e}^{iy(z-\tanh[\hat{m}+(a_0\xi+ax)\hat{\mu}])} \\
&= \frac{\sum_\xi \xi\, \mathrm{e}^{\beta_s m_0 \xi}}{2\cosh(\beta_s m_0)} \int_{-\infty}^{\infty} Dx\, \delta(z-\tanh[\hat{m}+(a_0\xi+ax)\hat{\mu}]).
\end{aligned}$$

Substituting the result into $\mathcal{M}_L$ and taking the limit of $L\to\infty$, we finally obtain

$$\begin{aligned}
\mathcal{M}(m,\mu) &= \lim_{L\to\infty} \mathcal{M}_L(m,\mu) = \int_{-\infty}^{\infty} dz\, \mathrm{sgn}(z) \lim_{L\to\infty} \mathcal{P}_L(z) = \int_{-\infty}^{\infty} dz\, \mathrm{sgn}(z)\mathcal{P}(z) \\
&= \frac{\sum_\xi \xi\, \mathrm{e}^{\beta_s m_0 \xi}}{2\cosh(\beta_s m_0)} \int_{-\infty}^{\infty} Dx\, \mathrm{sgn}[\hat{m}+(a_0\xi+ax)\hat{\mu}]
\end{aligned}$$

where we used the fact $\mathrm{sgn}(\tanh(x)) = \mathrm{sgn}(x)$. This result is nothing but equation (16). As parameters $\hat{m}$ and $\hat{\mu}$ are related to the order-parameters m and μ in equations (17)(18), overlap function $\mathcal{M}$ is influenced by quantum fluctuation through (17)(18) and the solution of equations (14)(15).

Acknowledgments

The present study was financially supported by *Grant-in-Aid Scientific Research on Priority Areas "Deepening and Expansion of Statistical Mechanical Informatics (DEX-SMI)" of The Ministry of Education, Culture, Sports, Science and Technology (MEXT)* No. 18079001 and *INSA (Indian National Science Academy) - JSPS (Japan Society of Promotion of Science) Bilateral Exchange Programme*. The author thanks Saha Institute of Nuclear Physics for their warm hospitality during his stay in India.

References

[1] Bishop C M 2006, *Pattern Recognition and Machine Learning*, (Singapore: Springer)
[2] Opper M and Saad D 2001 *Advanced Mean Field Methods: Theory and Practice* (Massachusetts: The MIT Press)
[3] Jordan M I 1998 *Learning in Graphical Models* (Massachusetts: The MIT Press)
[4] Landau D P and Binder K 2000 *A Guide to Monte Carlo Simulations in Statistical Physics* (Cambridge: Cambridge University Press)
[5] Mézard M, Parisi G and Virasoro M A 1987 *Spin Glass Theory and Beyond* (Singapore: World Scientific)
[6] Binder K and Young A P 1986 *Rev. Mod. Phys.* **58** 801
[7] Young A P 1998 *Spin Glass and Random Fields* (Singapore: World Scientific)
[8] Swendsen R H and Wang J S 1986 *Phys. Rev. Lett.* **57** 2607, Swendsen R H and Wang J S 1987 *Phys. Rev. Lett.* **58** 86
[9] Hukushima K and Nemoto K 1996 *J. Phys. Soc. Jpn.* **65** 1604
[10] Kirkpatrick S, Gelatt Jr C D and Vecchi MP (1983) *Science* **220** 671
[11] Geman S and Geman D 1984 *IEEE Trans. Pattern. Anal. and Mach. Intel.* **11** 721
[12] Kadowaki T and Nishimori H 1998 *Physical Review* E **58** 5355
[13] Farhi E, Goldstone J, Gutmann S, Lapan J, Lundgren A and Preda P 2001 *Science* **292** 472
[14] Morita S and Nishimori H 2006 *J. Phys. A* **39** 13903
[15] Suzuki S and Okada M 2005 *J. Phys. Soc. Jpn.* **74** 1649
[16] Santoro G E and Tosatti 2006 *J. Phys. A* **41** 209801
[17] Suzuki M 1976 *Prog. Theor. Phys.* **56** 1454
[18] de Oriveira M J and Chiappin J R N 1997 *Physica* A **238** 307
[19] Coolen A C C and Ruijgrok Th W 1988 *Phys. Rev.* A **38** 4253
[20] Hopfield J J 1982 *PNAS* **79** 2554
[21] Coolen A C C and Sherrington D 1994 *Phys. Rev. Lett.* **49** 1921
[22] Coolen A C C, Laughton S N and Sherrington D 1996 *Phys. Rev.* B **53** 8184
[23] Sherrington D and Kirkpatrick S 1975 *Phys. Rev. Lett.* **35** 1792
[24] Nishimori H 2001 *Statistical Physics of Spin Glasses and Information Processing: An Introduction* (Oxford: Oxford University Press)
[25] Tanaka K and Horiguchi T 1997 *IEICE* **J80-A-12** 2217 (in Japanese)
[26] Nishimori H and Wong K Y M 1999 *Phys. Rev.* E **60** 132
[27] Tanaka K 2002 *J. Phys. A: Math. Gen.* **35** R81
[28] Inoue J 2001 *Phys. Rev.* E **63** 046114
[29] Inoue J 2005 *Quantum Spin Glasses, Quantum Annealing, and Probabilistic Information Processing*, in *Quantum Annealing and Related Optimization Methods Lecture Notes in Physics* **679**, ed Das A and Chakrabarti B K (Berlin Heidelberg: Springer) p 259
[30] Chakrabarti K, Dutta A and Sen P 1996 *Quantum Ising Phases and Transitions in Transverse Ising Models* (Heidelberg: Springer)
[31] Sachdev S 1999 *Quantum Phase Transitions* (Cambridge: Cambridge University Press)
[32] Sourlas N 1989 *Nature* **339** 693
[33] Inoue J, Saika Y and Okada M 2009 *J. Phys: Conference Series* **143** 012019
[34] Suzuki M 1966 *J. Phys. Soc. Japan* **21** 2140
[35] Elliot R J, Pfeuty P and Wood C 1970 *Phys. Rev. lett.* **25** 443
[36] Feynman R P and Hibbs A R 1965 *Quantum Mechanics and Path Integrals* (New York: McGraw-Hill)
[37] Kleinert H 2009 *Path Integrals in Quantum Mechanics, Statistics, Polymer Physics, and Financial Markets* (Singapore: World Scientific)
[38] Ozeki T and Okada M 2003 *J. Phys. A* **36** 11011
[39] Inoue J and Tanaka K 2002 *Phys. Rev.* E **65** 016125
[40] Ma Y Q and Gong C D 1992 *Phys. Rev.* B **45** 793
[41] Nishimori H and Nonomura Y 1996 *J. Phys. Soc. Japan* **65** 3780
[42] Chandra A K, Inoue J and Chakrabarti B K 2010 *Phys. Rev.* E (to appear)

Solution space heterogeneity of the random K-satisfiability problem: Theory and simulations

Haijun Zhou[1,2]

[1]Key Laboratory of Frontiers in Theoretical Physics, Institute of Theoretical Physics, Chinese Academy of Sciences, Beijing 100190, China

[2]Kavli Institute for Theoretical Physics China (KITPC) at the Chinese Academy of Sciences, Beijing 100190, China

E-mail: zhouhj@itp.ac.cn

Abstract. The random K-satisfiability (K-SAT) problem is an important problem for studying typical-case complexity of NP-complete combinatorial satisfaction; it is also a representative model of finite-connectivity spin-glasses. In this paper we review our recent efforts on the solution space fine structures of the random K-SAT problem. A heterogeneity transition is predicted to occur in the solution space as the constraint density α reaches a critical value α_{cm}. This transition marks the emergency of exponentially many solution communities in the solution space. After the heterogeneity transition the solution space is still ergodic until α reaches a larger threshold value α_d, at which the solution communities disconnect from each other to become different solution clusters (ergodicity-breaking). The existence of solution communities in the solution space is confirmed by numerical simulations of solution space random walking, and the effect of solution space heterogeneity on a stochastic local search algorithm SEQSAT, which performs a random walk of single-spin flips, is investigated. The relevance of this work to glassy dynamics studies is briefly mentioned.

1. Introduction

A random K-Satisfiability (K-SAT) formula is constructed by adding M constraints on N variables [1]. Each variable i has a spin state $\sigma_i = \pm 1$, and each constraint a applies to K different variables that are randomly chosen from the whole set of variables. The energy of constraint a is expressed as

$$E_a = \prod_{i \in \partial a} \left[\frac{1 - J_a^i \sigma_i}{2} \right] , \tag{1}$$

where ∂a denotes the set of variables involved in constraint a (the size of ∂a is K), $J_a^i = \pm 1$ is the preferred spin state of constraint a on variable i. Each of the K preferred spin values J_a^i of a constraint a are randomly and independently assigned a value $+1$ or -1 with equal probability. The whole set $\{J_a^i\}$ of preferred spin values are then fixed, but the actual spin state σ_i of each variable i is allowed to change. The constraint energy E_a is zero if at least one of the variables $i \in \partial a$ takes the spin value $\sigma_i = J_a^i$, otherwise $E_a = 1$. Given a random K-SAT formula, the task is to construct at least one spin configuration $\vec{\sigma} \equiv \{\sigma_1, \sigma_2, \ldots, \sigma_N\}$ that satisfies all the constraint (i.e., makes all the constraint energy E_a to be zero), or to prove that no such solutions (i.e., satisfying spin configurations) exist.

When N is large, rigorous mathematical proofs (see, e.g., review article [2]) and numerical simulations [1] revealed that whether a random K-SAT formula is satisfiable or not depends on the constraint density α ($\equiv M/N$). As α increases beyond certain satisfiability threshold $\alpha_s(K)$, the probability of a randomly constructed K-SAT formula to be satisfiable quickly drops from being close to unity to being close to zero. Among the whole ensemble of random K-SAT formulas, the satisfiability of those instances with constraint density α in the vicinity of $\alpha_s(K)$ is the hardest to determine. This empirical observation has stimulated a lot of investigations.

The random K-SAT problem was intensively studied in the statistical physics community during the last decade [3–7]. The number of solutions for the random K-SAT problem as a function of constraint density was calculated in Ref. [3] by the replica method of spin-glass physics. Later the satisfiability threshold α_s as a function of K was calculated by the first-step replica-symmetry-breaking (1RSB) energetic cavity method [4, 8]. The evolution of the solution space of the random K-SAT problem with constraint density α was studied in Refs. [5, 9, 10] using the 1RSB entropic cavity method [11, 12]. Before the satisfiability threshold $\alpha_s(K)$ is reached, the solution space experiences a clustering transition at $\alpha = \alpha_d(K)$, where exponentially many solution clusters (Gibbs states) form and the ergodicity of the solution space is broken. This clustering transition is followed by a condensation transition at $\alpha = \alpha_c(K) \geq \alpha_d(K)$, where a sub-exponential number of solution clusters begin to dominate the solution space. Within a solution cluster, the spin states of a large fraction of variables start to be frozen to the same value as α exceeds certain threshold value α_f that may be different for different clusters. The freezing transition was investigated in Refs. [9, 13, 14]. Some of these phase transitions (i.e., the clustering and the condensation transition) were earlier found to occur in mean-field p-body-interaction spin glasses [15], with temperature T (instead of the constraint density) being the control parameter.

In the present paper, we review our recent efforts on the solution space fine structures of the random K-SAT problem. A heterogeneity transition is predicted to occur in the solution space as the constraint density α reaches a critical value $\alpha_{cm}(K)$ which is smaller than $\alpha_d(K)$ [16]. This transition marks the emergency of exponentially many solution communities in the solution space. For $\alpha_{cm} < \alpha < \alpha_d$, the heterogeneous solution space is ergodic; at $\alpha = \alpha_d$ the solution communities will turn into different solution clusters as an ergodicity-breaking transition occurs. The existence of solution communities in the solution space is confirmed by numerical simulations on single K-SAT formulas [17], and the effect of solution space heterogeneity on a stochastic local search algorithm `SEQSAT`, which performs a random walk of single-spin flips, is investigated [18]. Beyond the clustering transition point α_d, our numerical simulation results [17, 19] suggested that the individual solution clusters of the solution space also have rich internal structures.

The replica-symmetric cavity method is used in the next section to calculate the value of α_{cm} for the onset of solution space structural heterogeneity. Section 3 presents the data-clustering results on sampled solutions of single random K-SAT formulas; Section 4 reports the simulation results of the stochastic search algorithm `SEQSAT`. We conclude this work in Sec. 5 and point out some possible links with the phenomena of two-step relaxation and dynamical heterogeneity in supercooled liquids.

2. Solution space heterogeneity transition

Given a random K-SAT formula, the total energy $E(\vec{\sigma}) \equiv \sum_a E_a$ is equal to the number of violated constraints by the spin configuration $\vec{\sigma}$. The whole set $\mathcal{S}$ of spin configurations with $E(\vec{\sigma}) = 0$ form the solution space of this K-SAT formula. The similarity between any two

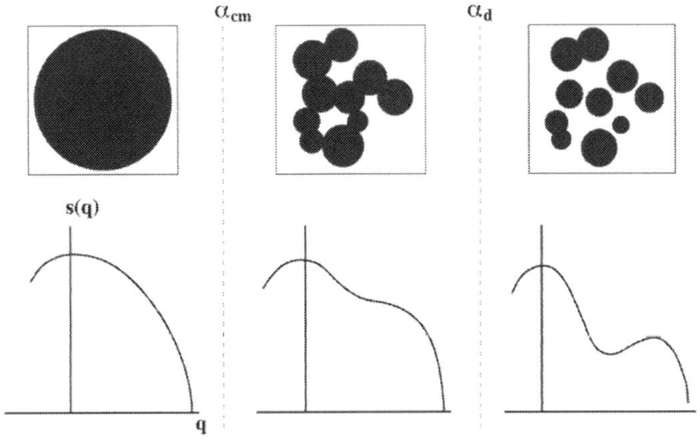

Figure 1. Detecting heterogeneity of the solution space by the shape of the entropy density function $s(q)$: At low constraint density α (left panel), $s(q)$ is a concave function of the solution-pair overlap q. Solution communities start to form as α exceeds a threshold value α_{cm} (middle panel) and the solution space becomes heterogeneous; this makes $s(q)$ to be non-concave. At the clustering transition point α_d (right panel), solution communities separate into different clusters, and $s(q)$ becomes non-monotonic. As α further increases, solution-pairs with intermediate overlap values may cease to exist and then $s(q) = -\infty$ for these intermediate q values.

solutions $\vec{\sigma}^1$ and $\vec{\sigma}^2$ of the space $\mathcal{S}$ can be measured by an overlap parameter defined as

$$q(\vec{\sigma}^1, \vec{\sigma}^2) = \frac{1}{N} \sum_{i=1}^{N} \sigma_i^1 \sigma_i^2 \ . \tag{2}$$

To characterize the statistical property of the solution space, we count the total number of solution-pairs with overlap value q and denote this number as $\mathcal{N}(q)$. For a random K-SAT formula of size $N \geq 1$ and constraint density $\alpha < \alpha_s(K)$, the size of $\mathcal{S}$ is exponential in N, and $\mathcal{N}(q)$ is also exponential in N. It is helpful to define an entropy density $s(q)$ as $s(q) \equiv (1/N)\ln[\mathcal{N}(q)]$.

The solution space $\mathcal{S}$ has a natural graphical representation (solution graph), in which each solution is denoted as a node and two nodes are connected by an edge if (and only if) the corresponding solutions are interchangeable by a single-spin flip. When α is less than the clustering transition point $\alpha_d(K)$, this solution graph has only a single connected component (solution cluster) that contains all (or almost all) of the solutions [5]. As is well known in social and biological network studies (see, e.g., Refs. [20, 21]), a connected network component may have rich internal community structures. A community of a graph refers to a subset of nodes of the graph that are much more densely inter-connected with each other than with the remaining nodes of the graph. If communities exist in the solution graph of a large random K-SAT formula, many subsets of solutions (solution communities) can then be extracted, with the property that each solution community contains a set of solutions that are much more similar with each other than with the other solutions. If such solution communities are abundant and of considerable sizes, they will be statistically relevant and will be manifested in the form of the entropy density function $s(q)$ (see Fig. 1).

We define a partition function $Z(x)$ as

$$Z(x) \equiv \sum_{\vec{\sigma}^1 \in \mathcal{S}} \sum_{\vec{\sigma}^2 \in \mathcal{S}} \exp\Big(Nxq(\vec{\sigma}^1, \vec{\sigma}^2)\Big) = \sum_{q} \exp\Big[N(s(q) + xq)\Big] \ , \tag{3}$$

Journal of Physics: Conference Series **233** (2010) 012011 doi:10.1088/1742-6596/233/1/012011

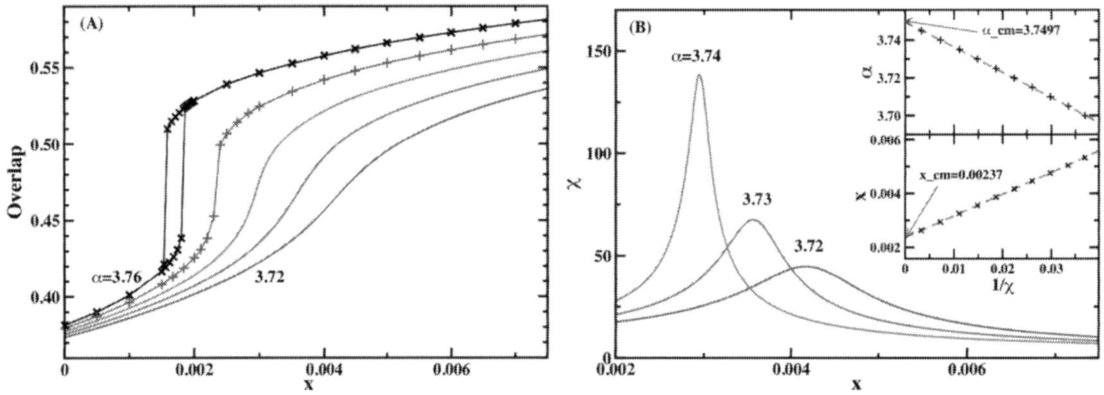

Figure 2. The mean overlap $\overline{q}(x)$ and the susceptibility $\chi(x)$ at constraint density value α for the random 3-SAT problem. In (A) α increases from 3.72 to 3.76 (right to left) with step size 0.01. The insets of (B) demonstrate that the peak value of χ diverges inverse linearly with α and x as the critical point ($\alpha_{cm} = 3.7497, x_{cm} = 0.0024$) is approached.

where x is an auxiliary binding field. The free entropy $\Phi(x)$ of the system is defined as $\Phi(x) \equiv \ln Z(x)$. In the thermodynamic limit of $N \to \infty$, the free entropy density $\phi(x) \equiv \Phi(x)/N$ is related to the entropy density $s(q)$ by

$$\phi(x) = \max_{q\in[-1,1]} [s(q) + xq] = s(\overline{q}(x)) + x\overline{q}(x) . \tag{4}$$

In Eq. (4), $\overline{q}(x)$ is the mean value of solution-solution overlaps under the binding field x, it is the argument value at which the function $s(q) + xq$ reaches the global maximum.

If the entropy density $s(q)$ is a concave function of $q \in [q_0, 1]$ (Fig. 1, left panel), where q_0 is the most probable solution-pair overlap value, then for each $x \geq 0$ there is only one mean overlap $\overline{q}$, and $\overline{q}(x)$ is a continuous function of x. On the other hand, if $s(q)$ is non-concave in $q \in [q_0, 1]$ (Fig. 1, middle and right panel), then the value of $\overline{q}(x)$ changes discontinuously at certain value of $x = x^* > 0$. We have exploited this correspondence between the non-concavity of $s(q)$ and the discontinuity of $\overline{q}(x)$ to determine the threshold constraint density α_{cm} at which $s(q)$ starts to be non-concave [16]. We regard $\alpha = \alpha_{cm}$ as the point at which the solution space of the random K-SAT problem transits into structural heterogeneity. This is because, as schematically shown in Fig. 1, at α_{cm} it starts to make sense to distinguish between intra-community overlap values and inter-community overlap values. As shown in Sec. 3, many solution communities can be identified in a heterogeneous solution space [17]. Each solution community contains a set of solutions which are more similar with each other than with the solutions of other communities. These differences of intra- and inter-community overlap values and the relative sparseness of solutions at the boundaries between solution communities cause the non-concavity of $s(q)$.

For the random K-SAT problem, we use the replica-symmetric cavity method of statistical mechanics [11] to calculate the mean overlap values $\overline{q}(x)$ at each value of $\alpha < \alpha_d(K)$. As the partition function Eq. (3) is a summation over pairs of solutions $(\vec{\sigma}^1, \vec{\sigma}^2)$, the state of each vertex is a pair of spins (σ, σ'). Details of this calculation can be found in Ref. [16] and here we cite the main results.

Figure 2A shows, for the random 3-SAT problem, the form of the function $\overline{q}(x)$ at several different α values. When $\alpha < \alpha_{cm}(3) \approx 3.75$, the mean overlap $\overline{q}$ increases with the binding field x smoothly, indicating that the entropy density function $s(q)$ is concave in shape (Fig. 1, left panel). The solution space $\mathcal{S}$ of the random 3-SAT problem is then regarded as homogeneous.

When $\alpha > \alpha_{cm}(3)$, there is a hysteresis loop in the $\overline{q}(x)$ curve as the binding field x increases and then decreases around certain threshold value x^*. This behavior is typical of a first-order phase-transition. At $x > x^*$ the partition function $Z(x)$ is contributed mainly by intra-community solution pairs, while at $x < x^*$ it is contributed mainly by inter-community solution pairs.

To determine precisely the critical value α_{cm}, we investigate the overlap susceptibility $\chi \equiv \mathrm{d}\overline{q}(x)/\mathrm{d}x$, which is a measures of the overlap fluctuations,

$$\chi(x) = \frac{1}{N}\sum_{i=1}^{N}\sum_{j=1}^{N}\Big[\langle\sigma_i^1\sigma_i^2\sigma_j^1\sigma_j^2\rangle - \langle\sigma_i^1\sigma_i^2\rangle\langle\sigma_j^1\sigma_j^2\rangle\Big] , \tag{5}$$

where $\langle\ldots\rangle$ means averaging over solution-pairs under the binding field x. From the divergence of the peak value of $\chi(x)$ as shown in Fig. 2B, we obtain that $\alpha_{cm}(3) = 3.7497$ for the random 3-SAT problem. This value is much below the value of $\alpha_d(3) = 3.87$ [5].

For the random 4-SAT problem, we find that $\alpha_{cm}(4) = 8.4746$. This value is again much below the clustering transition point $\alpha_d(4) = 9.38$ [5]. The difference $\alpha_d(K) - \alpha_{cm}(K)$ appears to be an increasing function of K.

3. Solution graph random walking

For single random K-SAT formulas, the solution space structural heterogeneity can also be detected by performing a long-time random walking in the corresponding solution graphs [17, 19, 22, 23]. The Hamming distance $d(\vec{\sigma}^1, \vec{\sigma}^2)$ between two solutions $\vec{\sigma}^1$ and $\vec{\sigma}^2$ of the solution space is defined as

$$d(\vec{\sigma}^1, \vec{\sigma}^2) = \sum_{i=1}^{N}\Big[1 - \delta(\sigma_i^1, \sigma_i^2)\Big] \tag{6}$$

where $\delta(\sigma, \sigma') = 1$ if $\sigma = \sigma'$ and $\delta(\sigma, \sigma') = 0$ if $\sigma = -\sigma'$. This distance counts the number of different spins between the two solutions. The Hamming distance $d(\vec{\sigma}^1, \vec{\sigma}^2)$ is related to the overlap $q(\vec{\sigma}^1, \vec{\sigma}^2)$ [Eq. (2)] through

$$q(\vec{\sigma}^1, \vec{\sigma}^2) = 1 - \frac{2d(\vec{\sigma}^1, \vec{\sigma}^2)}{N} . \tag{7}$$

In the solution graph of a satisfiable random K-SAT formula, a solution $\vec{\sigma}$ is linked to $k_{\vec{\sigma}}$ other solutions, all of which have unit Hamming distance with $\vec{\sigma}$. It was empirically found that the degrees $k_{\vec{\sigma}}$ of the solutions are narrowly distributed with a mean much less than N [17]. The solutions can therefore be regarded as equally important in terms of connectivity. However, the connection pattern of the solution graph can be highly heterogeneous. Even when the whole graph is connected, solutions may still form different communities such that the edge density of a community is much larger than that of the whole graph. The communities may even further organize into super-communities.

Consider two solutions $\vec{\sigma}^1$ and $\vec{\sigma}^2$ of the solution graph. The shortest-path length $l(\vec{\sigma}^1, \vec{\sigma}^2)$ between these two solutions in the solution graph satisfies the inequality $l(\vec{\sigma}^1, \vec{\sigma}^2) \geq d(\vec{\sigma}^1, \vec{\sigma}^2)$. If $\vec{\sigma}^1$ and $\vec{\sigma}^2$ belong to the same solution community, $l(\vec{\sigma}^1, \vec{\sigma}^2)$ may be equal to or just slightly greater than the Hamming distance $d(\vec{\sigma}^1, \vec{\sigma}^2)$. On the other hand, if $\vec{\sigma}^1$ and $\vec{\sigma}^2$ belong to two different solution communities, an extensive spin rearrangement may have to be made to change from $\vec{\sigma}^1$ to $\vec{\sigma}^2$ by single-spin flips, and then $l(\vec{\sigma}^1, \vec{\sigma}^2)$ is much greater than $d(\vec{\sigma}^1, \vec{\sigma}^2)$. If solutions are sampled by a random walking process at equal time interval Δt, the sampled solutions should contain useful information about the community structure of the solution space, with a resolution level depending on Δt. This is because that, when the edges are followed randomly by a random walker, the walker will be trapped in different communities most of the time, and the sampled solutions will then form different similarity groups.

Starting from an initial solution $\vec{\sigma}^0$, a sequences of solutions $(\vec{\sigma}^0, \vec{\sigma}^1, \ldots, \vec{\sigma}^l, \vec{\sigma}^{l+1}, \ldots)$ are generated by solution graph random walking. Two different random walking processes are used in the simulation. In the *unbiased* random walking process, the solution $\vec{\sigma}^{l+1}$ is a nearest neighbor of solution $\vec{\sigma}^l$ for $l = 0, 1, \ldots$ [17]. Under this simple dynamics, if the generated solution sequence is infinitely long, each solution $\vec{\sigma}$ in a connected component of the solution graph will appear in the sequence with frequency proportional to its connectivity $k_{\vec{\sigma}}$. On the other hand, in the *uniform* random walking process, with probability $k_{\vec{\sigma}^l}/N$, the solution $\vec{\sigma}^{l+1}$ $(l = 0, 1, \ldots)$ is a nearest neighbor of solution $\vec{\sigma}^l$, and with the remaining probability $1 - k_{\vec{\sigma}^l}/N$, solution $\vec{\sigma}^{l+1}$ is identical to $\vec{\sigma}^l$. Under this later dynamical rule, each solution in a connected component of the solution graph will have the same frequency to appear in the generated solution sequence. Both these two random walking processes were used to sample solutions, and the clustering analysis performed on these two sets of data gave qualitatively the same results [17]. The simulation results shown in Fig. 3 were obtained by the unbiased random walking process.

A set of solutions are sampled (with equal time interval Δt) from the generated long sequence of solutions for clustering analysis. (To avoid strong dependence on the input solution $\vec{\sigma}^0$, the initial part of the solution sequence was not used for sampling.) By calculating the overlap values between the sampled solutions, we obtain an overlap histogram as shown in Fig. 3. A hierarchical minimum-variance clustering analysis [24] (see also Ref. [25]) is performed on the sampled solutions. Initially each solution is regarded as a group, and the distance between two groups is the Hamming distance. At each step of the clustering, two groups C_a and C_b that have the smallest distance are merged into a single group C_c. The distance between the merged group C_c and another group C_d is calculated by

$$d(C_c, C_d) = \frac{(|C_a| + |C_d|)d(C_a, C_d) + (|C_b| + |C_d|)d(C_b, C_d) - |C_d|d(C_a, C_b)}{|C_c| + |C_d|}, \tag{8}$$

where $|C|$ is the number of solutions in group C. After the sampled solutions are listed in the order as reported by the minimal-variance clustering algorithm, the matrix of Hamming distances between these solutions are represented in a graphical form as shown in Fig. 3 (upper row).

For a random 4-SAT formula with $N = 20,000$ variables, the overlap histogram and the Hamming distance matrix of 1000 sampled solutions are shown in Fig. 3 for two different constraint density values $\alpha = 9.10$ and $\alpha = 9.22$ in the ergodic phase. At $\alpha = 9.10$, the Hamming distance matrix has a weak signature of the existence of many solution communities, and the overlap histogram is slightly non-concave. At $\alpha = 9.22$, the Hamming distance matrix has a very clear block structure and the overlap histogram is non-monotonic. These observations suggest that the explored solution spaces are heterogeneous both at $\alpha = 9.10$ and $\alpha = 9.22$; the entropic trapping effect of the solution communities becomes much stronger as α increases from 9.10 to 9.22. The simulation results are consistent with the analytical results of the preceding section, they are also in agreement with the expectation that, as $\alpha_d(4)$ is approached from below, a random walker becomes more and more trapped in a single solution community and finally becomes impossible to escape from this single community at $\alpha \geq \alpha_d(4)$.

We have obtained similar simulation results on single random 3-SAT formulas [17]. The solution space random walking simulations confirm that the solution space of the random K-SAT problem is already very heterogeneous before the clustering transition point $\alpha_d(K)$ is reached. The simulation results reported in Ref. [17] also suggested that, at $\alpha > \alpha_d(K)$, the single solution clusters of the solution space of the random K-SAT problem are themselves quite heterogeneous in internal structure, which may be the reason underlying the non-convergence of the belief-propagation iteration process within a single solution cluster [19].

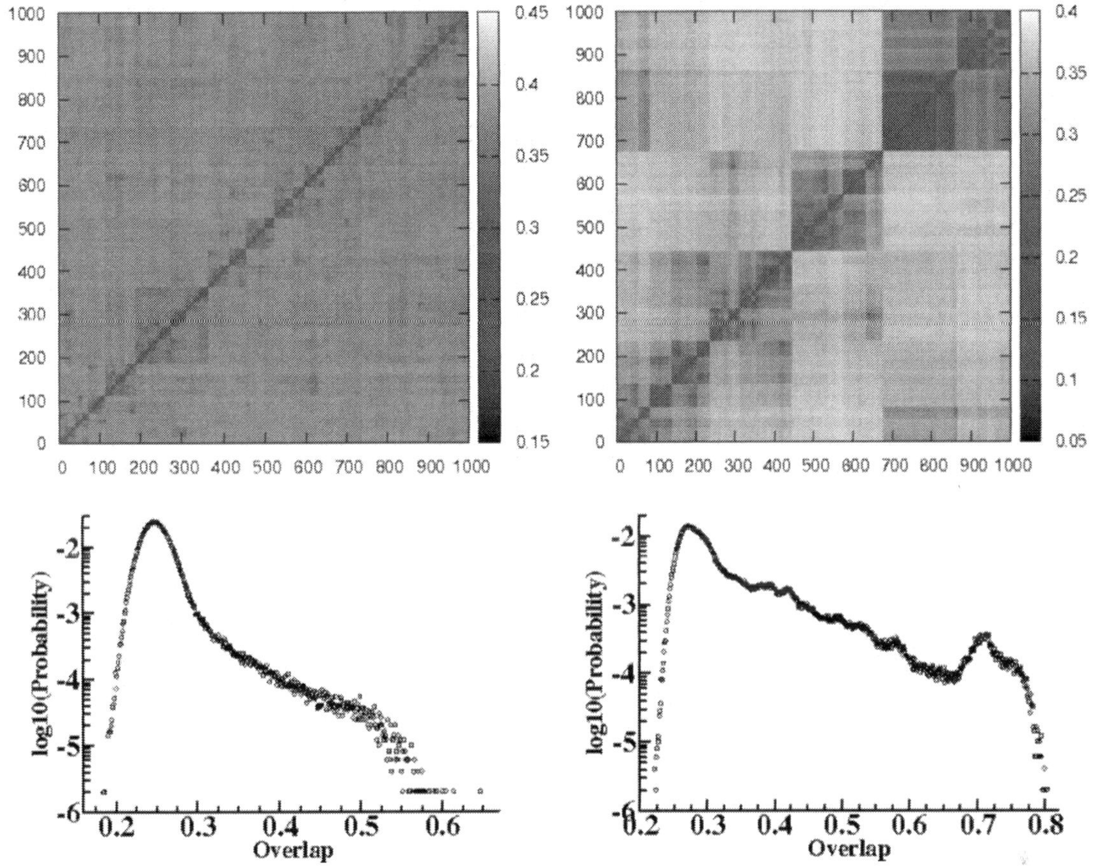

Figure 3. The matrix of Hamming distances of 1000 sampled solutions for a random 4-SAT formula of $N = 20,000$ variables, and the corresponding overlap distribution of these sampled solutions. The solutions were obtained by an unbiased random walking process (solution sampling began after running the random walking process for $3 \times 10^7 \times N$ steps starting from an initial solution, and two consecutively sampled solutions are separated by $\Delta t = 50,000 \times N$ random walking steps). The left panel corresponds to $\alpha = 9.10$ and the right panel to $\alpha = 9.22$. In the Hamming distance matrices, the solutions are ordered according to the results of the minimal-variance clustering analysis.

4. Glassy behavior of a stochastic search algorithm

After a slight change, the solution graph random walking process of the preceding section was turned into a stochastic local search algorithm [18]. This algorithm, referred to as `SEQSAT`, satisfies sequentially the constraints of a random K-SAT formula F. We denote by F_m the sub-formula containing the first m constraints of F. Suppose a configuration $\vec{\sigma}(t_m)$ that satisfies F_m is reached at time t_m. The $(m+1)$-th constraint of formula F is added to F_m to obtain the enlarged sub-formula F_{m+1}. Then, starting from $\vec{\sigma}(t_m)$, an unbiased random walk process is running on the solution graph of F_m until a spin configuration $\vec{\sigma}(t_{m+1})$ that satisfies F_{m+1} is first reached at time $t_{m+1} = t_m + \Delta t_m$ after $\Delta t_m \times N$ single-spin flips, N being the total number of variables in formula F. The waiting time of satisfying the $(m+1)$-th constraint is Δt_m (this time is zero if $\vec{\sigma}(t_m)$ already satisfies F_{m+1}). Starting from a completely random initial spin configuration $\vec{\sigma}(0)$ and an empty sub-formula F_0, every constraint of the formula F are satisfied

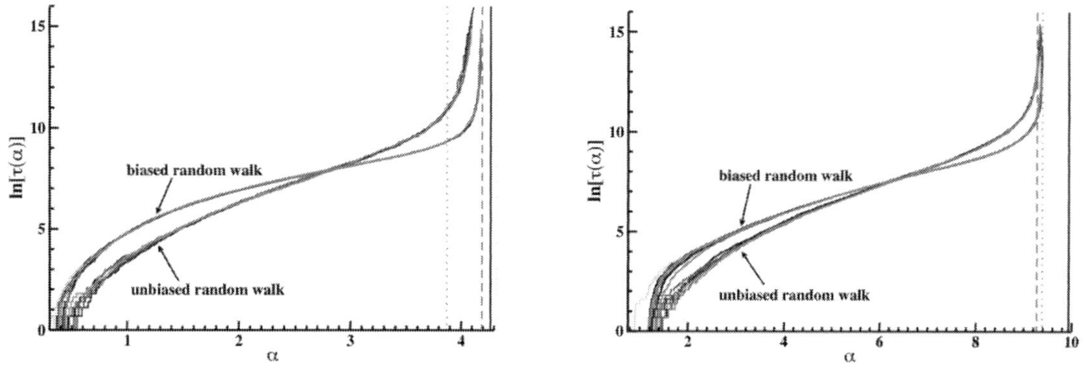

Figure 4. The total search time $\tau(\alpha)$ needed to satisfy the first $\alpha \times N$ constraints of a random K-SAT formula with $N = 100,000$ variables ($K = 3$, left panel; $K = 4$, right panel). The performances of an unbiased random-walk search process and a biased random-walk search process are compared. The red dashed lines correspond to the jamming transition points $\alpha_j(K)$ as predicted by a long-range-frustration mean-field theory [18, 26], the green dotted lines correspond to the clustering transition points $\alpha_d(K)$ [5], and the black solid lines correspond to the satisfiability threshold value $\alpha_s(K)$ [8].

by SEQSAT in this way. Notice that if a constraint was satisfied by SEQSAT, it remains to be satisfied as new constraints are added (energy barrier crossing is not allowed).

Figure 4 shows the simulation results of SEQSAT on a random 3-SAT formula and a random 4-SAT formula, $N = 10^5$. When the constraint density α of the satisfied sub-formula F_m is low, the waiting time needed to satisfy a constraint is very close to zero. As the heterogeneity transition point $\alpha_{cm}(K)$ is reached, however, the waiting time increases quickly and it starts to take more than $100 \times N$ single-spin flips to satisfy a constraint. The search process becomes more and more slow as α further increases. As α approaches another threshold value $\alpha_j(K)$ the waiting time is so long (exceeding $10^6 \times N$ single-spin flips) that SEQSAT essentially stops to satisfy a newly added constraint. The parameter $\alpha_j(K)$ is regarded as the jamming point of the random walk search algorithm. In the range of $\alpha_{cm}(K) < \alpha < \alpha_j(K)$ the SEQSAT is performing an increasingly viscous diffusion in the solution space of the random K-SAT formula. The simulation results of Fig. 4 and those reported in Ref. [18] for single random 5-SAT and 6-SAT formulas clearly demonstrate that, the solution space heterogeneity transition at $\alpha_{cm}(K)$ has significant dynamical consequences for stochastic local search processes.

For the random 3-SAT problem, the jamming point $\alpha_j(3)$ is significantly larger than the clustering transition point $\alpha_d(3)$. At $\alpha_d(3)$ the solution space of a large random 3-SAT formula is dominated by a sub-exponential number of solution clusters [5]. A dominating solution cluster also has internal community structures [17]. As α further increases, a dominating cluster shrinks in size, and it break into many sub-clusters. Figure 4 (left panel) indicates that, the random walker of SEQSAT is residing on one of the dominating clusters at $\alpha \approx \alpha_d(3)$ and it continue to be residing on one of the dominating sub-clusters of the visited cluster as α increases. If the spin values of a large fraction of variables become frozen in the residing solution cluster, a jamming transition then occurs. The value of $\alpha_j(3)$ is predicted to be 4.1897 by a long-range frustration mean-field theory [18, 26], in agreement with the simulation results.

For the random K-SAT problem with $K \geq 4$, the simulation results shown in Fig. 4 (right panel) and in Ref. [18] suggest that the jamming transition point $\alpha_j(K)$ is identical to or very close to the clustering transition point $\alpha_d(K)$. We notice that for the random K-SAT problem with $K \geq 4$, at $\alpha = \alpha_d(K)$, the union of an exponential number of small solution clusters is contributing predominantly to the solution space [5]. It is expected that, as α approaches $\alpha_d(K)$ from below, the ergodic solution space of a large K-SAT formula is dominated by an exponential number of small solution communities. The solutions reached by SEQSAT probably are residing on one of these small communities. As $\alpha \to \alpha_d(K)$, each of these small (but statistically relevant) solution communities probably contains a large fraction of variables that are almost frozen.

Figure 4 also shows the performances of a biased random walk search algorithm. The biased random walk process differs from the unbiased random walk process in that, in each single-spin flip, a variable that is flippable but not yet being flipped is flipped with priority [18]. This biased random walk SEQSAT algorithm was implemented with the hope of escaping from a solution community more quickly. For the random 3-SAT and 4-SAT problem, the biased SEQSAT algorithm indeed is more efficient than the unbiased algorithm. But for the random K-SAT problem with $K \geq 5$, the biased algorithm diverges earlier than the unbiased algorithm [18].

5. Conclusion and discussions

The solution space structure of the random K-SAT problem evolves with the constraint density α. Several qualitative transitions occurs in the solution space as α becomes relatively large. We demonstrated, both theoretically and by computer simulations, that the first qualitative structural transition is the heterogeneity transition at $\alpha = \alpha_{cm}(K)$, where exponentially many solution communities start to form in the (still ergodic) solution space. The dynamic behavior of a stochastic search algorithm was investigated. This simple algorithm SEQSAT constructs satisfying spin configurations for a single K-SAT formula by performing a random walk of single-spin flips. Due to the entropic trapping effect of solution communities, the solution space random walking process starts to be very viscous as α goes beyond $\alpha_{cm}(K)$. SEQSAT is able to find solutions for a random K-SAT formula with constraint density less than a threshold value $\alpha_j(K)$. For $K = 3$, the jamming point $\alpha_j(3) \approx 4.19$ is larger than the solution space clustering transition point $\alpha_d(3)$. But for $K \geq 4$, it appears that $\alpha_j(K)$ is very close to $\alpha_d(K)$.

When the constraint density α of a large random K-SAT formula is in the region of $\alpha_{cm}(K) < \alpha < \alpha_d(K)$, the solution space of the formula is heterogeneous but ergodic. If a random walking process is running on the solution space starting from an initial solution, a two-step relaxation behavior can be observed [17], corresponding to a quick slipping into a solution community, a relatively long wandering within this community, and finally the viscous diffusing in the whole solution space. Similar two-step relaxation behaviors were observed in glassy dynamics studies of supercooled liquids (see, e.g., the review article [27]) both in the non-activated dynamics region and in the activated dynamics region.

The existence of solution communities in the solution space may lead to a phenomenon that is similar to the dynamical heterogeneity of supercooled liquids (see, e.g., review articles [28, 29]). For a random walking process on a heterogeneous solution space, if the observation time is less than the typical relaxation time of escaping from the solution communities, one may find that the spin values of a large fraction of variables change only very infrequently among the visited solutions, while the spin values of the remaining variables are flipped much more frequently. The variables of the K-SAT formula then divides into an active group and an inactive group in this observation time window. One may further observe that the active variables are clustered into many distantly separated sub-groups, each of which containing variables that are very close to each other [30]. Numerical simulations need to be performed to investigate in more detail the solution space random walking processes.

The constraint density α of the random K-SAT problem corresponds to the particle density ρ in supercooled liquid. In supercooled liquids, another important control parameter is the temperature T. We can also introduce a positive temperature T to the random K-SAT problem and investigate how the configuration space evolves with T. It is anticipated that, as T is lowered to a threshold value $T_{cm}(K, \alpha)$ the configuration space will also experience a heterogeneity transition [16]. Detailed theoretical and simulation results will be reported in a later work. For the fully-connected p-spin spherical model of glasses, it has already been shown [32, 33] that a quantity called the Franz-Parisi potential has a change in its concavity property as T is lowered (see [34] for more recent extended studies). Another related problem, the weight space property of the Ising perceptron, was studied in [35]. For this fully connected model, a change of concavity property was also predicted for its characteristic function.

The solution space heterogeneity transition also exists in the random K-XORSAT problem [18]. It may be a general feature of the configuration spaces of spin glass models on finite-connectivity random graphs.

Acknowledgments
The author thanks Kang Li, Hui Ma and Ying Zeng for collaborations. This work was partially supported by the National Science Foundation of China (Grant numbers 10774150 and 10834014) and the China 973-Program (Grant number 2007CB935903).

References

[1] Kirkpatrick S and Selman B 1994 *Science* **264** 1297–1301
[2] Achlioptas D 2001 *Theor. Comput. Sci.* **265** 159–185
[3] Monasson R and Zecchina R 1996 *Phys. Rev. Lett.* **76** 3881–3885
[4] Mézard M, Parisi G and Zecchina R 2002 *Science* **297** 812–815
[5] Krzakala F, Montanari A, Ricci-Tersenghi F, Semerjian G and Zdeborova L 2007 *Proc. Natl. Acad. Sci. USA* **104** 10318–10323
[6] Hartmann A K and Weigt W 2005 *Phase Transitions in Combinatorial Optimization Problems* (Weinheim, Germany: Wiley-VCH)
[7] Mezard M and Montanari A 2009 *Information, Physics, and Computation* (New York, USA: Oxford Univ. Press)
[8] Mertens S, Mézard M and Zecchina R 2006 *Rand. Struct. Algorithms* **28** 340–373
[9] Montanari A, Ricci-Tersenghi F and Semerjian G 2008 *J. Stat. Mech.: Theor. Exper.* P04004
[10] Zhou H 2008 *Phys. Rev. E* **77** 066102
[11] Mézard M and Parisi G 2001 *Eur. Phys. J. B* **20** 217–233
[12] Mézard M and Montanari A 2006 *J. Stat. Phys.* **124** 1317–1350
[13] Semerjian G 2008 *J. Stat. Phys.* **130** 251–293
[14] Ardelius J and Zdeborova L 2008 *Phys. Rev. E* **78** 040101(R)
[15] Gardner E 1985 *Nucl. Phys. B* **257** **[FS14]** 747–765
[16] Zhou H 2009 Criticality and heterogeneity in the solution space of random constraint satisfaction problems arXiv:0911.4328 (Int. J. Mod. Phys. B, in press)
[17] Zhou H and Ma H 2009 *Phys. Rev. E* **80** 066108
[18] Zhou H 2010 *Eur. Phys. J. B* **73** 617–624
[19] Li K, Ma H and Zhou H 2009 *Phys. Rev. E* **79** 031102
[20] Wasserman S and Faust K 1994 *Social Network Analysis: Methods and Applications* (New York: Cambridge University Press)
[21] Girvan M and Newman M E J 2002 *Proc. Natl. Acad. Sci. USA* **99** 7821–7826
[22] Zhou H 2003 *Phys. Rev. E* **67** 041908
[23] Zhou H 2003 *Phys. Rev. E* **67** 061901
[24] Jain A K and Dubes R C 1988 *Algorithms for Clustering Data* (Englewood Cliffs, NJ, USA: Prentice-Hall)
[25] Barthel W and Hartmann A K 2004 *Phys. Rev. E* **70** 066120
[26] Zhou H 2005 *New J. Phys.* **7** 123
[27] Cavagna A 2009 *Phys. Report* **476** 51–124
[28] Ediger M D 2000 *Annu. Rev. Phys. Chem.* **51** 99–128
[29] Glotzer S C 2000 *J. Non-Cryst. Solids* **274** 342–355
[30] A random K-SAT formula can be represented by a bipartite graph (see, e.g., Ref. [31]). In this bipartite graph, a circular node represents a variable and a square node represents a constraint, an edge between a variable node i and a constraint node a means that variable i is involved in constraint a. In this bipartite graph, the distance $d(i, j)$ between two variable nodes i and j is defined as the number of constraint nodes

on a shortest-length path between i and j. For example, if i and j both are constrained by the same constraint, then $d(i, j) = 1$; if i and j are not constrained by the same constraint but i and a variable k are involved in constraint a and j and k are involved in constraint b, then $d(i, j) = 2$.

[31] Kschischang F R, Frey B J and Loeliger H A 2001 *IEEE Trans. Infor. Theor.* **47** 498–519

[32] Franz S and Parisi G 1995 *J. de Physique I* **5** 1401–1415

[33] Franz S and Parisi G 1997 *Phys. Rev. Lett.* **79** 2486–2489

[34] Zdeborova L and Krzakala F 2010 Generalization of the cavity method for adiabatic evolution of Gibbs states arXiv:1003.2748

[35] Obuchi T and Kabashima Y 2009 *J. Stat. Mech.: Theor. Exp.* P12014

International Workshop on Statistical-Mechanical Informatics 2010 IOP Publishing
Journal of Physics: Conference Series **233** (2010) 012012 doi:10.1088/1742-6596/233/1/012012

Two models of spin glasses — Ising versus Heisenberg

Hikaru Kawamura
Department of Earth and Space Science, Faculty of Science, Osaka University, Toyonaka 560-0043, Japan

E-mail: kawamura@ess.sci.osaka-u.ac.jp

Abstract. Brief review is given on recent numerical research of the ordering of two typical models of spin glasses (SGs), the three-dimensional (3D) Ising SG and the 3D Heisenberg SG models. Particular attention is paid to the questions of whether there is a thermodynamic transition in zero field, what are the associated critical properties, what is the nature of the ordered state, particularly of a possible replica-symmetry breaking, and whether there is a thermodynamic transition in applied fields. The properties of the two models are contrasted, and possible relation to experiments is discussed.

1. Introduction

Spin glasses are the type of random magnets in which both ferromagnetic and antiferromagnetic interactions coexist and compete, thereby giving rise to the effects of frustration and quenched randomness. Spin glasses (SGs) have been regarded as a typical example of complex systems, and their ordering properties have been studied quite extensively, either experimentally, analytically or numerically [1]. Experimentally, we now have fairly convincing evidence that SG magnets exhibit an equilibrium phase transition at a finite temperature into the glassy ordered state. This has been particularly well established in canonical SGs which represent a class of dilute transition-metal alloys soluted in the noble-metal host.

In theoretical studies of the ordering of SGs, a statistical model called the Edwards-Anderson (EA) model has been widely used [2]. In this model, spins are located on a certain regular lattice, say, on a three-dimensional (3D) simple cubic lattice, and are assumed to interact via the random exchange interaction taking either ferromagnetic or antiferromagnetic sign. One crucially important ingredient of the EA model is its spin symmetry. In its original version as proposed by Edwards and Anderson, the spin was assumed to be the Heisenberg one with three spin components. In fact, this is expected to be a good approximation for most of real SG materials, particularly for canonical SGs which possess relatively weak magnetic anisotropy. Another well-studied SG model is the Ising EA model with only one spin component. Although an ideal Ising SG is relative scarce in real SG materials, it has been widely used in numerous SG simulations performed in the past because of its simplicity. A naive expectation might be that the weak magnetic anisotropy inherent to real magnets eventually causes a crossover from the isotropic Heisenberg behavior to the anisotropic Ising behavior even in real SG materials [3], so that the Ising EA model corresponding to the strongly anisotropic limit is enough to describe the asymptotic behavior of weakly anisotropic SG magnets. Although such a naive expectation sounds plausible from the standard view of critical phenomena, the actual situation is not necessarily so simple as we shall see below.

Obviously, it is also crucially important to fully understand the ordering properties of the Heisenberg EA model, in particular, whether the Heisenberg EA model behaves similarly or differently from the Ising EA model. Indeed, via recent extensive studies, it now becomes increasing clear that the 3D Heisenberg EA model exhibits an intriguing ordering behavior quite different from that of its Ising counterpart. Among others, a special degree of freedom called a "chirality", totally absent in the corresponding Ising model, might play a crucial role in the ordering process of the 3D Heisenberg EA model [4], which might be crucially important in the ordering of real weakly anisotropic SG materials [4, 5].

In the present article, we wish to review the recent studies on the ordering properties of the two typical models of SGs, *i.e.*, the 3D Ising EA model versus the 3D Heisenberg EA model, focusing on their similarities and differences. Indeed, although the recent numerical studies have indicated that both the Ising EA model and the Heisenberg EA model exhibit a finite-temperature transition, the nature of the ordering is often in sharp contrast with each other. In the present article, we wish to focus on the following four issues.

(a) Is there a thermodynamic transition in zero field ?
(b) If so, what are the associated critical properties ?
(c) What is the nature of the ordered state ? Does it exhibit a replica-symmetry breaking (RSB) ? If yes, what type ?
(d) Is there a thermodynamic transition in applied magnetic fields ?

In the following, we shall examine the ordering properties of the 3D Ising SG in section 2, and those of the 3D Heisenberg SG in section 3. Section 4 is devoted to summary and discussion.

2. The Ising spin glass

In this section, we review the ordering properties of the 3D Ising EA model. The Hamiltonian of the 3D Ising EA model is given by,

$$\mathcal{H} = -\sum_{<ij>} J_{ij} S_i S_j - H \sum_i S_i \quad , \tag{1}$$

where $S_i = \pm 1$ is an Ising spin variable located at the i-th site of a 3D simple-cubic lattice, H is an external field intensity, and the $< ij >$ sum is taken over all nearest-neighbor pairs on the lattice. Periodic boundary conditions are applied in all directions. The nearest-neighbor couplings J_{ij} are quenched random variables with zero mean and standard deviation unity. For the form of the J_{ij} distribution, either the Gaussian distribution or the binary distribution ($\pm J$ distribution) is widely used.

The ordering properties of the 3D Ising EA model has been studied quite extensively for years as a standard model of SGs since the beginning of the SG research in 1970's. Note that the model has a finite-range (nearest-neighbor) interaction so that, in contrast to the mean-field-type (or an infinite-ranged) model such as the Sherrington-Kirkpatrick (SK) model [6], the correlation effect is taken into account in the model. In spite of its apparent simplicity, however, the ordering properties of the 3D Ising SG model turns out to be highly nontrivial and hard to be elucidated. So far, rigorous analytical information on this model has been scarce, and most of the information we now have for this model is necessarily a numerical one. Nevertheless, after more than 30 years of extensive research, we now have several established (or at least almost established) knowledge about the ordering properties of the 3D Ising EA model. We summarize some of them below, together with several open issues.

2.1. Ordering in zero field

In earlier studies, the issue of whether the 3D Ising EA model exhibits a finite-temperature transition or not remained controversial. Monte Carlo (MC) simulations on the model made by Ogielski and by Bhatt and Young in 1985 changed this situation, giving strong support to the occurrence of a finite-temperature SG transition [7, 8]. Since then, many numerical works have been made, all of which have agreed in that the 3D EA model exhibits a thermodynamic SG transition at a finite temperature [9, 10, 11, 12, 13, 14, 15, 16, 17, 18, 19]. In particular, the correlation-length ratio and the Binder ratio played an important role in identifying the SG transition point. In figure 1, the correlation-length ratio ξ_L/L, defined by the finite-size SG correlation length ξ_L divided by the linear size of the system L, calculated by Katzgraber *et al* for the $\pm J$ Ising EA model of $L \leq 32$ is shown, together with the Binder ratio calculated by the same authors [16]. The correlation-length ratio and the Binder ratio are dimensionless quantities so that the data of various L should be scale-invariant and exhibits a crossing behavior at the respective SG and chiral-glass transition points. Indeed, clear fan-out of the correlation-length ratio and the Binder ratio is observed at $\beta^{-1} = T \simeq 1.12$, indicating the occurrence of a finite-temperature SG transition.

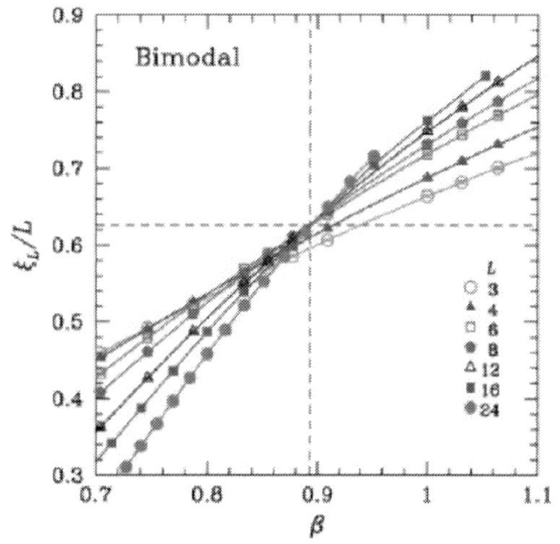

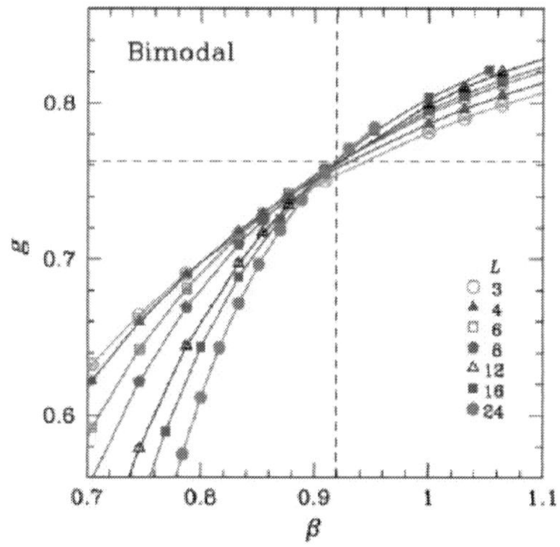

Figure 1. (Color online) The temperature and size dependence of the correlation-length ratio (upper figure), and the Binder ratio (lower figure), of the 3D Ising EA model with the binary ($\pm J$) coupling. β is the inverse temperature. Reprinted figures with permission from [H. G. Katzgraber, M. Körner and A. P. Young, Phys. Rev. B**73** 224432 (2006)]. Copyright (2006) by the American Physical Society. http://prb.aps.org/abstract/PRB/v73/i22/e224432

By contrast, reliable estimates of the associated SG exponents are more difficult. Concerning the critical-point-decay exponent (or the anomalous dimension) η, various simulations have given nearly a common value of $\eta \simeq -0.38 \sim -0.40$. By contrast, concerning the correlation-length exponent ν, smaller values around $\nu \simeq 1.2 - 1.3$ reported by earlier simulations [7, 8] have been revised by later simulations to significantly larger values. Indeed, most recent MC simulations have yielded $\nu \simeq 2.5 \sim 2.7$, together with $\eta \simeq -0.38 \sim -0.40$ [18, 19]. The SG susceptibility exponent γ is given by $\gamma \simeq 6.0 \sim 6.5$. As an example, we show in figure 2 the finite-size scaling plot of the correlation-length ratio of the $\pm J$ Ising EA model reported by Campbell *et al*, which yielded $\nu = 2.72(8)$ [18].

Although a good experimental realization of the Ising SG has been rather scarce, at least

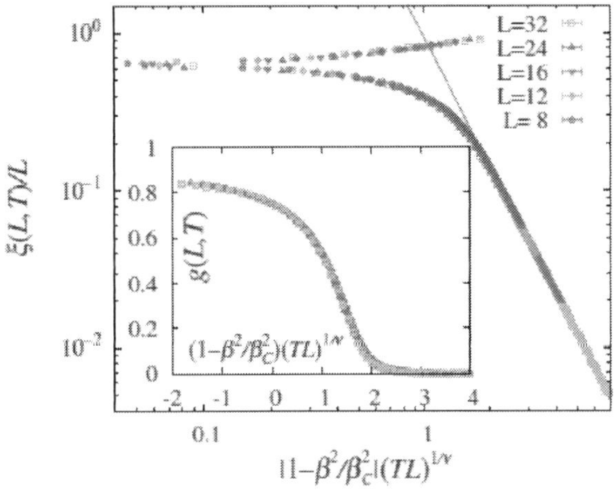

Figure 2. (Color online) Finite-size scaling plot of the correlation-length ratio of the 3D Ising EA model with the binary ($\pm J$) coupling. The exponent ν is taken to be 2.72. Taken from [I. A. Campbell, K. Hukushima, and H. Takayama: Phys. Rev. Lett. **97** (2006) 117202].

one example, an insulating SG magnet $FeMnTiO_3$, has been thoroughly studied. Indeed, the experimental exponent values measured for $FeMnTiO_3$ turns out to be roughly consistent with the numerical values of the 3D Ising SG [20, 5].

Meanwhile, in many of real SG magnets including canonical SGs, magnetic anisotropy is much weaker than the isotropic exchange interaction so that they might be described as a weakly anisotropic Heisenberg-like SG. It should be stressed that, in contrast to a naive expectation mentioned above, the experimentally determined SG critical exponents of such weakly anisotropic Heisenberg-like SGs largely deviate from the exponent values of the 3D Ising EA model. Indeed, for typical canonical SG materials, consistent experimental estimates are now available thanks to careful experimental measurements [21, 22, 23, 24, 25, 26, 27]. The exponents determined by various authors for canonical SGs indeed come close to each other, yielding the values $\beta \simeq 1$, $\gamma \simeq 2.2-2.3$, $\nu \simeq 1.3-1.4$ and $\eta \simeq 0.4-0.5$, which are clearly at odd with the exponent values of the 3D Ising EA model. The experimental ν and γ are about half and the sign of η is reversed.

To elucidate the origin of the discrepancy observed between the critical properties of experimental Heisenberg-like SG magnets and of the 3D Ising SG is one of important issues left in the SG research.

2.2. Possible replica-symmetry breaking

One of hot issues in the SG research has been concerned with the nature of the SG ordered state: In particular, whether the SG ordered state spontaneously breaks a replica symmetry or not [1]. Two typical views have been common. One is a droplet picture, which claims that the SG ordered state is a "disguised ferromagnet" without a spontaneous RSB [28]. The other is a hierarchical RSB picture inspired by the exact solution of the mean-field model, which claims that the SG ordered state is intrinsically more complex accompanied with a hierarchical or full RSB where the phase space is hierarchically organized in the SG ordered state [29]. Hot debate has continued over years concerning which view applies to the ordered state of real SG magnets.

For more quantitative discussion, it is convenient to introduce an "overlap" variable q, which

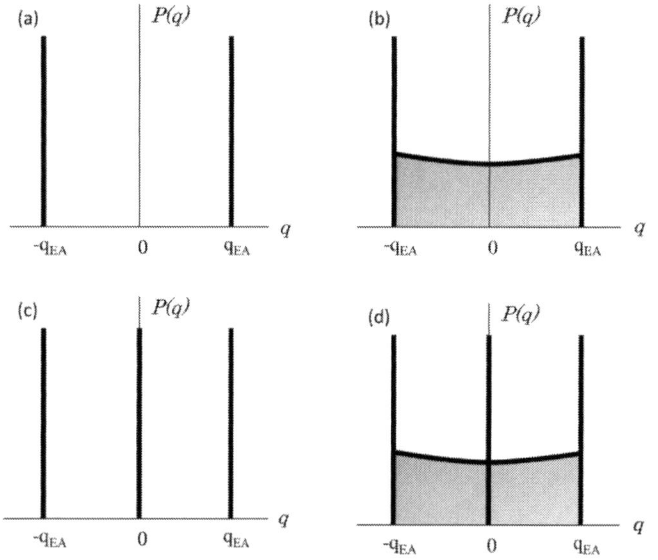

Figure 3. (Color online) Typical patterns of the overlap distribution function $P(q)$ in the thermodynamic limit. In the case (a) expected in a droplet theory, $P(q)$ consists of just two delta-function peaks at $q = \pm q_{EA}$. In the case of (b) expected in a hierarchical or full RSB picture, $P(q)$ possesses an additional plateau part connecting the two delta-function peaks at $q = \pm q_{EA}$. In the case of (c) expected in a one-step RSB picture, $P(q)$ possesses a central peak located at $q = 0$ in addition to two delta-function peaks at $q = \pm q_{EA}$. The case (d) is a combination of (b) and (c).

is defined by

$$q = \frac{1}{N} \sum_i S_i^{(\alpha)} S_i^{(\beta)}, \tag{2}$$

where $S_i^{(\alpha)}$ represents the i-th spin variable of the "replica" α and the summation is taken over all N spins of the system. Replicas α and β mean here the two independent copies of the system with the same realization of quenched randomness. One can then consider the distribution function of the overlap variable $P(q)$

$$P(q') = [< \delta(q - q') >], \tag{3}$$

where $< \cdots >$ represents a thermal average and $[\cdots]$ an average over the quenched disorder (configurational average). Some typical forms of $P(q)$ in the thermodynamic limit is illustrated in figure 3.

The droplet picture claims that the overlap distribution describing the SG ordered state to be a trivial one in the thermodynamic limit consisting of just two delta functions located at $q = q_{EA}$ and at $q = -q_{EA}$. It means that the SG ordered state consists of unique pure state and its symmetry counterpart, irrespective of its apparent complexity in real-space spin pattern.

In the hierarchical RSB picture, by contrast, $P(q)$ exhibits a continuous plateau part spanning between the two delta-function peaks at $q = \pm q_{EA}$. It means that the phase space is divided into infinitely many pure states organized in a hierarchical manner, each of which is separated by infinitely high free-energy barrier.

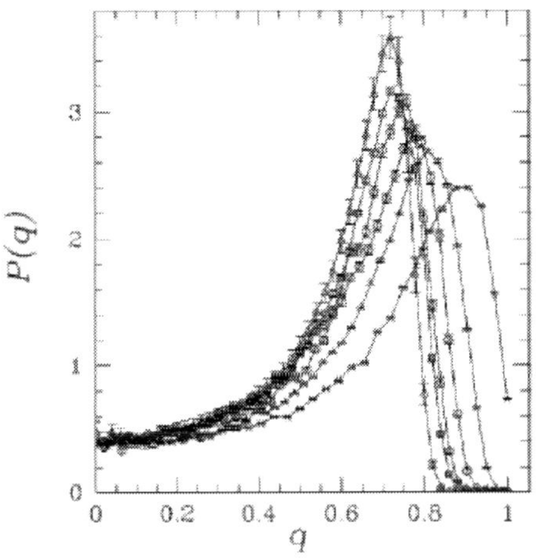

Figure 4. The overlap distribution function $P(q)$ in the spin-glass ordered state of the 3D Ising EA model with the Gaussian coupling. The temperature is $T = 0.7$ where T_g was estimated as $T_g = 0.95 \pm 0.04$. Only the positive side of q is shown. The sizes are $L = 4, 6, 8, 10, 12$ and 16. Reprinted figure with permission from [E. Marinari, G. Parisi and J.J. Ruiz-Lorenzo, Phys. Rev. **B58** 14852 (1998)]. Copyright (1998) by the American Physical Society. http://prb.aps.org/abstract/PRB/v58/i22/p14852_1

Other types of RSB have also been known. One well-known example might be a one-step RSB, in which $P(q)$ possesses a central δ-function peak at $q = 0$ in addition to the self-overlap peaks at $q = \pm q_{EA}$. In this case, the phase space is divided into many components, but all of them, except for itself and its symmetry partner, are completely dissimilar or orthogonal. It is realized, *e.g.*, in the ordered states of the mean-field Potts glass or of the mean-field p-spin model [1, 30, 31]. In the past, such a one-step RSB has often been discussed in the context of the structural-glass problem rather than SG problem. The combination of the full RSB and the one-step RSB is also possible in certain models.

Many numerical studies have focused on the 3D EA model to solve this issue of RSB, asking whether this model exhibits a hierarchical RSB or no RSB. The controversy has not yet been solved completely. However, most of the recent studies have indicated that the overlap distribution $P(q)$ persistently exhibits a nontrivial component suggesting the occurrence of a hierarchical-type RSB [11, 32, 33]. As a typical example, we show in figure 4 the overlap distribution $P(q)$ of the 3D Gaussian Ising EA model of various lattice sizes at a temperature $T = 0.7$ which is well below the estimated SG transition temperature $T_g = 0.95 \pm 0.04$ [11].

2.3. Ordering in finite fields

In contrast to the situation in zero field, the question of the existence/nonexistence of an equilibrium transition under external fields has remained quite controversial for years and not yet been completely settled. In fact, the question is closely related to the one discussed in the previous subsection, *i.e.*, that of an RSB. For example, if the SG ordered state of the 3D Ising EA model does not accompany an RSB, the model is likely to exhibit no finite-temperature transition under finite fields since applied fields destroy the global Z_2 spin-inversion symmetry, a symmetry broken at the zero-field transition. If, on the other hand, the SG ordered state of the model accompanies an RSB, a finite-temperature transition associated with RSB may become possible even in the situation where no global symmetry is left under fields, though it does not immediately necessitate the occurrence of an in-field transition.

Although there is no completely consensus among researchers, most of recent numerical studies point to the absence of an equilibrium transition under finite fields [34, 35]. Such numerical observations seem consistent with the experimental observation on $FeMnTiO_3$: For this compound, Mattsson *et al* indicated that there was no in-field phase transition [36], in sharp contrast to the zero-field case. Meanwhile, the absence of an equilibrium transition in finite fields seem in apparent contrast to the possible occurrence of a hierarchical-type RSB observed in zero field. Further study is required to clarify this point.

3. The Heisenberg spin glass

In the present section, we deal with the ordering of the 3D Heisenberg EA model. The Hamiltonian of the 3D Heisenberg EA model is given by,

$$\mathcal{H} = -\sum_{<ij>} J_{ij}\vec{S}_i \cdot \vec{S}_j - H\sum_i S_i^z \ , \tag{4}$$

where $\vec{S}_i = (S_i^x, S_i^y, S_i^z)$ is now a three-component unit vector located at the i-th site of a 3D simple-cubic lattice, H is an external field intensity applied along the z-direction, and the $< ij >$ sum is taken over all nearest-neighbor pairs. Periodic boundary conditions are applied in all directions. The nearest-neighbor couplings J_{ij} are quenched random variables with zero mean and standard deviation unity. As in the Ising case, either the Gaussian distribution or the binary distribution ($\pm J$ distribution) is widely used for the form of the J_{ij} distribution.

3.1. Ordering in zero field

The ordering properties of the 3D Heisenberg EA model have long been studied [37, 38, 39, 40, 41, 42, 43, 44, 45, 46, 47, 48, 49, 50, 51, 52, 53, 54, 55, 56, 57, 58, 59, 60, 61, 62, 63], and are hotly debated even now. Earlier numerical simulations on the model suggested in common that the model exhibited only a zero-temperature transition [37, 38, 39, 40, 41], in apparent contrast to experiments. Common attitude in the community at that time was to invoke the weak magnetic anisotropy inherent to real materials to explain this apparent discrepancy with experiments, assuming that the weak anisotropy caused a rapid crossover from the $T_g = 0$ Heisenberg behavior to the $T_g > 0$ Ising behavior [3]. In fact, however, the situation was not quite satisfactory as already discussed above, in view of the fact that the experimental exponent values measured for canonical SGs are actually far from the Ising SG values, and no clear sign of Heisenberg-to-Ising crossover has been observed in experiments.

In 1992, the present author suggested that the 3D Heisenberg EA model might exhibit a finite-temperature transition *in its chiral sector* [42]. Chirality is a multispin variable representing the sense or the handedness of the noncollinear or noncoplanar structures induced by frustration, *i.e.*, whether the frustration-induced noncollinear or noncoplanar spin structure is right- or left-handed. The scalar chirality χ is defined by the product of three neighboring spins by

$$\chi = \vec{S}_i \cdot \vec{S}_j \times \vec{S}_k. \tag{5}$$

It has subsequently been suggested that, in the ordering of the 3D Heisenberg SG, the chirality was "decoupled" from the spin, the chiral-glass order taking place at a temperature higher than the SG order, $T_{CG} > T_{SG}$ [45, 46, 57, 61, 62]. Based on such a spin-chirality decoupling picture of the 3D isotropic Heisenberg SG, a chirality scenario of experimental SG transition was proposed by the author [42, 61]: According to this scenario, the chirality is a hidden order parameter of real SG transition.

The chirality scenario consists of the two parts, *i.e.*, the "spin-chirality decoupling" part for a completely isotropic system and the "spin-chirality recoupling" part for a weakly anisotropic system [4]. The first part, the spin-chirality decoupling, is a key ingredient of the scenario. It claims that the fully isotropic 3D Heisenberg SG exhibits a peculiar two-step ordering process, in which the systems exhibits, with decreasing the temperature, first the glass ordering of the chirality at a finite temperature $T = T_{CG}$ spontaneously breaking only a discrete Z_2 symmetry with preserving the continuous $SO(3)$ symmetry, and at a lower temperature $T = T_{SG} < T_{CG}$ exhibits the glass ordering of the spin itself fully breaking both the Z_2 and $SO(3)$ symmetries. The higher transition at $T = T_{CG}$ associated with the discrete Z_2 symmetry breaking is called the "chiral-glass transition", while the intermediate phase between T_{CG} and T_{SG} where only the Z_2 symmetry is broken with preserving the continuous $SO(3)$ symmetry is called the "chiral-glass state".

The second part of the chirality scenario concerns the role of the weak anisotropy which inevitably exists in real SG magnets to certain extent. The chirality scenario claims that the weak random magnetic anisotropy, which reduces the Hamiltonian symmetry from $Z_2 \times SO(3)$ to only Z_2, weakly "mixes" the chirality to the spin sector, and the chiral-glass transition hidden in the chiral sector in fully isotropic system is 'revealed' in the spin sector in weakly anisotropic system. In this scenario, the chiral-glass transition of the fully isotropic system, not the SG transition of the isotropic system, dictates the SG of real weakly anisotropic SGs.

The chirality scenario is capable of explaining several long-standing puzzles concerning the experimental SG transition in a natural way, *e.g.*, the origin of the non-Ising critical exponents experimentally observed in canonical SGs, and remains to be an attractive hypothesis in consistently explaining various experimental observations for canonical SGs [4]. In recent numerical studies of the 3D Heisenberg EA model, although consensus now seems to appear that the 3D Heisenberg SG indeed exhibits a finite-temperature transition of some sort [57, 61, 58, 59, 62, 63], the nature of the transition, especially whether the model really exhibits the spin-chirality decoupling, is still under hot debate [47, 48, 49, 52, 53, 54, 56, 58, 59, 63].

Thus, the issue of whether the spin-chirality decoupling really occurs in the 3D Heisenberg SG has remained controversial in spite of the potential importance in understanding the nature of experimental SG ordering. While several numerical results in favor of the occurrence of the spin-chirality decoupling were reported [45, 46, 50, 51, 55, 57, 62, 63], a simultaneous spin and chirality transition without the spin-chirality decoupling was claimed in other works [47, 48, 49, 52, 53, 54, 56, 58, 59, 63]. The recent simulation of Ref.[62], however, has provided a fairly strong numerical support for the occurrence of the spin-chirality decoupling. This calculation indicates that the SG transition point T_{SG} is located about $10\% \sim 15\%$ below the chiral-glass transition point T_{CG}, which we shall briefly describe below.

In figure 5, we show the recent MC data of the correlation-length ratios for the chirality ξ_{CG}/L (a), and for the spin ξ_{SG}/L (b), of the isotropic 3D Heisenberg EA model with random Gaussian coupling [62]. The system contains total $N = L^3$ spins with L ranging from 6 to 32. For further details, refer to Ref.[62]. As can be seen from the figure, while the chiral ξ_{CG}/L curves cross at temperatures which are only weakly L-dependent, the spin ξ_{SG}/L curves cross at progressively lower temperatures as L increases.

As an other indicator of the transition, we show in figure 6 the Binder ratios for the chirality (a), and for the spin (b). The Binder ratios are also dimensionless, and are expected to exhibit a scale-invariant behavior at the respective chiral-glass and SG transition points. As can be seen from the figure, the chiral Binder ratio g_{CG} exhibits a negative dip which deepens with increasing L. The data of different L cross on the *negative* side of g_{CG} unlike the correlation-length ratio. These features indicate a finite-temperature transition in the chiral sector.

To estimate the bulk chiral-glass and SG transition temperatures quantitatively, we plot in figure 7 the crossing temperature of ξ_{CG}/L and ξ_{SG}/L for pairs of successive L values versus $1/L_{av}$, where L_{av} is a mean of the two sizes, together with the dip temperature T_{dip} and the crossing temperature T_{cross} of the chiral Binder ratio g_{CG}. The data show a near-linear $1/L_{av}$-

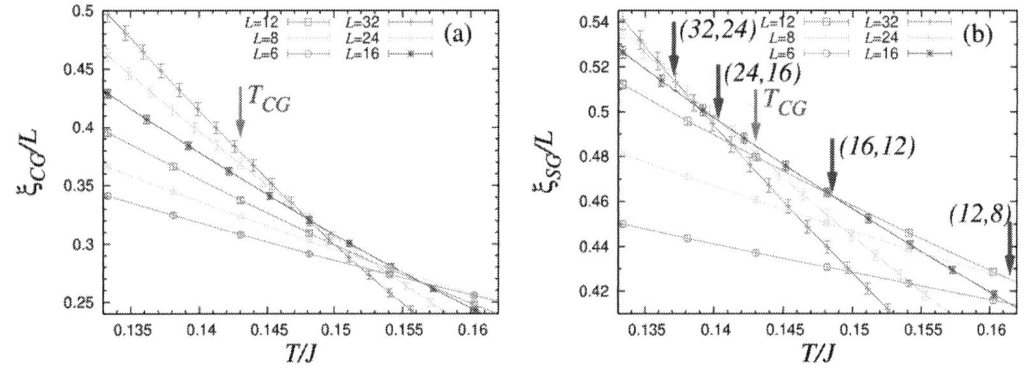

Figure 5. (Color online) The temperature and size dependence of the correlation-length ratio for the chirality (a), and for the spin (b). The arrow indicates the bulk chiral-glass transition point.

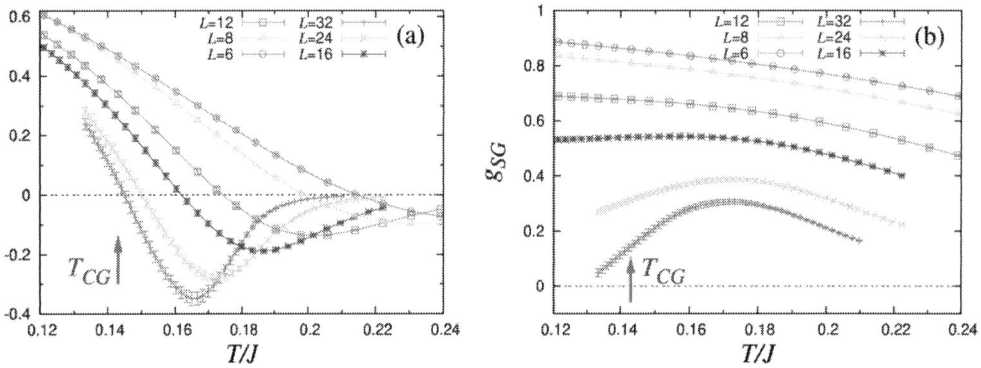

Figure 6. (Color online) The temperature and size dependence of the Binder ratio for the chirality (a), and for the spin (b). The arrow indicates the bulk chiral-glass transition point.

dependence. The chiral crossing temperatures of ξ_{CG}/L and of g_{CG} exhibits a weaker size dependence than the spin crossing temperature, and are extrapolated to $T_{CG} = 0.143 \pm 0.004$. The spin crossing temperature exhibits a stronger size dependence, which is extrapolated to $T_{SG} = 0.125 \pm 0.006$. Hence, T_{SG} is lower than T_{CG} by about $10\% \sim 15\%$.

Various chiral-glass exponents have been estimated via the standard finite-size scaling analysis. As an example, we show in figure 8 the finite-size scaling plot of the chiral correlation-length ratio ξ_{CG}/L in which the effect of the leading correction-to-scaling term has been taken into account. The exponents determined in this way from various physical quantities yield $\nu_{CG} = 1.5 \pm 0.2$ and $\eta_{CG} = 0.6 \pm 0.2$, *etc.*, which differ significantly from the standard 3D Ising SG values, $\nu \simeq 2.5 \sim 2.7$ and $\eta \simeq -0.38 \sim -0.40$ [18, 19]. The results indicate that the chiral-glass transition belongs to a universality class different from that of the 3D Ising SG. Possible long-range and/or many-body nature of the chirality-chirality interaction might be the cause of this difference.

Concerning the critical properties associated with the SG transition, which is likely to be located below the chiral-glass transition, the available information is still rather limited. The exponent η_{SG} was reported to be $\eta_{SG} \simeq -0.3$ in Ref.[62], while the exponent ν_{SG} to be $\nu_{SG} \simeq 1.5$

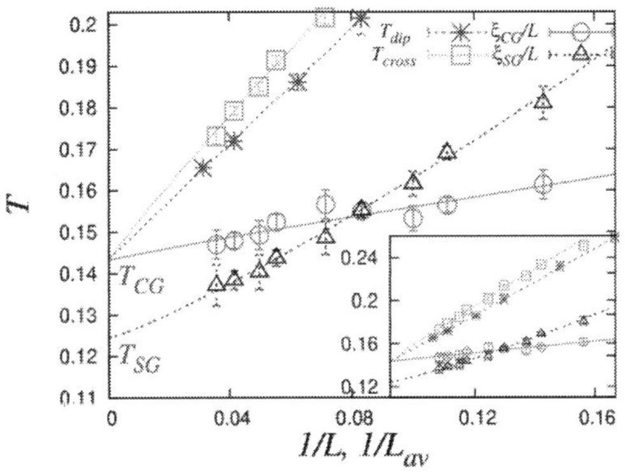

Figure 7. (Color online) The (inverse) size dependence of the crossing temperatures of ξ_{CG}/L and ξ_{SG}/L, the dip temperature T_{dip} and the crossing temperature T_{cross} of g_{CG}. The inset exhibits a wider range.

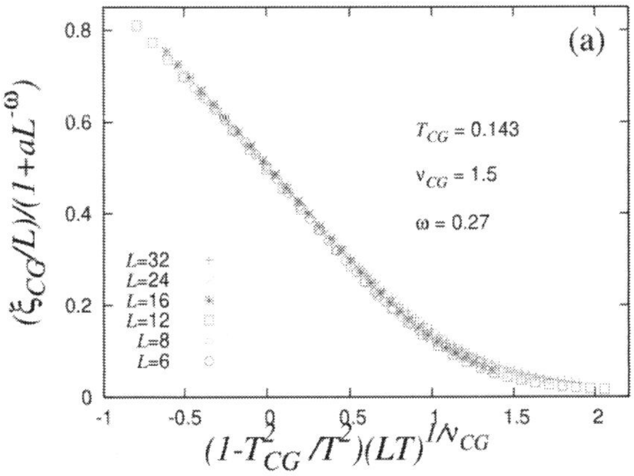

Figure 8. (Color online) Finite-size-scaling plot of the chiral-glass correlation-length ratio ξ_{CG}/L, where the correction-to-scaling effect is taken into account. The chiral-glass transition temperature and the leading correction-to-scaling exponents are $T_{CG} = 0.143$ and $\omega = 0.27$. The best fit is obtained with $\nu_{CG} = 1.5$.

in Ref.[63].

3.2. Possible replica-symmetry breaking

In this subsection, we discuss the issue of the nature of the chiral-glass ordered state, *i.e.*, whether the chiral-glass state exhibits an RSB, and if so, what type. In this connection, it should be noticed that the form of the chiral Binder ratio g_{CG} shown in figure 6, which exhibits a prominent negative dip, is quite peculiar. In fact, this form of the Binder ratio resembles the one of the system exhibiting a one-step RSB [30, 31].

In figure 9, we show the chiral-overlap distribution $P(q_\chi)$ in the chiral-glass phase calculated in Ref.[57] for the 3D Heisenberg SG model with the binary coupling. Here, the chiral overlap q_χ is defined by $q_\chi = \frac{1}{3N}\sum_{i\mu}\chi_{i\mu}^{(1)}\chi_{i\mu}^{(2)}$, where (1) and (2) indicate two copies (replicas) of the system. The local chirality variable $\chi_{i\mu}$ is defined at the ith site and in the μ-direction ($\mu = x, y, z$) by $\chi_{i\mu} = \vec{S}_{i-e_\mu}\cdot(\vec{S}_i\times\vec{S}_{i+e_\mu})$ where e_μ is a unit lattice vector along the μ-direction. The calculated

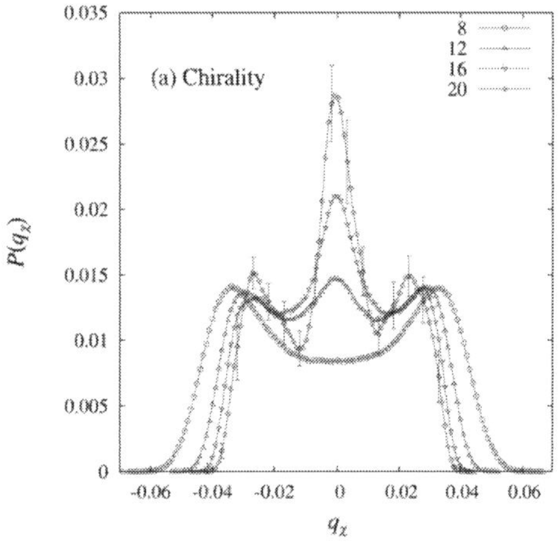

Figure 9. The overlap distribution function for the chirality of the 3D Heisenberg spin glass with the $\pm J$ binary coupling. The temperature is $T = 0.15$ below the chiral-glass transition temperature of this model $T = T_{CG} \simeq 0.19$. [From K. Hukushima and H. Kawamura, Phys. Rev. B**72** (2005), 144416.]

$P(q_\chi)$ exhibits besides symmetric side peaks located at $q_\chi = \pm q_\chi^{\rm EA}$ corresponding to the long-range chiral-glass order, which grow with increasing L, it also exhibits a prominent central peak at $q_\chi = 0$, which also grows with increasing L. The existence of such a pronounced central peak is a characteristic feature of the system exhibiting a one-step-like RSB, never seen in the Ising SG. The data strongly suggest that the chiral-glass ordered state exhibits a one-step-like RSB [46, 57, 62], in quite contrast to the one of the 3D Ising SG. Indeed, recent off-equilibrium experiments of the fluctuation-dissipation ratio of the Heisenberg SG magnet, thiospinel $CdCr_{1.7}In_{0.3}S_4$, seems consistent with such an expectation [4, 64]. Recent MC also indicates that the chiral-glass ordered state is non-self-averaging [57, 62].

3.3. Ordering in finite fields

In this subsection, we discuss the ordering of the 3D Heisenberg SG under finite fields [50]. When one applies a field to the fully isotropic Heisenberg SG, the symmetry of the Hamiltonian reduces from $Z_2 \times SO(3)$ in zero field to $Z_2 \times SO(2)$ under fields, where Z_2 refers to the chiral degeneracy associated with a spin-reflection operation (solely in spin space, not in real space) with respect to an arbitrary plane in spin space including the magnetic-field axis, while $SO(2)$ refers to the continuous degeneracy associated with a spin-rotation operation (in spin space, not in real space) around the magnetic-field axis in spin space.

Since the Z_2 chiral symmetry characterized by the sign of the scalar chirality remains under magnetic fields, the chiral-glass transition is expected to persist under magnetic fields. Of course, applied fields change the symmetry, but lower only the continuous part from $SO(3)$ to $SO(2)$. Since the continuous part has already been decoupled from the discrete Z_2 part, a natural expectation here would be that the chiral-glass transition persists even under fields essentially of the same type as the zero-field one. In particular, the chiral-glass transition line under fields should be a regular function of the filed intensity H. Since there is a trivial $H \leftrightarrow -H$ symmetry, the chiral-glass transition temperature under fields should behave for weak fields as $T_{CG}(H) \approx T_{CG}(0) - cH^2 \cdots$ (c is a constant). In fact, this yields a transition line resembling the so-called GT line of the mean-field model [65], $|T_{CG}(0) - T_{CG}(H)| \propto H^{1/2}$, although the origin of the exponent $1/2$ is entirely different: Here, $1/2$ is just of regular origin, whereas the

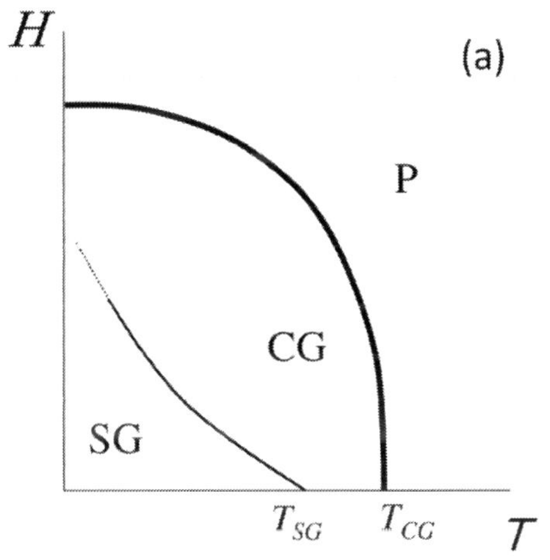

Figure 10. Typical phase diagram of the 3D isotropic Heisenberg SG in the magnetic field (H) - temperature (T) plane, expected from the chirality scenario. 'P', 'SG' and 'CG' represent the paramagnetic, spin-glass and chiral-glass phases, respectively.

exponent 1/2 in the mean-field model cannot be regarded as of regular origin. The expected phase diagram expected from the chirality scenario is sketched in figure 10.

The ordering associated with the continuous part should also occur under finite fields. The in-field transition line associated with the continuous $SO(2)$ symmetry breaking should be a continuation of the $SO(3)$ breaking SG transition in zero field. Since the broken symmetry is different in zero and finite fields, *i.e.*, $SO(3)$ versus $SO(2)$, the SG transition line at low fields should exhibit a singular form, $H \propto |T_{SG}(0) - T_{SG}(H)|^{\phi/2}$. The exponent ϕ is not yet precisely determined, but may roughly be estimated as $\phi = \beta_{SG} + \gamma_{SG} \approx 4$. This $SO(2)$ breaking transition line is also sketched in figure 10.

In-field ordering properties of the isotropic 3D Heisenberg SG were studied by Imagawa and the author also by means of MC simulations, which successfully demonstrates some of the feature of the phase diagram mentioned above [50, 55]. The obtained magnetic phase diagram is also consistent with the experimental one for canonical SGs [5, 66, 67]. In figure 11, we show the magnetic phase diagram of the 3D isotropic Heisenberg EA model with binary coupling as determined by MC simulation [50]. (Note that the SG transition line was not examined in this work). Another interesting observation from MC is that the SG ordered state turns out to be quite robust against applied magnetic fields as can be seen from figure 11 [50, 55]. It appears to be stable up to fields as large as $25k_BT_{SG}(H = 0)$. This might be understandable if one notices that the coupling between the chirality and magnetic fields might be rather weak, since magnetic fields couple directly to the spin via the Zeeman term, only indirectly to the chirality.

4. Summary and discussion

In this article, I reviewed the ordering properties of the two representative models of SGs, the 3D Ising EA model and the 3D Heisenberg EA model. Particular attention has been paid to i)

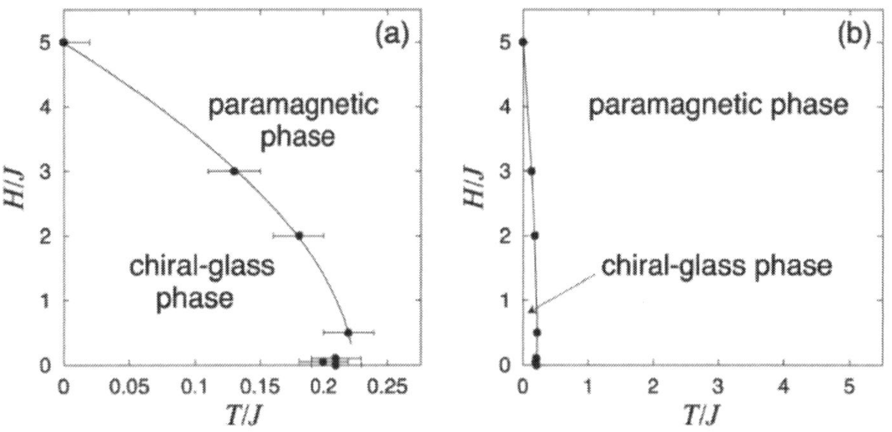

Figure 11. Magnetic phase diagram of the 3D isotropic Heisenberg SG in the magnetic field (H) - temperature (T) plane. Note that the energy scales of the magnetic field H and the temperature T are mutually different in (a), while they are taken to be common in (b). Taken from [D. Imagawa and H. Kawamura, J. Phys. Soc. Jpn. **71** (2002) 127].

whether the model exhibits a finite-temperature transition in zero field or not, and if it does, what are the critical properties of the transition; (ii) what is the nature of the ordered state, whether the model exhibits an RSB, and if yes, what type; (iii) whether the model exhibits a finite-temperature transition in finite fields.

Concerning the zero-field ordering, both models, *i.e.*, the 3D Ising and Heisenberg EA models, exhibit a finite-temperature transition. A single SG transition occurs in the Ising EA model, whereas successive chiral-glass and SG transitions are likely to occur in the Heisenberg EA model, though there still remains controversy about the occurrence of such spin-chirality decoupling. Concerning the associated critical properties, recent numerical estimates yield $\nu_{SG} = 2.5 \sim 2.7$ and $\eta_{SG} = -0.38 \sim -0.40$ for the 3D Ising SG, while, for the chiral-glass critical properties of the 3D Heisenberg SG, recent numerical estimates yield $\nu_{CG} \sim 1.5$ and $\eta_{CG} \sim 0.6$. For the SG exponents of the 3D Heisenberg SG at its SG transition, which is likely to be located below the chiral-glass transition, recent numerical calculation suggests $\nu_{SG} \sim 1.5$ and $\eta_{SG} \sim -0.3$. One sees from these results that the universality classes associated with the spin ordering of the Ising SG, the chiral and the spin orderings of the isotropic Heisenberg SG are all different from each other. Numerically observed critical properties of the 3D Ising SG are roughly consistent with the experimental values of the Ising-like SG magnet $FeTiO_3$. Numerically observed chiral-glass critical properties of the 3D Heisenberg SG are close to those experimentally observed for canonical SGs. This coincidence supports the chirality scenario of Refs.[42] which claims that the SG critical properties of real weakly anisotropic Heisenberg-like SG magnets such as canonical SGs should be equal to the chiral-glass (*not* SG !) critical properties of the fully isotropic Heisenberg SG.

Concerning the issue of the RSB in the ordered state, numerical simulations on the 3D Ising SG point to the occurrence of a hierarchical-type RSB like the one in the infinite-range SK model, although the situation has remained somewhat controversial. For the 3D Heisenberg SG, recent simulations suggest that the ordered state might exhibit a one-step-like RSB, the one quite different from that of the Ising SG or of the mean-field SK model. According to the chirality scenario, such a one-step-like RSB should be realized in real weakly anisotropic SG magnets.

Concerning the in-field ordering properties, recent numerical simulations suggest that the 3D Ising EA model does not exhibit a finite-temperature transition in magnetic fields, which seems consistent with the experiment on the Ising-like SG magnet $FeMnTiO_3$. By contrast, the 3D Heisenberg EA model exhibits a finite-temperature transition in magnetic fields associated with its chiral degrees of freedom, consistently with experimental observations for canonical SGs.

Thus, the ordering properties of the 3D Ising SG and the Heisenberg SG are quite different from each other, each of which might provide a useful reference in attacking various problems related to SGs, not only of random and/or frustrated magnets but also of many systems including glassy materials, complex systems and even information problem.

Acknowledgments

The author is thankful to I.A. Campbell, D.X. Viet, K. Hukushima, D. Imagawa for the collaboration and many useful discussion, and to T. Taniguchi, H. Yoshino, T. Okubo, M. Gingras, E. Vincent, M. Ocio, M. Picco and H. Takayama for discussion. This study was supported by Grant-in-Aid for Scientific Research on Priority Areas "Novel States of Matter Induced by Frustration" (19052006). We thank ISSP, Tokyo University and YITP, Kyoto University for providing us with the CPU time.

[1] For reviews on spin glasses, see *e. g.*, (a) Binder K and Young A P 1986 *Rev. Mod. Phys.* **58** 801; (b) Fischer K H and Hertz J A 1991 *Spin Glasses* (Cambridge: Cambridge University Press); (c) Mydosh J A 1993 *Spin Glasses* (London: Taylor & Francis); (d) Young A P (*ed.*) 2004 *Spin glasses and random fields* (Singapore: World Scientific); (e) Kawashima N and Rieger H 2004 in *Frustrated Spin Systems*, ed. H.T. Diep (Singapore: World Scientific)
[2] Edwards S F and Anderson P W 1975 *J. Phys. F: Met. Phys.* **5** 965
[3] Bray A J, Moore M A and Young A P 1986 *Phys. Rev. Lett.* **56** 2641
[4] For a review of the chirality in spin glasses, see Kawamura H 2010 *J. Phys. Soc. Jpn.* **79** 011007
[5] For a review of the experimental Heisenberg-like spin-glass ordering and the chirality, see Campbell I A and Petit D C M C 2010 *J. Phys. Soc. Jpn.* **79** 011006
[6] Sherrington D and Kirkpatrick S 1975 *Phys. Rev. Lett.* **35** 1972
[7] Ogielski A T and Morgenstern I 1985 *Phys. Rev. Lett.* **54** (1985) 928; Ogielski A T 1985 *Phys. Rev.* B **32** 7384
[8] R. N. Bhatt R N and Young A P 1985 *Phys. Rev. Lett.* **54** 924; 1988 *Phys. Rev.* B **37** 5606
[9] Kawashima N and Young A P 1996 *Phys. Rev.* B **53** R484
[10] Bernardi L W, Prakash S and Campbell I A 1996 *Phys. Rev. Lett.* **77** 2798
[11] Marinari E, Parisi G and Ruiz-Lorenzo J J 1988 *Phys. Rev.* B **58** 14852
[12] Palassini M and Caracciolo S 1999 *Phys. Rev. Lett.* **82** 5128
[13] Mari P O and Campbell I A 1999 *Phys. Rev.* E **59** 2653
[14] Ballesteros H G, Cruz A, Fernandez L A, Martin-Mayor V, Pech J, Ruiz-Lorenzo J J, Tarancon A, Tellez P, Ullod C L and Ungil C 2000 *Phys. Rev.* B **62** 14237
[15] Mari P O and Campbell I A 2002 *Phys. Rev.* B **65** 184409
[16] Katzgraber H G, Körner M and Young A P 2006 *Phys. Rev.* B **73** 224432
[17] Jorg T 2006 *Phys. Rev.* B**73** 224431
[18] Campbell I A, Hukushima K, and Takayama H 2006 *Phys. Rev. Lett.* **97** 117202
[19] Hasenbusch M, Pellissetto A, and Vicari E 2008 *J. Stat. Mech.: Theory Exp.* L02001; 2008 *Phys. Rev.* B **78** 214205
[20] Gunnarsson K, Svedlindh P, Nordblad P, Lundgren L, Aruga H and Ito A 1991 *Phys. Rev.* B **43** 8199
[21] Simpson M 1979 *J. Phys.* F **9** 1377
[22] de Courtenay N, Bouchiat H, Herdequint H and Fert A 1986 *J. Phys. (France)* **47** 1507
[23] Bouchiat H 1986 *J. Phys. (France)* **47** 71
[24] Lévy L P and Ogielsky A T 1986 *Phys. Rev. Lett.* **57** 3288
[25] Coles B R and Williams G 1988 *J. Phys. F* **18** 1279
[26] Fert A, de Courtenay N and Bouchiat H 1988 *J. Phys. (France)* **49** 1173
[27] Taniguchi T and Miyako Y 1988 *J. Phys. Soc. Jpn.* **57** 3520
[28] Fisher D S and Huse D A 1986 *Phys. Rev. Lett.* **56** 416; 1988 *Phys. Rev.* B **38** 373, 386
[29] Parisi G 1979 Phys. Rev. Lett. **43** 1754; 1980 *J. Phys. A* **13** L115, 1101, 1187; 1980 *Phil. Mag.* B **41** 677; 1983 *Phys. Rev. Lett.* **50** 1946

[30] Hukushima K and Kawamura H 2000 *Phys. Rev.* E **62** 3360
[31] Picco M, Ritort F and Sales M 2001 *Euro Phys. J.* B **19** 565
[32] Katzgraber H, Palassini M and Young A P 2001 *Phys. Rev.* B **63** 184422
[33] Katzgraber H and Young A P 2002 *Phys. Rev.* B **65** 214402
[34] Young A P and Katzgraber H G 2004 *Phys. Rev. Lett.* **93** 207203
[35] Jorg T, Katzgraber H G, and Krzakala F 2008 *Phys. Rev. Lett.* **100** 19720
[36] Mattsson J, Jonsson T, Nordblad P, ArugaKatori H and Ito A 1995 *Phys. Rev. Lett.* **74** 4305
[37] Banavar J R and Cieplak M 1982 *Phys. Rev. Lett.* **48** 832; Cieplak M K and Banavar J R 1984 *Phys. Rev.* B **29** 469
[38] McMillan W L 1985 *Phys. Rev.* B. **31** 342
[39] Olive J A, Young A P and Sherrington D 1986 *Phys. Rev.* B **34** 6341
[40] Matsubara F, Iyota T and Inawashiro S 1991 *Phys. Rev. Lett.* **67** 1458
[41] Yoshino H and Takayama H 1993 *Europhys. Lett.* **22** 631
[42] Kawamura H 1992 *Phys. Rev. Lett.* **68** 3785
[43] Kawamura H 1995 *J. Phys. Soc. Jpn.* **64** 26
[44] Kawamura H 1996 *Int. J. Mod. Phys.* **7** 345
[45] Kawamura H 1998 *Phys. Rev. Lett.* **80** 5421
[46] Hukushima K and Kawamura H 2000 *Phys. Rev.* E **61** R1008
[47] Matsubara F, Endoh S and Shirakura T 2000 *J. Phys. Soc. Jpn.* **69** 1927
[48] Endoh S, Matsubara F and Shrakura T 2001 *J. Phys. Soc. Jpn.* **70** 1543
[49] Matsubara F, Shirakura T and Endoh S 2001 *Phys. Rev.* B **64** 092412
[50] Kawamura H and Imagawa D 2001 *Phys. Rev. Lett.* **87** 207203; Imagawa D and Kawamura H 2002 *J. Phys. Soc. Jpn.* **71** 127
[51] Matsumoto M, Hukushima K and Takayama H 2002 *Phys. Rev.* B. **66** 104404
[52] Nakamura T and Endoh S 2002 *J. Phys. Soc. Jpn.* **71** 2113
[53] Lee L W and Young A P 2003 *Phys. Rev. Lett.* **90** 227203
[54] Berthier L and Young A P 2004 *Phys. Rev.* B **69** 184423
[55] Imagawa D and Kawamura H 2004 *Phys. Rev. Lett.* **92** 077204; 2004 *Phys. Rev.* B **70** 144412
[56] Picco M and Ritort F 2005 *Phys. Rev.* B **71** 100406
[57] Hukushima K and Kawamura H 2005 *Phys. Rev.* B **72** 144416
[58] Campos I, Cotallo-Aban M, Martin-Mayor V, Perez-Gaviro S and Tarancon A 2006 *Phys. Rev. Lett.* **97** 217204
[59] Lee L W and Young A P 2007 *Phys. Rev.* B **76** 024405
[60] Campbell I and Kawamura H 2007 *Phys. Rev. Lett.* **99** 019701
[61] Kawamura H 2007 *J. Mag. Mag. Mater.* **310** 1487
[62] Viet D X and Kawamura H 2009 *Phys. Rev. Lett.* **102** 027202; 2009 *Phys. Rev.* B **80** 064118
[63] Fernandez L A, Martin-Mayor V, Perez-Gaviro S, Tarancon A and Young A P, *Phys. Rev.* B **80** 024422
[64] Hérisson D and Ocio M 2002 *Phys. Rev. Lett.* **88** 257202; 2004 *Eur. Phys. J.* B **40** 283
[65] Gabay M and Toulouse G 1981 *Phys. Rev. Lett.* **47** 201
[66] Petit D, Fruchter L and Campbell I A 1999 *Phys. Rev. Lett.* **83** 5130
[67] Petit D, Fruchter L, and Campbell I A 2002 *Phys. Rev. Lett.* **88** 207106

Journal of Physics: Conference Series **233** (2010) 012013 doi:10.1088/1742-6596/233/1/012013

Statistical performance analysis by loopy belief propagation in Bayesian image modeling

Kazuyuki Tanaka†, Shun Kataoka†† and Muneki Yasuda†††

Graduate School of Information Sciences, Tohoku University, 6-3-09 Aramaki-aza-aoba, Aoba-ku, Sendai 980-8579, Japan

E-mail: †kazu@smapip.is.tohoku.ac.jp, ††xkataoka@smapip.is.tohoku.ac.jp, †††muneki@smapip.is.tohoku.ac.jp

Abstract. The mathematical structures of loopy belief propagation are reviewed for Bayesian image modeling from the standpoint of statistical mechanical informatics. We propose some schemes for evaluating the statistical performance of probabilistic binary image restoration. The schemes are constructed by means of the LBP, which is known as the Bethe approximation in statistical mechanics. We show some results of numerical experiments obtained by using the LBP algorithm as well as the statistical performance analysis for the probabilistic image restorations.

1. Introduction

Bayesian analysis and statistical-mechanical informatics are two of many useful approaches in probabilistic information processing systems [1, 2]. Bayesian analysis gives us a lot of statistical models for treating massive and realistic data. Statistical-mechanical informatics provides us with powerful algorithms including the Monte Carlo methods and the advanced mean-field methods. Particularly, the loopy belief propagation (LBP) is equivalent to the Bethe approximation in statistical mechanics [3]. Moreover, we can analyze the statistical properties of probabilistic information processing systems by using the spin glass theory [1, 2]. This fact is also regarded as one of advantages of statistical-mechanical informatics.

Image processing is one of the important problems in the computer sciences. It is defined on a finite size square lattice that consists of many pixels. The state at each pixel corresponds to light intensity and is expressed in terms of integers. When we consider binary image processing, the state for light intensity can be regarded as the Ising spin. Bayesian modeling for image processing is often systematically achieved as a posterior probabilistic model by defining a conditional probability for the data generating processes as well as prior probabilistic model. Although such probabilistic models usually involve a high computational complexity, they are often reduced to classical spin systems with spatially-non-uniform external fields and nearest-neighbour interactions. Some useful approximate algorithms have been proposed by applying advanced mean-field methods, including the LBP [4].

Other successful applications of statistical-mechanical informatics are statistical performance analysis in terms of the spin glass theory [1, 2]. The random average in the spin glass theory corresponds to the sample average with respect to data. In Bayesian modeling, data are often treated as random external fields or random interactions in the spin glass theory. In calculating statistical performances, we have to take averages with respect to all the possible data. These calculations can be regarded as the random average with respect to all the possible external random fields or random interactions. Statistical analysis has been achieved in low density parity check (LDPC) codes [5] and code division multiple access (CDMA) decoding [6] in information and communication technology by using the replica method. In Bayesian image analysis, Nishimori and Wong have calculated the statistical performance in binary image restoration problems by using the replica method [7]. However, their approach was based on spin systems with non uniform external fields on the complete graph as a prior probabilistic model, whereas we would consider spin systems on regular graphs, for example, on the square lattice.

We propose new schemes for evaluating the statistical performance of binary image restoration based on Bayesian statistics. The main idea has been published in [8]. The schemes are constructed by means of the LBP, which is known as the Bethe approximation in statistical mechanics. They are reduced to problems involving the solution of simultaneous integral equations for distributions of messages in the LBP. Our schemes are proposed for two cases. One of them is for the case in which the original image is given explicitly. The other one is for the case in which only the prior probability for the original image is given. In section 2, we explain the Bayesian modeling and the LBP algorithm in binary image analysis. In section 3, we give some explicit schemes of statistical performance analysis by means of the LBP. In each section, we show some results of the numerical experiments using the LBP algorithm and the statistical performance analyses for the probabilistic restoration of binary images. In section 4 we give some concluding remarks.

2. Bayesian Image Analysis and Loopy Belief Propagation

In this section, we explain the Bayesian modeling and the LBP algorithm in the binary image analysis. We show some results of numerical experiments using the LBP algorithm.

We consider an image on a square lattice $V \equiv \{i | i = 1, 2, \cdots, |V|\}$ such that each pixel takes one of the binary levels -1 and $+1$ corresponding to black and white, respectively. The intensities at pixel i in the original image and the degraded image are regarded as random variables denoted by F_i and G_i, respectively. Then the random fields of intensities in the original image and the degraded image are represented by $\boldsymbol{F} \equiv (F_1, F_2, \cdots, F_{|V|})$ and $\boldsymbol{G} \equiv (G_1, G_2, \cdots, G_{|V|})$, respectively. The actual original image and degraded image are denoted by $\boldsymbol{f} = (f_1, f_2, \cdots, f_{|V|})$ and $\boldsymbol{g} = (g_1, g_2, \cdots, g_{|V|})$, respectively.

In the present paper, it is assumed that the degraded image $\boldsymbol{g}$ is generated from the original image $\boldsymbol{f}$ by introducing additive white Gaussian noise with average 0 and variance σ^2. The conditional probability for the degradation process is given by

$$\Pr\{\boldsymbol{G} = \boldsymbol{g} | \boldsymbol{F} = \boldsymbol{f}, \sigma\} = \Big(\sqrt{\frac{1}{2\pi\sigma^2}}\Big)^{|V|} \prod_{i \in V} \exp\Big(-\frac{1}{2\sigma^2}(f_i - g_i)^2\Big). \tag{1}$$

Moreover, the prior probability of an original image $\boldsymbol{f}$ is assumed to be

$$\Pr\{\boldsymbol{F} = \boldsymbol{f} | \alpha\} = \frac{1}{\mathcal{Z}(\alpha)} \exp\Big(-\frac{1}{2}\alpha \sum_{\{i,j\} \in E} (f_i - f_j)^2\Big), \tag{2}$$

where $\mathcal{Z}(\alpha)$ is a normalization constant and E is the set of all the nearest-neighbour pairs of pixels, $\{i, j\}$, on the square lattice V.

By substituting equations (1) and (2) into the Bayes formula, we obtain

$$\Pr\{\boldsymbol{F} = \boldsymbol{f} | \boldsymbol{G} = \boldsymbol{g}, \alpha, \sigma\} = \frac{1}{\mathcal{Z}(\boldsymbol{g}, \alpha, \sigma)}$$

$$\times\exp\Big(-\frac{1}{2\sigma^2}\sum_{i\in V}(f_i-g_i)^2-\frac{1}{2}\alpha\sum_{\{i,j\}\in E}(f_i-f_j)^2\Big), \tag{3}$$

where $\mathcal{Z}(\boldsymbol{g},\alpha,\sigma)$ is a normalization constant. By using the LBP, we obtain the estimate of the original image $\widehat{\boldsymbol{f}}(\boldsymbol{g},\alpha,\sigma)=(\widehat{f}_1(\boldsymbol{g},\alpha,\sigma),\widehat{f}_2(\boldsymbol{g},\alpha,\sigma),\cdots,\widehat{f}_{|V|}(\boldsymbol{g},\alpha,\sigma))$ as follows:

$$\widehat{f}_i(\boldsymbol{g},\alpha,\sigma)=\mathrm{sign}\Big(\tanh(\frac{g_i}{\sigma^2}+\sum_{k\in\partial i}\lambda_{k\to i})\Big)=\mathrm{sign}\Big(\frac{g_i}{\sigma^2}+\sum_{k\in\partial i}\lambda_{k\to i}\Big)\ (i\in V), \tag{4}$$

$$\lambda_{j\to i}=\tanh^{-1}\Big(\tanh(\alpha)\tanh(\frac{g_j}{\sigma^2}+\sum_{k\in\partial j\backslash\{i\}}\lambda_{k\to j})\Big)\ (i\in V, j\in\partial i). \tag{5}$$

Here $\partial i\equiv\{j|\{i,j\}\in E\}$ is the set of all the nearest-neighbour pixels of the i-th pixel.

One of the examples of probabilistic image restorations by using the LBP is shown in figure 1. The original image $\boldsymbol{f}$ in figure 1(a) is randomly generated by means of the MCMC for the prior probability $\Pr\{\boldsymbol{F}=\boldsymbol{f}|\alpha=\alpha^*\}$ in equation (2) on the square lattice (V,E) with $|V|=128\times128$ and the periodic boundary conditions. Here α^* corresponds to the true value for α that the original image $\boldsymbol{f}$ is generated. In figure 1(a), we set $\alpha^*=0.465$ The degraded image $\boldsymbol{g}$ in figure 1(b) is randomly generated by adding the white Gaussian noise with average 0 and the variance $\sigma^2=1$ to the original image $\boldsymbol{f}$. The restored image $\widehat{\boldsymbol{f}}=\widehat{f}(\boldsymbol{g},\alpha^*,\sigma)$ in figure 1(c) is obtained by using the LBP algorithm in equations (4)-(5).

We introduce the mean square error (MSE) $D(\boldsymbol{f},\widehat{\boldsymbol{f}})$ between the original image $\boldsymbol{f}$ and the restored image $\widehat{\boldsymbol{f}}$:

$$D(\boldsymbol{f},\widehat{\boldsymbol{f}})\equiv\frac{1}{|V|}\sum_{i\in V}\Big(f_i-\widehat{f}_i\Big)^2, \tag{6}$$

and the improvement of signal to noise ratio (dB):

$$\Delta_{\mathrm{SNR}}\equiv10\log_{10}\left(\frac{D(\boldsymbol{f},\boldsymbol{g})}{D(\boldsymbol{f},\widehat{\boldsymbol{f}})}\right)\ (\mathrm{dB}). \tag{7}$$

The values of $D(\boldsymbol{f},\boldsymbol{g})$, $D(\boldsymbol{f},\widehat{\boldsymbol{f}})$ and Δ_{SNR} in figure 1 are shown in table 1.

Table 1. Estimates of hyperparameters α which is estimated from the original image $\boldsymbol{f}$ by using the maximum likelihood estimation (9). The estimates obtained by using the LBP in equations (10)-(11) and the Onsager solution in equation (12) are denoted by notations $\widehat{\alpha}_{\mathrm{LBP}}$ and $\widehat{\alpha}_{\mathrm{Onsager}}$, respectively.

	$D(\boldsymbol{f},\boldsymbol{g})$	$D(\boldsymbol{f},\widehat{\boldsymbol{f}})$	Δ_{SNR} (dB)
figure 1	0.9868	0.1892	7.2562

In the statistics, the maximum likelihood (ML) method is usually used for estimating the value of hyperparameter α from the given original image $\boldsymbol{f}$. In the framework, we regard the prior probability $\Pr\{\boldsymbol{F}=\boldsymbol{f}|\alpha\}$ as a likelihood function for the hyperparameter α and determine the estimate $\widehat{\alpha}$ of hyperparameter α so as to maximize $\Pr\{\boldsymbol{F}=\boldsymbol{f}|\alpha\}$ with respect to α as follows:

$$\widehat{\alpha}=\underset{\alpha}{\mathrm{argmax}}\Pr\{\boldsymbol{F}=\boldsymbol{f}|\alpha\}. \tag{8}$$

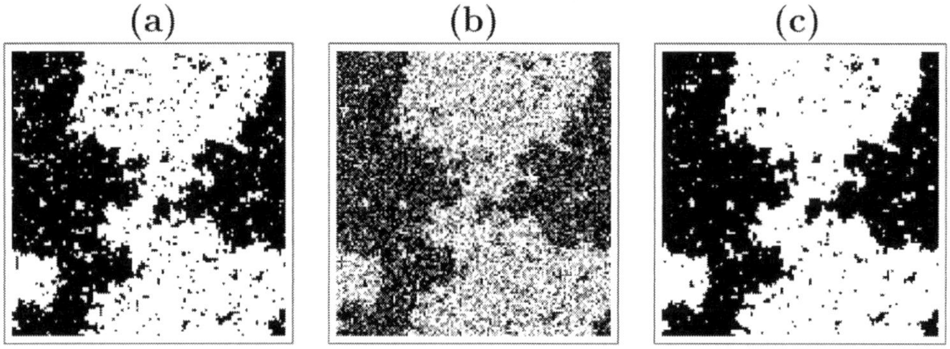

Figure 1. Image restorations by using the LBP. The original image $\boldsymbol{f}$ is randomly generated by means of the MCMC for the prior probability $\Pr\{\boldsymbol{F}=\boldsymbol{f}|\alpha=\alpha^*\}$ in equation (2). We set that $\alpha^*=0.465$. The degraded image $\boldsymbol{g}$ is randomly generated by adding the white Gaussian noise with average 0 and the variance $\sigma^2=1$ to the original image $\boldsymbol{f}$. The restored image $\widehat{\boldsymbol{f}}=\widehat{f}(\boldsymbol{g},\alpha^*,\sigma)$ is obtained by using the LBP algorithm in equations (4)-(5). (a) Original image $\boldsymbol{f}$ ($|V|=128\times128$). (b) Degraded image $\boldsymbol{g}$. (c) Restored image $\widehat{\boldsymbol{f}}=\widehat{\boldsymbol{f}}(\boldsymbol{g},\alpha^*,\sigma)$.

The extremum condition $\frac{\partial}{\partial\alpha}\Pr\{\boldsymbol{F}=\boldsymbol{f}|\alpha\}=0$ can be reduced to

$$\frac{1}{|E|}\sum_{\{i,j\}\in E}\sum_{\boldsymbol{z}}(z_i-z_j)^2\Pr\{\boldsymbol{F}=\boldsymbol{z}|\widehat{\alpha}\}=\frac{1}{|E|}\sum_{\{i,j\}\in E}(f_i-f_j)^2. \tag{9}$$

In the LBP, the left-hand side of equation (9) can be approximately expressed as follows:

$$\begin{aligned}\frac{1}{|E|}\sum_{\{i,j\}\in E}\sum_{\boldsymbol{z}}(z_i-z_j)^2\Pr\{\boldsymbol{F}=\boldsymbol{z}|\alpha\}&=\frac{1}{|E|}\sum_{\{i,j\}\in E}\sum_{z_i=\pm1}\sum_{z_j=\pm1}(z_i-z_j)^2\Pr\{F_i=z_i,F_j=z_j|\alpha\}\\&\simeq\frac{1}{|E|}\sum_{\{i,j\}\in E}\frac{4e^{-\alpha}\cosh\Big(\sum_{k\in\partial i\backslash\{j\}}\mu_{k\to i}-\sum_{l\in\partial j\backslash\{i\}}\mu_{l\to j}\Big)}{e^{\alpha}\cosh\Big(\sum_{k\in\partial i\backslash\{j\}}\mu_{k\to i}+\sum_{l\in\partial j\backslash\{i\}}\mu_{l\to j}\Big)+e^{-\alpha}\cosh\Big(\sum_{k\in\partial i\backslash\{j\}}\mu_{k\to i}-\sum_{l\in\partial j\backslash\{i\}}\mu_{l\to j}\Big)},\end{aligned} \tag{10}$$

In this equation, the quantities $\mu_{j\to i}$ ($i\in V$, $j\in\partial i$) correspond to the messages in the LBP. They are determined by the following simultaneous fixed point equations:

$$\mu_{j\to i}=\tanh^{-1}\Big(\tanh(\alpha)\tanh\big(\sum_{k\in\partial j\backslash\{i\}}\mu_{k\to j}\big)\Big)\ \ (i\in V,j\in\partial i). \tag{11}$$

They correspond to the message passing rules of the LBP. The values of $\frac{1}{|E|}\sum_{\{i,j\}\in E}\sum_{\boldsymbol{z}}(z_i-z_j)^2\Pr\{\boldsymbol{F}=\boldsymbol{z}|\alpha\}$ by using the LBP for $|V|=128\times128$ can be expressed in terms of thick lines in figure 2 for the cases of the periodic boundary conditions and of the free boundary conditions.

On the other hand, the quantity $\frac{1}{|E|}\sum_{\{i,j\}\in E}\sum_{\boldsymbol{z}}(z_i-z_j)^2\Pr\{\boldsymbol{F}=\boldsymbol{z}|\alpha\}$ can be calculated by means of the MCMC method. The values obtained by using the MCMC method for $|V|=128\times128$ are shown as open circles in figure 2. Moreover, the left-hand side of equation (9) for the infinite size limit $|V|\to+\infty$, can be exactly expressed as follows:

$$\lim_{|V|\to+\infty}\frac{1}{|E|}\sum_{\{i,j\}\in E}\sum_{\boldsymbol{z}}(z_i-z_j)^2\Pr\{\boldsymbol{F}=\boldsymbol{z}|\alpha\}$$

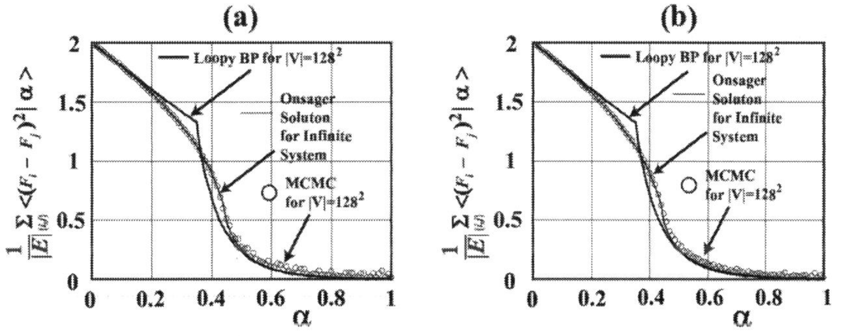

Figure 2. $\frac{1}{|E|}\sum_{\{i,j\}\in E}\langle (F_i-F_j)^2|\alpha\rangle = \frac{1}{|E|}\sum_{\{i,j\}\in E}\sum_{\boldsymbol{z}}(z_i-z_j)^2\Pr\{\boldsymbol{F}=\boldsymbol{z}|\alpha\}$, obtained by the LBP and by the MCMC method for the square lattice with $|V|=128\times128$. The values of the LBP are shown by thick lines and the values of MCMC method are by open circles. The values obtained by the Onsager solutions for $|V|\to+\infty$ are shown by thin lines. (a) Periodic boundary conditions. (b) Free boundary conditions.

$$= 2\left\{1-\alpha\coth(2\alpha)\left(1+\frac{2}{\pi}(2\tanh^2(2\alpha)-1)\int_0^{\pi/2}\sqrt{1-\Big(\frac{2\tanh(2\alpha)}{\cosh(2\alpha)}\Big)^2\sin(\theta)d\theta}\right)\right\}. \quad (12)$$

This expression is derived by using the Onsager solution for the two-dimensional Ising model. The values obtained by the Onsager solutions for $|V|\to+\infty$ are shown by thin lines. They show that the results for the infinite size system are close to the ones for the finite size system with $|V|=128\times128$.

Estimates of hyperparameters α are estimated from the original image $\boldsymbol{f}$ by using the ML method. We can apply the LBP method (10)-(11) to solve the deterministic equation (9) for $\widehat{\alpha}$ in the ML method. The maximum likelihood estimates for α obtained by the LBP and the ML methods are denoted by $\widehat{\alpha}_{\rm LBP}$. The estimates obtained by combining the Onsager solution in equation (12) with the deterministic equation (9) for $\widehat{\alpha}$ in the ML method are denoted by $\widehat{\alpha}_{\rm Onsager}$. The estimated values of $\widehat{\alpha}_{\rm LBP}$ and $\widehat{\alpha}_{\rm Onsager}$ for the original image $\boldsymbol{f}$ in figure 1 are shown in table 2.

Table 2. Estimates of hyperparameters α which are estimated from the original image $\boldsymbol{f}$ by using the maximum likelihood estimation (9). The estimates obtained by using the LBP in equations (10)-(11) and the Onsager solution in equation (12) are denoted by notations $\widehat{\alpha}_{\rm LBP}$ and $\widehat{\alpha}_{\rm Onsager}$, respectively. The value of the right-hand side of equation (9), $\frac{1}{|E|}\sum_{\{i,j\}\in E}(f_i-f_j)^2$, for each original image $\boldsymbol{f}$ is also given.

$\boldsymbol{f}$	α^*	$\widehat{\alpha}_{\rm LBP}$	$\widehat{\alpha}_{\rm Onsager}$	$\frac{1}{\lvert E\rvert}\sum_{\{i,j\}\in E}(f_i-f_j)^2$
figure 1(a)	0.4650	0.4343	0.4530	0.4624
figure 3(a)	—	0.4171	0.4416	0.5629
figure 4(a)	—	0.3991	0.4309	0.6925

Moreover, we show some numerical experiments for binary natural images $\boldsymbol{f}$ given in figures 3(a) and 4(a). The estimates of $\widehat{\alpha}_{\rm LBP}$ and $\widehat{\alpha}_{\rm Onsager}$ for these binary images $\boldsymbol{f}$ are also given in table 2. The degraded images $\boldsymbol{g}$ in figures 3(b) and 4(b) are randomly generated by adding the white Gaussian noise with average 0 and the variance $\sigma^2 = 1$ to the original images $\boldsymbol{f}$, respectively. The restored images $\widehat{\boldsymbol{f}} = \widehat{f}(\boldsymbol{g}, \widehat{\alpha}_{\rm LBP}, \sigma)$ in figures 3(c) and 4(c) are obtained by using the LBP algorithm in equations (4)-(5). The values of $D(\boldsymbol{f}, \boldsymbol{g})$, $D(\boldsymbol{f}, \widehat{\boldsymbol{f}})$ and $\Delta_{\rm SNR}$ for the results in figures 3 and 4 are shown in table 3.

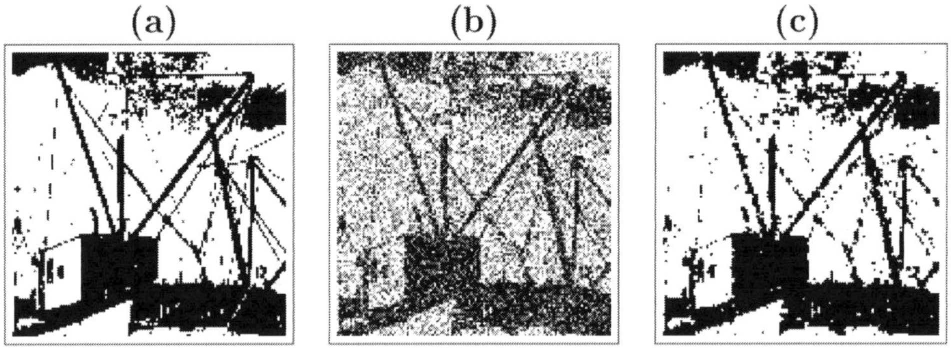

Figure 3. Image restorations by using the LBP. The original image $\boldsymbol{f}$ is given on the square lattice (V, E) with $|V| = 128\times128$. The degraded image $\boldsymbol{g}$ is randomly generated by adding the white Gaussian noise with average 0 and the variance $\sigma^2 = 1$ to the original image $\boldsymbol{f}$. The restored image $\widehat{\boldsymbol{f}} = \widehat{\boldsymbol{f}}(\boldsymbol{g}, \widehat{\alpha}_{\rm LBP}, \sigma)$ is obtained by using the LBP algorithm in equations (4)-(5) for $\alpha = \widehat{\alpha}_{\rm LBP}$. (a) Original image $\boldsymbol{f}$. (b) Degraded image $\boldsymbol{g}$. (c) Restored image $\widehat{\boldsymbol{f}} = \widehat{\boldsymbol{f}}(\boldsymbol{g}, \widehat{\alpha}_{\rm LBP}, \sigma)$.

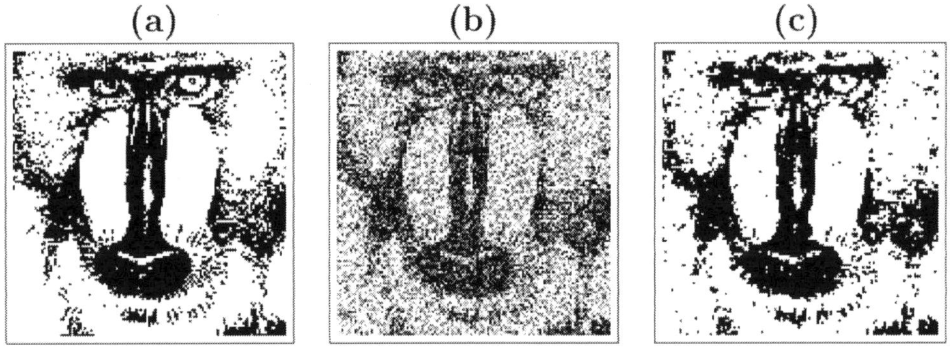

Figure 4. Image restorations by using the LBP. The original image $\boldsymbol{f}$ is on the square lattice (V, E) with $|V| = 128\times128$. The degraded image $\boldsymbol{g}$ is randomly generated by adding the white Gaussian noise with the average 0 and the variance $\sigma^2 = 1$ to the original image $\boldsymbol{f}$. The restored image $\widehat{\boldsymbol{f}} = \widehat{\boldsymbol{f}}(\boldsymbol{g}, \widehat{\alpha}_{\rm LBP}, \sigma)$ is obtained by using the LBP algorithm in equations (4)-(5). (a) Original image $\boldsymbol{f}$. (b) Degraded image $\boldsymbol{g}$. (c) Restored image $\widehat{\boldsymbol{f}} = \widehat{\boldsymbol{f}}(\boldsymbol{g}, \widehat{\alpha}_{\rm LBP}, \sigma)$.

3. Statistical Performance Analysis

In this section, we give some explicit schemes of statistical performance analyses by means of the LBP. We show some results of the numerical experiments of the LBP algorithm and statistical performance analyses for probabilistic restorations of binary images.

Table 3. Estimates of hyperparameters α which is estimated from the original image $\boldsymbol{f}$ by using the maximum likelihood estimation (9). The estimates obtained by using the LBP in equations (10)-(11) and the Onsager solution in equation (12) are denoted by notations $\widehat{\alpha}_{\rm LBP}$ and $\widehat{\alpha}_{\rm Onsager}$, respectively.

	$D(\boldsymbol{f},\boldsymbol{g})$	$D(\boldsymbol{f},\widehat{\boldsymbol{f}})$	$\Delta_{\rm SNR}$ (dB)
figure 3	0.9868	0.2288	5.9873
figure 4	0.9868	0.2881	4.5816

Now, we introduce the statistical average of the MSE between the original image $\boldsymbol{f}$ and the restored image $\widehat{\boldsymbol{f}}(\boldsymbol{g},\alpha,\sigma)$, defined by

$$\mathcal{D}(\alpha,\sigma|\alpha^*,\sigma^*) \equiv \sum_{\boldsymbol{f}}\int D(\boldsymbol{f},\widehat{\boldsymbol{f}}(\boldsymbol{g},\alpha,\sigma))\Pr\{\boldsymbol{G}=\boldsymbol{g}|\boldsymbol{F}=\boldsymbol{f},\sigma^*\}\Pr\{\boldsymbol{F}=\boldsymbol{f}|\alpha^*\}d\boldsymbol{g}. \quad (13)$$

If we fix the values of α^* and σ^*, the statistical average is taken over all the original images $\boldsymbol{f}$ according to the prior probability $\Pr\{\boldsymbol{F}|\alpha^*\}$ in equation (2) and all the degraded images $\boldsymbol{g}$ generated by adding the white Gaussian noise with average 0 and variance σ^{*2} to each original image $\boldsymbol{f}$. By using the LBP, we obtain the statistical performance of the restored images $\widehat{\boldsymbol{f}}(\boldsymbol{g},\alpha,\sigma)$ that are generated by using the LBP algorithm in binary image restorations as follows:

$$\begin{aligned}\mathcal{D}(\alpha,\sigma|\alpha^*,\sigma^*) &\simeq \frac{1}{|V|}\sum_{i\in V}\sum_{f_i=\pm1}\int_{-\infty}^{+\infty}dg_i\Big(\prod_{j\in\partial i}\int_{-\infty}^{+\infty}d\lambda_{j\to i}\Big)\Big(f_i-\mathrm{sign}\big(\frac{g_i}{\sigma^2}+\sum_{j\in\partial i}\lambda_{j\to i}\big)\Big)^2\\ &\quad\times\Big(\prod_{j\in\partial i}\rho_{j\to i}(\lambda_{j\to i})\Big)\Pr\{G_i=g_i|\alpha^*,\sigma^*\}. \end{aligned}\quad (14)$$

Here $\rho_{j\to i}(\lambda_{j\to i})$ ($j\in\partial i$, $i\in V$) are the distributions of messages $\lambda_{j\to i}$ and are determined by the following simultaneous integral equations:

$$\begin{aligned}\rho_{j\to i}(\lambda_{j\to i}) &= \int_{-\infty}^{+\infty}dg_j\Big(\prod_{k\in\partial j\backslash\{i\}}\int_{-\infty}^{+\infty}d\lambda_{k\to j}\Big)\\ &\quad\times\delta\Big[\lambda_{j\to i}-\tanh^{-1}\Big\{\tanh(\alpha)\tanh\big(\frac{g_j}{\sigma^2}+\sum_{l\in\partial j\backslash\{i\}}\lambda_{l\to j}\big)\Big\}\Big]\\ &\quad\times\Big(\prod_{k\in\partial j\backslash\{i\}}\rho_{k\to j}(\lambda_{k\to j})\Big)\Pr\{G_j=g_j|\alpha^*,\sigma^*\}.\end{aligned}\quad (15)$$

The marginal probability $\Pr\{G_i=g_i|\alpha^*,\sigma^*\}$ in equations (14) and (15) can be given by

$$\Pr\{G_i=g_i|\alpha^*,\sigma^*\}=\sqrt{\frac{1}{2\pi\sigma^{*2}}}\sum_{\boldsymbol{f}}\exp\Big(-\frac{1}{2\sigma^{*2}}(f_i-g_i)^2\Big)\Pr\{\boldsymbol{F}=\boldsymbol{f}|\alpha^*\}. \quad (16)$$

It is also computed by using the LBP algorithm as follows:

$$\Pr\{G_i=g_i|\alpha^*,\sigma^*\}\simeq\frac{\Big(\prod_{\{j,k\}\in E}\mathcal{Z}_{\{j,k\},i}(g_i,\alpha^*,\sigma^*)\Big)\Big(\prod_{j\in V}\mathcal{Z}_{j,i}(g_i,\alpha^*,\sigma^*)^{1-|\partial j|}\Big)}{\Big(\prod_{\{j,k\}\in E}\mathcal{Z}_{\{j,k\},i}(g_i,\alpha^*,+\infty)\Big)\Big(\prod_{j\in V}\mathcal{Z}_{j,i}(g_i,\alpha^*,+\infty)^{1-|\partial j|}\Big)}, \quad (17)$$

where

$$\mathcal{Z}_{j,i}(g_i,\alpha,\sigma) = 2\cosh\Big(\frac{g_i}{\sigma^2}\delta_{i,j} + \sum_{l\in\partial j}\kappa_{l\to j,i}\Big), \tag{18}$$

$$\begin{aligned}\mathcal{Z}_{\{j,k\},i}(g_i,\alpha,\sigma) &= 2e^{\alpha}\cosh\Big(\frac{g_i}{\sigma^2}\delta_{i,j} + \frac{g_i}{\sigma^2}\delta_{i,k} + \sum_{l\in\partial j\backslash\{k\}}\kappa_{l\to j,i} + \sum_{l\in\partial k\backslash\{j\}}\kappa_{l\to k,i}\Big)\\ &+2e^{-\alpha}\cosh\Big(\frac{g_i}{\sigma^2}\delta_{i,j} - \frac{g_i}{\sigma^2}\delta_{i,k} + \sum_{l\in\partial j\backslash\{k\}}\kappa_{l\to j,i} - \sum_{l\in\partial k\backslash\{j\}}\kappa_{l\to k,i}\Big). \end{aligned}\tag{19}$$

The messages $\{\kappa_{j\to k,i}|i,j\in V,\ k\in\partial j\}$ can be determined by solving the following simultaneous equations:

$$\kappa_{k\to j,i} = \tanh^{-1}\Big(\tanh(\alpha)\tanh(\frac{g_i}{\sigma^2}\delta_{i,k} + \sum_{l\in\partial k\backslash\{j\}}\kappa_{l\to k,i})\Big)\ \ (i\in V, k\in V, j\in\partial k). \tag{20}$$

We give the results of our numerical experiments in figure 5. In those experiments, we set $\alpha^* = 0.465$ and $\sigma^* = 1$. Figure 5 show the α-dependence of the statistical performance $\mathcal{D}(\alpha,\sigma^*|\alpha^*,\sigma^*)$. It is shown that the minimum point is at $\alpha = \alpha^*$. The solid circles are the sample averages of $D(\boldsymbol{f},\widehat{\boldsymbol{f}}(\boldsymbol{g},\alpha,\sigma^*))$ obtained by the LBP algorithm and the bars are their 99% confidence intervals. The sample averages and the confidence intervals are calculated from forty-nine pairs of the original image $\boldsymbol{f}$ and the degraded image $\boldsymbol{g}$. Seven original images are generated by according to the prior probability $\Pr\{\boldsymbol{F}=\boldsymbol{f}|\alpha^*\}$ and seven degraded images $\boldsymbol{g}$ are generated from each original image $\boldsymbol{f}$ by the additive white Gaussian noise with the average 0 and the variance $\sigma^{*2} = 1$.

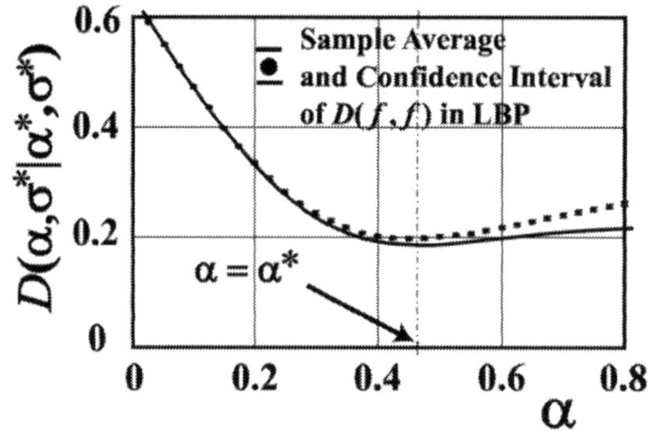

Figure 5. α-dependence of the statistical performance $\mathcal{D}(\alpha,\sigma^*|\alpha^*,\sigma^*)$ ($\alpha^*=0.465$, $\sigma^*=1$). The thick line is the statistical performance that is calculated by using equations (14)-(15) and equations (17)-(20). The solid circles are the sample averages of $D(\boldsymbol{f},\widehat{\boldsymbol{f}}(\boldsymbol{g},\alpha,\sigma^*))$ obtained by the LBP algorithm and the bars are their 99% confidence intervals. The sample averages and the confidence intervals are calculated from forty-nine pairs of the original image $\boldsymbol{f}$ and the degraded image $\boldsymbol{g}$.

Instead of equation (13), we can consider another statistical average defined by

$$\mathcal{D}(\alpha,\sigma|\boldsymbol{f},\sigma^*) \equiv \int D(\boldsymbol{f},\widehat{\boldsymbol{f}}(\boldsymbol{g},\alpha,\sigma))\Pr\{\boldsymbol{G}=\boldsymbol{g}|\boldsymbol{F}=\boldsymbol{f},\sigma^*\}d\boldsymbol{g}. \tag{21}$$

If we set an original image $\boldsymbol{f}$, the statistical average is taken over all the degraded images $\boldsymbol{g}$ generated by adding the white Gaussian noise with average 0 and variance σ^{*2} to the fixed original image $\boldsymbol{f}$. By using the Bethe approximation for random spin systems, we obtain the statistical performance for the restored images obtained by the LBP algorithm. The deterministic equations for $\mathcal{D}(\alpha,\sigma|\boldsymbol{f},\sigma^*)$ are derived by replacing $\Pr\{G_i=g_i|\alpha^*,\sigma^*\}$ with

$$\Pr\{G_i=g_i|F_i=f_i,\sigma^*\}=\sqrt{\frac{1}{2\pi\sigma^{*2}}}\exp\Big(-\frac{1}{2\sigma^{*2}}(f_i-g_i)^2\Big) \tag{22}$$

in equations (14)-(15). The statistical performance of the restored images $\widehat{\boldsymbol{f}}(\boldsymbol{g},\alpha,\sigma)$ that are generated by using the LBP algorithm in binary image restorations is derived as follows:

$$\begin{aligned}\mathcal{D}(\alpha,\sigma|\boldsymbol{f},\sigma^*) \;\simeq\; & \frac{1}{|V|}\sum_{i\in V}\int_{-\infty}^{+\infty}dg_i\Big(\prod_{j\in\partial i}\int_{-\infty}^{+\infty}d\lambda_{j\to i}\Big)\Big(f_i-\mathrm{sign}\big(\frac{g_i}{\sigma^2}+\sum_{j\in\partial i}\lambda_{j\to i}\big)\Big)^2 \\ & \times\Big(\prod_{j\in\partial i}\rho_{j\to i}(\lambda_{j\to i})\Big)\Pr\{G_i=g_i|F_i=f_i,\sigma^*\}. \end{aligned} \tag{23}$$

Here $\rho_{j\to i}(\lambda_{j\to i})$ $(j\in\partial i,\ i\in V)$ are the distributions of messages $\lambda_{j\to i}$ and are determined by the following simultaneous integral equations:

$$\begin{aligned}\rho_{j\to i}(\lambda_{j\to i}) \;=\; & \int_{-\infty}^{+\infty}dg_j\Big(\prod_{k\in\partial j\backslash\{i\}}\int_{-\infty}^{+\infty}d\lambda_{k\to j}\Big) \\ & \times\delta\Big[\lambda_{j\to i}-\tanh^{-1}\Big\{\tanh(\alpha)\tanh\big(\frac{g_j}{\sigma^2}+\sum_{l\in\partial j\backslash\{i\}}\lambda_{l\to j}\big)\Big\}\Big] \\ & \times\Big(\prod_{k\in\partial j\backslash\{i\}}\rho_{k\to j}(\lambda_{k\to j})\Big)\Pr\{G_j=g_j|F_j=f_j,\sigma^*\}. \end{aligned} \tag{24}$$

When we set the original image $\boldsymbol{f}$ to the one in figures 1(a), 3(a) and 4(a), α-dependence of the statistical performance $\mathcal{D}(\alpha,\sigma^*|\boldsymbol{f},\sigma^*)$ for $\sigma^*=1$ is shown in figure 6(a), (b) and (c), respectively. The solid circles are the sample averages of $D(\boldsymbol{f},\widehat{\boldsymbol{f}}(\boldsymbol{g},\alpha,\sigma^*))$ obtained by the LBP algorithm and the bars are their 99% confidence intervals. The sample averages and the confidence intervals. are calculated from forty-nine pairs $(\boldsymbol{f},\boldsymbol{g})$. Seven degraded images $\boldsymbol{g}$ are generated from each original image $\boldsymbol{f}$ by the additive white Gaussian noise with the average 0 and the variance $\sigma^{*2}=1$.

In figure 5, we see that the difference between the statistical performance $\mathcal{D}(\alpha,\sigma^*|\alpha^*,\sigma^*)$ and the results of numerical experiments using the LBP is not negligible in the region of large α. On the other hand, in figure 6, we see that the difference between $\mathcal{D}(\alpha,\sigma^*|\boldsymbol{f},\sigma^*)$ and the results of numerical experiments by the LBP is small in the whole region of $\alpha\in(0,1)$. Although, in the scheme of calculation of $\mathcal{D}(\alpha,\sigma^*|\boldsymbol{f},\sigma^*)$, $\Pr\{G_i=g_i|F_i=f_i,\sigma^*\}$ is derived as the explicit expression in equation (22), we have to calculate $\Pr\{G_i=g_i|\alpha^*,\sigma^*\}$ by using the LBP in the scheme of calculation of $\mathcal{D}(\alpha,\sigma^*|\alpha^*,\sigma^*)$. We conclude that the LBP procedure in calculating $\Pr\{G_i=g_i|\alpha^*,\sigma^*\}$ yields the reduction of the accuracy in $\mathcal{D}(\alpha,\sigma^*|\boldsymbol{f},\sigma^*)$.

In the numerical experiments of the LBP for figure 4(a), the confidence intervals are larger than the ones for figures 1(a) and 3(a). The prior probability $\Pr\{\boldsymbol{F}=\boldsymbol{f}|\alpha\}$ yields the second order phase transition in the infinite system $|V|\to+\infty$. The transition point is given as $\alpha=\frac{1}{2}\mathrm{arcsinh}(1)=0.4406...$ in the Onsager solution. In table 2, we remark that $\widehat{\alpha}_{\mathrm{Onsager}}$ for figure 4(a) is a little smaller than the transition point, although the corresponding values for figures 1(a) and 3(a) are larger than the transition point. We suggest that this is one of the reasons why the confidence intervals for figure 4(a) are larger than for the other cases.

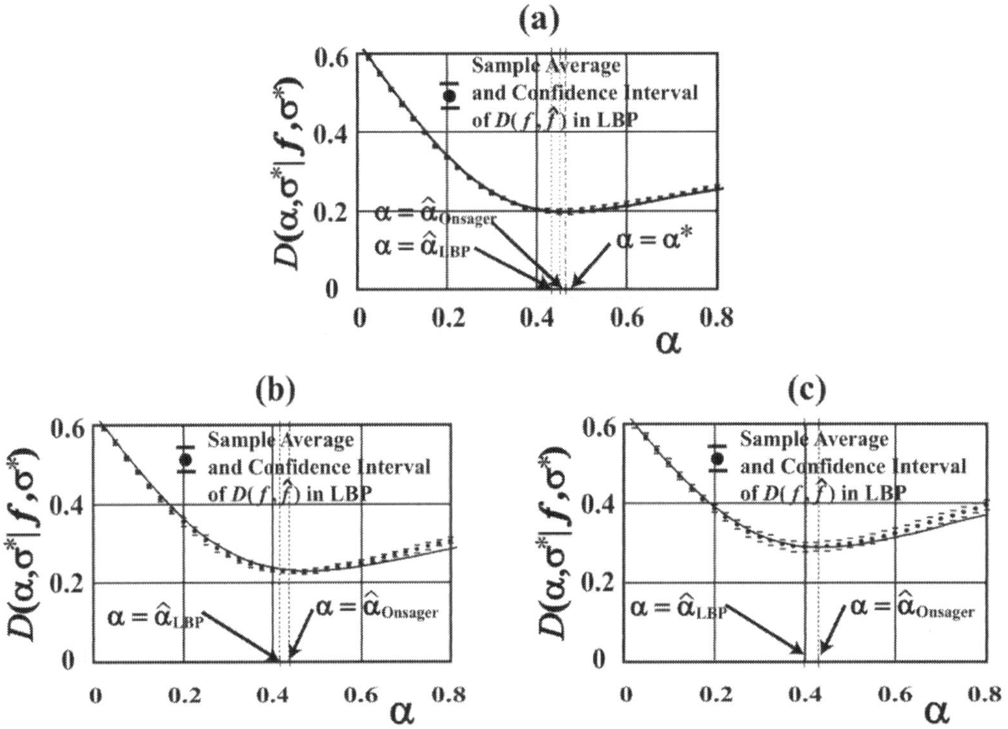

Figure 6. α-dependence of the statistical performance $\mathcal{D}(\alpha, \sigma^*|\boldsymbol{f}, \sigma^*)$ ($\alpha^* = 0.465$, $\sigma^* = 1$). The thick line is the statistical performance that is calculated by using equations (23)-(24) with equation (22). The solid circles are the sample averages of $D(\boldsymbol{f}, \widehat{\boldsymbol{f}}(\boldsymbol{g}, \alpha, \sigma^*))$ obtained by the LBP algorithm and the bars are their 99% confidence intervals. The sample averages and the confidence intervals are calculated from seven degraded images $\boldsymbol{g}$ that are generated by the additive white Gaussian noise with the average 0 and the variance σ^{*2}. The graphs of (a), (b) and (c) are calculated for the original images $\boldsymbol{f}$ in figures 1(a), 3(a) and 4(a), respectively.

4. Concluding Remarks

In the present paper, we have surveyed the Bayesian image modeling and have explained the practical algorithm for estimating the original image from each given degraded image based on the LBP method. We considered two kinds of statistical performances when the Ising model is adopted as the prior probabilistic model in the Bayesian image analysis. The algorithms for these statistical performances have been proposed in [8]. The algorithms of performance analyses have been constructed by using the LBP method and have been reduced to the computations of simultaneous integral equations for distributions of message in LBP.

The proposed algorithms give us approximate values of the statistical performances and are constructed on the square lattice. Our algorithms are applicable also to the statistical analyses for Bayesian modeling defined on more generalized graphs including hypergraphs, non-sparse random graphs and so on. This is one of future problems.

In section 2 of the present paper, we have discussed the estimation of hyperparameter α from the original image $\boldsymbol{f}$. The estimated value of α is very close to each optimal point for the statistical performance as shown in figure 6. Now we have to estimate the hyperparameters α and σ in the posterior probability from only one given degraded image without using the original image. The basic scheme and the practical algorithm for binary image restorations has been given in [4]. The statistical analysis for the hyperparameter estimation is one of interesting problems. In the Bayesian image modeling based on the expectation-maximization (EM) algorithm and the LBP as well as the generalized belief propagation for the Gaussian graphical model, we have investigated the statistical analysis of the hyperparameter estimation.

As the next step for practical applications, we have to consider multi-valued image modeling. Inoue and Carlucci [10], adopted Q-Ising models with infinite-range interactions as prior probabilities, and investigated the statistical performance by using the replica method. Inoue and Tanaka [11], extended it to the statistical analysis of the EM algorithm for the image restorations based on Q-Ising models with infinite-range interactions as prior probabilities. However these analyses have been done not for the square lattice but for the complete graph. In our work in progress, we are extending the framework to the Q-Ising model on the square lattice.

Acknowledgements
This work was partly supported by the Grants-In-Aid (No.18079002) for Scientific Research from the Ministry of Education, Culture, Sports, Science and Technology of Japan.

References

[1] Nishimori H 2001 *Statistical Physics of Spin Glass and Information Processing —Introduction—* (New York: Oxford University Press)
[2] Mézard M and Montanari A 2009 *Information, Physics and Computation* (New York: Oxford University Press)
[3] Kabashima Y and Saad D 1998 *Europhysics Letters* **44** 668
[4] Tanaka K 2002 *J. Phys. A: Math. Gen.* **35** R81
[5] Kabashima Y and Saad D 2004 *J. Phys. A: Math. Gen.* **37** R1
[6] Tanaka T 2002 *IEEE Trans. on Information Theory* **48** 2888
[7] Nishimori H and Wong K Y M 1999 *Phys. Rev.* E **60** 132
[8] Kataoka S, Yasuda M and Tanaka K 2010 *J. Phys. Soc. Jpn* **79** 025001
[9] Tanaka K and Titterington D M 2007 *J. Phys. A: Math. Theor.* **40** 11285
[10] Inoue J and Carlucci D M 2001 *Phys. Rev.* E **64** 036121
[11] Inoue J and Tanaka K 2003 *J. Phys. A: Math. Gen.* **36** 10997

International Workshop on Statistical-Mechanical Informatics 2010 IOP Publishing
Journal of Physics: Conference Series **233** (2010) 012014 doi:10.1088/1742-6596/233/1/012014

Asymptotic learning curve and renormalizable condition in statistical learning theory

Sumio Watanabe
P&I Lab., Tokyo Institute of Technology
4259 Nagatsuta, Midoriku, Yokohama 226-8503, Japan
E-mail: swatanab(at)pi.titech.ac.jp

Abstract. Bayes statistics and statistical physics have the common mathematical structure, where the log likelihood function corresponds to the random Hamiltonian. Recently, it was discovered that the asymptotic learning curves in Bayes estimation are subject to a universal law, even if the log likelihood function can not be approximated by any quadratic form. However, it is left unknown what mathematical property ensures such a universal law. In this paper, we define a renormalizable condition of the statistical estimation problem, and show that, under such a condition, the asymptotic learning curves are ensured to be subject to the universal law, even if the true distribution is unrealizable and singular for a statistical model. Also we study a nonrenormalizable case, in which the learning curves have the different asymptotic behaviors from the universal law.

1. Introduction

In recent studies, it was pointed out that Bayes statistics and statistical physics have the common mathematical structure, where the log likelihood function plays the same role as the random Hamiltonian, and the Bayes posterior distribution can be understood as the Boltzmann distribution. However, there are some differences between them. In statistical learning theory, the random Hamiltonian can not be necessarily approximated by any quadratic form because the Hessian matrix of the log likelihood function can be singular [1]. For example, artificial neural networks [2], normal mixtures [3], reduced rank regressions [4], Bayes networks [5], binomial mixtures, Boltzmann machines, and hidden Markov models are singular models.

The statistical properties of such models have been left unknown in statistics and information science, because it was difficult to analyze a singular likelihood function [1, 6]. Recently, new statistical learning theory has been established based on algebraic geometry, by which it was proved that the generalization and training errors are subject to a universal law, even if the statistical model does not satisfy the regularity condition [7, 8, 9, 10, 11, 12]. However, it is not yet clarified what mathematical properties ensure that such a universal law holds, therefore it is unknown the range of statistical problems which are subject to the universal law.

In this paper, we define a renormalizable condition of a statistical problem. The renormalizable condition requires that the variance function of the random Hamiltonian is bounded by the average one. We show that, if a statistical problem is renormalizable, then the algebraic geometrical method can be successfully applied, resulting that the learning curves are subject to the universal law. Also we show that, if it is not renormalizable, then the large fluctuation of the random Hamiltonian prevents the system from obeying to the universal law in general.

2. Bayes Learning Theory

Let $q(x)$ be a probability density function on N dimensional real Euclidean space $\mathbb{R}^N$. The training samples and the testing sample are respectively defined by random variables $X_1, X_2, ..., X_n$ and X, which are independently subject to the same probability distribution $q(x)dx$.

A statistical model is defined as a probability density function $p(x|w)$ of $x \in \mathbb{R}^N$ for a given parameter $w \in W \subset \mathbb{R}^d$, where W is a set of all parameters. In Bayes estimation, we prepare a probability density function $\varphi(w)$ on W. Although $\varphi(w)$ is called a prior distribution, it does not necessary represent an *a priori* knowledge of the parameter, in general.

For a given function $F(w)$ on W, its expectation value $\langle F(w)\rangle$ with respect to the posterior distribution is defined by

$$\langle F(w)\rangle = \frac{\displaystyle\int F(w) \prod_{i=1}^{n} p(X_i|w)^{\beta}\, \varphi(w)dw}{\displaystyle\int \prod_{i=1}^{n} p(X_i|w)^{\beta}\, \varphi(w)dw},$$

where $0 < \beta < \infty$ is the inverse temperature. The case $\beta = 1$ is most important because it corresponds to the strict Bayes estimation. The Bayes predictive distribution is defined by

$$p^*(x) = \langle p(x|w)\rangle.$$

In Bayes learning theory, the following random variables play an important role. The Bayes generalization loss B_g, the Bayes training loss B_t, the Gibbs generalization loss G_g, and the Gibbs training loss G_t are respectively defined by

$$B_g = -E_X[\log\langle p(X|w)\rangle], \quad (1)$$

$$B_t = -\frac{1}{n}\sum_{i=1}^{n} \log\langle p(X_i|w)\rangle, \quad (2)$$

$$G_g = -\Big\langle E_X[\log p(X|w)]\Big\rangle, \quad (3)$$

$$G_t = -\Big\langle \frac{1}{n}\sum_{i=1}^{n} \log p(X_i|w)\Big\rangle, \quad (4)$$

where $E_X[\ ]$ shows the expectation value over X. Let us introduce two random variables by

$$Y_g = E_X\Big[\langle(\log p(X|w))^2\rangle - \langle \log p(X|w)\rangle^2\Big], \quad (5)$$

$$Y_t = \frac{1}{n}\sum_{i=1}^{n}\Big\{\Big\langle(\log p(X_i|w))^2\Big\rangle - \Big\langle \log p(X_i|w)\Big\rangle^2\Big\}, \quad (6)$$

where $V_g = nY_g$ and $V_t = nY_t$ are referred to as the *functional variances* [9, 10]. In this paper, we study the expectation values of these six random variables, which are called *Bayes observables*. The log loss function $L(w)$ and the entropy S are respectively defined by

$$L(w) = -E_X[\log p(X|w)],$$
$$S = -E_X[\log q(X)].$$

Note that $L(w) = S + D(q||p_w)$, where $D(q||p_w)$ is the relative entropy or Kullback-Leibler distance defined by

$$D(q||p_w) = \int q(x) \log \frac{q(x)}{p(x|w)} dx.$$

Therefore, always $L(w) \geq S$. Moreover, $L(w) = S$ if and only if $p(x|w) = q(x)$. In this paper, we assume that there exists a parameter $w_0 \in W$ which minimizes $L(w)$,

$$L(w_0) = \min_{w \in W} L(w).$$

Note that such w_0 is not unique in general, because the map $w \mapsto p(x|w)$ is not one-to-one in general. We assume that, for an arbitrary w that satisfies $L(w) = L(w_0)$, $p(x|w)$ is the same probability density function. Let $p_0(x)$ be such a unique probability density function. For simplicity, we use notation $L_0 = -E_X[\log p_0(X)]$.

Definition. If $q(x) = p_0(x)$, then $q(x)$ is said to be *realizable* by $p(x|w)$, if otherwise it is said to be *unrealizable*.

Definition. If the set $W_0 = \{w \in W; p_0(x) = p(x|w)\}$ consists of a single point w_0 and if the Hessian matrix $J \equiv \nabla\nabla L(w_0)$ is strictly positive definite, then $q(x)$ is said to be *regular* for $p(x|w)$. If otherwise, then $q(x)$ is said to be *singular* for $p(x|w)$.

Bayes learning theory was studied in realizable and regular cases [13, 14, 15], realizable and singular cases [7, 9, 10], and unrealizable and regular cases [11]. In such cases, it was proved that there exists a universal relation between the generalization and training errors. In this paper, we mainly study unrealizable and singular cases.

3. Generating Function of Statistical Learning

The log density ratio function $f(x, w)$ and the log likelihood ratio function $H_n(w)$ are respectively defined by

$$\begin{aligned} f(x, w) &= \log \frac{p_0(x)}{p(x|w)}, \\ H_n(w) &= \frac{1}{n} \sum_{i=1}^{n} f(X_i, w), \end{aligned}$$

where $nH_n(w)$ is referred to as the random Hamiltonian. In this paper, we introduce the generating function of Bayes learning theory by

$$F_n(\alpha) = E\Big[-\log \int \exp(-\alpha f(X, w) - \beta n H_n(w))\varphi(w) dw\Big],$$

where $E[\ \ ]$ shows the expectation value over $X_1, X_2, .., X_n$ and X. Then, by the definitions eq.(1)-eq.(6) and by using the fact that $\log p_0(x)$ is a constant function of w, it immediately follows that

$$\begin{aligned} E[B_g] &= L_0 + F_n(1) - F_n(0), & (7) \\ E[B_t] &= L_0 + F_{n-1}(1+\beta) - F_{n-1}(\beta), & (8) \\ E[G_g] &= L_0 + F'_n(0), & (9) \\ E[G_t] &= L_0 + F'_{n-1}(\beta), & (10) \\ E[Y_g] &= -F''_n(0), & (11) \\ E[Y_t] &= -F''_{n-1}(\beta). & (12) \end{aligned}$$

These equations show that $F_n(\alpha)$ determines the behaviors of average Bayes observables [7, 14, 15]. In order to analyze these values, we need assumptions.

Definition. If there exists a constant $\gamma > 0$ such that

$$\lim_{n\to\infty} \sup_{0\le\alpha\le 1+\beta} |F_n^{(3)}(\alpha)| n^{\gamma} = 0, \tag{13}$$

$$\lim_{n\to\infty} |F_n'(0) - F_{n-1}'(0)| n^{\gamma} = 0, \tag{14}$$

$$\lim_{n\to\infty} |F_n''(0) - F_{n-1}''(0)| n^{\gamma} = 0, \tag{15}$$

then the generating function is said to satisfy the *conditions of learnability* with index γ.

Let us assume that the conditions of learnability are satisfied. Then, by using Taylor expansions of $F_n(\alpha)$, $F_n'(\alpha)$, and $F_n''(\alpha)$, it follows that

$$E[B_g] = L_0 + F_n'(0) + \frac{1}{2}F_n''(0) + o(\frac{1}{n^\gamma}), \tag{16}$$

$$E[B_t] = L_0 + F_n'(0) + \frac{2\beta + 1}{2}F_n''(0) + o(\frac{1}{n^\gamma}), \tag{17}$$

$$E[G_g] = L_0 + F_n'(0), \tag{18}$$

$$E[G_t] = L_0 + F_n'(0) + \beta F_n''(0) + o(\frac{1}{n^\gamma}), \tag{19}$$

$$E[Y_g] = -F_n''(0), \tag{20}$$

$$E[Y_t] = -F_n''(0) + o(\frac{1}{n^\gamma}). \tag{21}$$

Therefore, we obtain the *equations of states in statistical learning*,

$$E[B_g] = E[B_t] + \beta E[Y_t] + o(\frac{1}{n^\gamma}), \tag{22}$$

$$E[G_g] = E[G_t] + \beta E[Y_t] + o(\frac{1}{n^\gamma}). \tag{23}$$

That is to say, if the conditions of learnability are satisfied, then the equations of states hold. Minimization of both $E[B_g]$ and $E[G_g]$ is one of the main purposes of statistical estimation, however, they need the expectation value over the testing sample $E_X[\ ]$, hence they cannot be calculated directly from training samples. On the other hand, B_t, G_t, and Y_t can be calculated from only training samples without any direct information about $q(x)$. In other words, the equations of states show that $E[B_g]$ and $E[G_g]$ can be estimated from training samples, therefore $B_t+\beta Y_t$ and $G_t+\beta Y_t$ are information criteria which show how appropriate the set $(p(x|w), \varphi(w))$ is. In fact, they are equal to AIC [16] if $q(x)$ is realizable by and regular for $p(x|w)$. If $q(x)$ is unrealizable by or singular for $p(x|w)$, then AIC is not equal to the asymptotic generalization error, whereas $B_t + \beta Y_t$ and $G_t + \beta Y_t$ are. Hence they are called *widely applicable information criteria* (WAIC) [9, 10, 12].

4. Renormalizable Case

Let us define the renormalizability.

Definition. Let $W_\epsilon = \{w \in W; D(p_0||p_w) \le \epsilon\}$. If there exist $A > 0$ and $\epsilon > 0$ such that

$$w \in W_\epsilon \Longrightarrow L(w) - L_0 \ge A\ D(p_0||p_w),$$

then the pair $(q(x), p(x|w))$ is said to be *renormalizable*. If otherwise, *nonrenormalizable*.

It is easy to show that, if $q(x)$ is regular for $p(x|w)$, then $(q(x), p(x|w))$ is renormalizable. In fact, $D(p_0||p_w)$ is smaller than some quadratic form of $w - w_0$ in the neighborhood of unique w_0 and $L(w) - L_0$ has a positive definite Hessian matrix. Also, it is trivial to show that, if $q(x)$ is realizable by $p(x|w)$, then $(q(x), p(x|w))$ is renormalizable. In fact, since $q(x) = p_0(x)$, $L(w) - L_0 = D(p_0||p_w)$. However, if $q(x)$ is unrealizable by and singular for $p(x|w)$, then $(q(x), p(x|w))$ may be renormalizable or nonrenormalizable.

In this section, we study the renomalizable case, and show that the conditions of learnability hold with index $\gamma = 1$ and that the Bayes observables are subject to the universal law.

We assume that $L(w)$ is an analytic function of $w \in W$ and that $w \mapsto f(x, w)$ is a function-valued analytic function. Since $\int p_0(x)dx = \int p_w(x)dx = 1$,

$$D(p_0||p_w) = \int p_0(x)(f(x,w) + e^{-f(x,w)} - 1)dx.$$

There exists a constant $B > 0$ such that

$$\frac{t + e^{-t} - 1}{t^2} \geq B \quad (|t| < \epsilon).$$

By combining this inequality with the renormalizability, it follows that

$$L(w) - L_0 \geq AB \int p_0(x) f(x,w)^2 dx. \tag{24}$$

Since $L(w) - L_0$ is an analytic function, we can apply resolution of singularities [17, 19] to $L(w) - L_0$, and obtain the following result. There exist both a real d-dimensional analytic manifold $\mathcal{M}$ and a real analytic map $g : \mathcal{M} \to W$ such that, in each local coordinate of $\mathcal{M}$,

$$\begin{aligned} L(g(u)) - L_0 &= u^{2k} \equiv \prod_{j=1}^{d} u_j^{2k_j}, \\ |g'(u)|\varphi(g(u)) &= b(u)u^h \equiv b(u)\prod_{j=1}^{d} u_j^{h_j}, \end{aligned}$$

where $k = (k_1, k_2, ..., k_d)$ and $h = (h_1, h_2, ..., h_d)$ are multiple indeces made of nonnegative integers, $|g'(u)|$ is the Jacobian determinant of the map $w = g(u)$, and $b(u) > 0$. Then, by using eq.(24), $f(x, g(u))^2$ can be divided by u^{2k}, in other words, $f(x, g(u))/u^k$ is a well-defined analytic function. In fact, if $f(x, g(u))$ can not be divided by u^{2k}, then eq.(24) does not hold. Hence, there exists a function-valued analytic function $a(x, u)$ such that

$$f(x, g(u)) = a(x, u)u^k.$$

Moreover, from $L(w) - L_0 = E_X[f(X, w)]$, we have $E_X[a(X, u)] = u^k$. Remark that both renormalizability and resolution theorem are necessary to prove the existence of $a(x, u)$. Let us define an empirical process on $\mathcal{M}$,

$$\xi_n(u) = \frac{1}{\sqrt{n}} \sum_{i=1}^{n} \{a(X_i, u) - u^k\}.$$

Then the probability distribution of $\xi_n(u)$ converges to that of the gaussian process $\xi(u)$, which is uniquely determined by its average and covariance [9, 18],

$$\begin{aligned} E_\xi[\xi(u)] &= 0, \\ E_\xi[\xi(u)\xi(u')] &= E_X[a(X,u)a(X,u')] - E_X[a(X,u)]E_X[a(X,u')], \end{aligned}$$

where $E_\xi[\ ]$ shows the expectation value over the gaussian process $\xi(u)$. Moreover, the gaussian process $\xi(u)$ can be represented by

$$\xi(u) = \sum_{j=1}^{\infty} c_j(u) g_j$$

where $\{g_j\}$ are independent random variables and each g_j is subject to the standard normal distribution. Then

$$E_\xi[\xi(u)\xi(u')] = \sum_{j=1}^{\infty} c_j(u)c_j(u').$$

The random Hamiltonian is rewritten as

$$nH_n(g(u)) = nu^{2k} - \sqrt{n}u^k\xi_n(u).$$

To study the generatining function $F_n(\alpha)$, we need the asymptotic behavior of

$$Z_n(s) = \int f(x,w)^s \exp(-\beta n H_n(w))\varphi(w)dw,$$

where $s \geq 0$ is a real value. For example,

$$F_n'(0) = E\Big[\frac{Z_n(1)}{Z_n(0)}\Big], \tag{25}$$

$$F_n''(0) = -E\Big[\frac{Z_n(2)}{Z_n(0)}\Big] + E\Big[\frac{Z_n(1)}{Z_n(0)}\Big]^2. \tag{26}$$

Then by using the function $w = g(u)$,

$$\begin{aligned} Z_n(s) &= \sum_\alpha \int du\, a(x,u)^s u^{sk+h} \exp(-\beta n u^{2k} + \beta\sqrt{n}u^k\xi_n(u))b_\alpha(u) \\ &= \sum_\alpha \int_0^\infty dt \int du\, \frac{1}{n}\,\delta\Big(\frac{t}{n} - u^{2k}\Big)\, a(x,u)^s u^{sk+h} \exp(-\beta t + \beta\sqrt{t}\xi_n(u))b_\alpha(u), \end{aligned}$$

where $\sum_\alpha$ shows the sum over all local coordinates and $b_\alpha(u) \geq 0$ satisfies $\sum_\alpha b_\alpha(u) = b(u)$. By using the asymptotic expansion of the Shcwartz distribution $\delta(t/n - u^{2k})$ for $n \to \infty$ [7, 9, 10, 20, 21, 22, 25, 26], there exists a Schwartz distribution $D_\alpha(u)$ such that

$$\sum_\alpha \frac{1}{n}\,\delta\Big(\frac{t}{n} - u^{2k}\Big)\, u^{sk+h}\, b_\alpha(u) \cong \frac{(\log n)^{m-1}}{n^{\lambda+s/2}}\, t^{\lambda-1+s/2}\Big(\sum_{\alpha^*} D_{\alpha^*}(u)\Big),$$

where $\lambda > 0$ is the *log canonical threshold* defined by

$$\lambda = \min_\alpha \min_{j=1}^{d} \Big(\frac{h_j+1}{2k_j}\Big),$$

and m is the maximum number of j which attains the above minimum. Also $\sum_{\alpha^*}$ shows the sum over all local coordinates that attain the above minimum and the support of $D_{\alpha^*}(u)$ is contained in the set $\{u \in \mathcal{M}; L(g(u)) - L_0 = 0\}$. Hence

$$Z_n(s) \cong \frac{(\log n)^{m-1}}{n^{\lambda+s/2}}\Big(\int \mathcal{D}(u,t) t^{s/2} \exp(\beta\sqrt{t}\xi(u))\Big).$$

where $\int \mathcal{D}(u,t)$ is defined by the integration over the manifold,

$$\int \mathcal{D}(u,t) = \sum_{\alpha^*}\int_0^\infty dt \int du D_{\alpha^*}(u)\, t^{\lambda-1}\, \exp(-\beta t).$$

Let us define

$$\hat{Z}(q,r,s) = \int \mathcal{D}(u,t)\, \xi(u)^q\, t^{r/2}\, a(x,u)^s\, \exp(\beta\sqrt{t}\xi(u)).$$

Then

$$Z_n(s) \cong \frac{(\log n)^{m-1}}{n^{\lambda+s/2}}\hat{Z}(0,s,s). \tag{27}$$

Firstly, since $E_X[a(X,u)] = u^k$,

$$E_X[\hat{Z}(0,1,1)] = \hat{Z}(0,2,0).$$

Secondly, by using the partial integration of t

$$\int_0^\infty dt\; t^\lambda e^{-\beta t+\beta\sqrt{t}\xi(u)} = \frac{\lambda}{\beta}\int_0^\infty dt\; t^{\lambda-1}e^{-\beta t+\beta\sqrt{t}\xi(u)} + \frac{1}{2}\int_0^\infty dt\; t^{\lambda-1/2}\xi(u)e^{-\beta t+\beta\sqrt{t}\xi(u)},$$

it follows that

$$\hat{Z}(0,2,0) = \frac{\lambda}{\beta}\hat{Z}(0,0,0) + \frac{1}{2}\hat{Z}(1,1,0).$$

And lastly, by using the partial integration over the gaussian process $\xi(u)$,

$$\begin{aligned}
E_\xi\Big[\frac{\hat{Z}(1,1,0)}{\hat{Z}(0,0,0)}\Big] &= E_\xi\Big[\int \mathcal{D}(u,t)\Big(\sum_{j=1}^\infty c_j(u)g_j\Big)\frac{t^{1/2}\exp(\beta\sqrt{t}\xi(u))}{\int \mathcal{D}(u',t')\exp(\beta\sqrt{t'}\xi(u'))}\Big] \\
&= E_\xi\Big[\int \mathcal{D}(u,t)\Big(\sum_{j=1}^\infty c_j(u)\frac{\partial}{\partial g_j}\Big)\frac{t^{1/2}\exp(\beta\sqrt{t}\xi(u))}{\int \mathcal{D}(u',t')\exp(\beta\sqrt{t'}\xi(u'))}\Big] \\
&= \beta E_X E_\xi\Big[\frac{\hat{Z}(0,2,2)}{\hat{Z}(0,0,0)}\Big] - \beta E_X E_\xi\Big[\frac{\hat{Z}(0,1,1)}{\hat{Z}(0,0,0)}\Big]^2,
\end{aligned} \tag{28}$$

where we used $E_\xi[\xi(u)\xi(u')] = E_X[a(X,u)a(X,u')]$ on the set $\{u; L(g(u)) - L_0 = 0\}$. Let us define the constant 2ν by the right hand side of eq.(28), where ν is referred to as the *singular fluctuation*. Then by using eqs.(25),(26),(27),

$$\begin{aligned}
F_n'(0) &\cong (\frac{\lambda}{\beta}+\nu)\cdot\frac{1}{n}, \\
F_n''(0) &\cong -\frac{2\nu}{\beta}\cdot\frac{1}{n}.
\end{aligned}$$

Therefore, we obtained the *universal law* of Bayes observables,

$$E[B_g] \cong L_0 + (\frac{\lambda - \nu}{\beta} + \nu)\frac{1}{n}, \quad (29)$$
$$E[B_t] \cong L_0 + (\frac{\lambda - \nu}{\beta} - \nu)\frac{1}{n}, \quad (30)$$
$$E[G_g] \cong L_0 + (\frac{\lambda}{\beta} + \nu)\frac{1}{n}, \quad (31)$$
$$E[G_t] \cong L_0 + (\frac{\lambda}{\beta} - \nu)\frac{1}{n}, \quad (32)$$
$$E[Y_g] \cong E[Y_t] \cong \frac{2\nu}{\beta} \cdot \frac{1}{n}. \quad (33)$$

In this case, we can prove that the conditions of learnability with index $\gamma = 1$ are satisfied by the same way as [9, 10]. Hence, equations of states hold with $\gamma = 1$.

5. Nonrenormalizable Case

In this section, we study a nonrenormalizable case. It is still difficult to clarify the general nonrenormalizable case. Hence, in this section, we show that there exists a simple example in which the Bayes observables do not satisfy the universal law.

$$q(x, y) = \frac{1}{2\pi}\exp(-\frac{1}{2}(x^2 + y^2)), \quad (34)$$
$$p(x, y|a) = \frac{1}{2\pi}\exp(-\frac{1}{2}\{(x-a)^2 + (y - \sqrt{a^4 - a^2 + 1})^2\}), \quad (35)$$

where $a \in \mathbb{R}^1$ is a parameter. Then the relative entropy is

$$D(q||p_a) = \int q(x, y)\log\frac{q(x, y)}{p(x, y|a)}dxdy = \frac{1}{2}(a^4 + 1).$$

Hence $D(q||p_a)$ is minimized at $a = 0$, and $L_0 = \log(2\pi) + 3/2$. The Hessian is given by $\partial_a^2 D(q||p_a)|_{a=0} = 0$. Therefore $q(x)$ is unrealizable by and singular for $p(x|a)$. The log density ratio function is

$$f(x, a) = -ax - h(a)y + \frac{a^4}{2},$$

where $h(a) = \sqrt{a^4 - a^2 + 1} - 1$ is a real analytic function, and

$$D(p_0||p_a) = \frac{a^4}{2} - h(a).$$

Note that $D(p_0||p_a) \cong a^2/2$ in the neighborhood of $a = 0$. On the other hand, $L(a) - L_0 = a^4/2$, resulting that $(q(x), p(x|a))$ is not renormalizable. The random Hamiltonian is

$$nH_n(a) = \frac{n\,a^4}{2} - \sqrt{n}\,a\,\xi_1 - \sqrt{n}\,h(a)\,\xi_2,$$

where

$$\xi_1 = \frac{1}{\sqrt{n}}\sum_{i=1}^{n} X_i, \quad \xi_2 = \frac{1}{\sqrt{n}}\sum_{i=1}^{n} Y_i$$

are independently subject to the standard normal distribution.

$$\begin{aligned} nH_n(a)' &= 2\,n\,a^3 - \sqrt{n}\,\xi_1 - \sqrt{n}\,h(a)'\,\xi_2, \\ nH_n(a)'' &= 6\,n\,a^2 - \sqrt{n}\,h(a)''\,\xi_2. \end{aligned}$$

The parameter a that minimizes $nH_n(a)'$ is denoted by a^*. Since

$$a^* = \left(\frac{\xi_1}{kn}\right)^{1/3} + o_p(\frac{1}{n^{1/3}}),$$

the main order term of $nH_n(a)$ is given by

$$\begin{aligned} nH_n(a) &= \frac{1}{2}nH_n(a^*)(a-a^*)^2 + nH_n(a^*) \\ &= \frac{1}{2}C_n(a-D_n)^2 - \frac{1}{2}C_nD_n^2, \end{aligned}$$

where

$$\begin{aligned} C_n &= 6n(\xi_1/2\sqrt{n})^{2/3}, \\ D_n &= (\xi_1/2\sqrt{n})^{1/3}. \end{aligned}$$

Therefore, by using $E[(\xi_1)^{\mu-1}] = 2^{\mu/2}\Gamma(\mu/2)/\sqrt{2\pi}$,

$$\begin{aligned} F_n'(0) &= \frac{Q}{2}\cdot\frac{1}{n^{2/3}}, \\ F_n''(0) &= -\frac{2Q}{\beta}\cdot\frac{1}{n^{2/3}}, \end{aligned}$$

where

$$Q = \frac{2^{7/6}}{\sqrt{2\pi}}\Gamma(\frac{7}{6}).$$

The asymptotic behaviors of Bayes observables are different from the universal law,

$$E[B_g] \cong L_0 + (\frac{1}{2} - \frac{1}{\beta})\cdot\frac{Q}{n^{2/3}}, \quad (36)$$

$$E[B_t] \cong L_0 - (\frac{3}{2} + \frac{1}{\beta})\cdot\frac{Q}{n^{2/3}}, \quad (37)$$

$$E[G_g] \cong L_0 + \frac{Q}{2}\cdot\frac{1}{n^{2/3}}, \quad (38)$$

$$E[G_t] \cong L_0 - \frac{3Q}{2}\cdot\frac{1}{n^{2/3}}, \quad (39)$$

$$E[Y_g] \cong E[Y_t] \cong \frac{2Q}{\beta}\cdot\frac{1}{n^{2/3}}. \quad (40)$$

Also in this case, the conditions of learnability are satisfied with index $2/3$, hence the equations of states hold with $\gamma = 2/3$, however,

$$E[V_t] = nE[W_t] \cong n^{1/3}$$

does not converge to the constant. It seems that both renormalizable and nonrenormalizable statistical problems satisfy the more general universal law.

6. Discussion

In this section, let us discuss three points, birational invariants, renormalizability, and Bayes observables as random variables.

6.1. Birational Invariants

In section 4, we proved that, in the renormalizable case, the asymptotic learning curves are determined by λ and ν, which are defined by using resolution of singularities. Let us study the mathematical properties of them. For a given analytic function, $L(w) - L_0$, there exist infinitely many desingularization pairs $(\mathcal{M}, g)$. If a value defined by using $(\mathcal{M}, g)$ does not depend on the choice of $(\mathcal{M}, g)$, then it is called a *birational invariant.*

Firstly, as is shown in [7, 9], the value $(-\lambda)$ is equal to the largest pole of the *zeta function* on $\mathbb{C}$ obtained by the analytic continuation of

$$\zeta(z) = \int (L(w) - L_0)^z \varphi(w) dw \quad (\mathrm{Re}(z) > 0).$$

Therefore, λ is a birational invariant. This value is well known in algebraic geometry and algebraic analysis, which shows the relative relation of the pair of two algebraic varieties (W, W_0) [17, 20, 23, 24, 25, 26].

Secondly, the value ν is characterized by

$$\nu = \lim_{n\to\infty} \frac{\beta}{2} E\Big[\frac{1}{n}\sum_{i=1}^{n}\Big\{\Big\langle(\log p(X_i|w))^2\Big\rangle - \Big\langle\log p(X_i|w)\Big\rangle^2\Big\}\Big].$$

Hence ν is also a birational invariant.

It was clarified by [11] that, if a true distribution is unrealizableby and regular for a parametric model, then

$$\begin{aligned}\lambda &= d/2,\\ \nu &= \mathrm{tr}(IJ^{-1})/2,\end{aligned}$$

where I and J are respectively $d \times d$ matrices defined by

$$\begin{aligned}I &= E_X[\nabla \log p(x|w_0) \nabla \log p(x|w_0)],\\ J &= \nabla^2 L(w_0).\end{aligned}$$

For singular and realizable cases, λ was calculated in [3, 4, 5, 10], whereas ν is unknown.

6.2. Renormalizability

Let us discuss the renormalizable condition.

Firstly, we study the renormalizable condition from the physical point of view. In physics, a set of functions $\{f_n(x); n = 1, 2, ...,\}$ is sometimes called renormalizable if there exists some rescaling transform by which a universal law is discovered. For example, if there exist both a set (a, b) and a function $f^*(x)$ such that

$$\lim_{n\to\infty} n^a f_n(n^b x) \to f^*(x),$$

then such a system is called renormalizable. In this paper, we have shown that, if $(q(x), p(x|w))$ is renormalizable, then the Boltzmann distribution satisfies the convergence in law,

$$\frac{n^\lambda}{(\log n)^{m-1}} \exp(-n\beta H_n(g(u))) \to \int_0^\infty t^{\lambda-1} \exp(-n\beta t + \beta\sqrt{t}\,\xi(u)) dt,$$

when n tends to infinity, where $\xi(u)$ is a gaussian process defined by the central limit theorem of the functional space. If $(q(x), p(x|w))$ does not satisfy the renormalizable condition, then such a rescaling transform does not exist in general. The expectation and the variance of

$$nH_n(w) = \sum_{i=1}^{n} f(X_i, w)$$

are respectively given by

$$E[nH_n(w)] = nE_X[f(X,w)], \quad V[nH_n(w)] = nV_X[f(X,w)].$$

Because $E_X[f(X,w)] = L(w) - L_0 \geq 0$ and $V_X[f(X,w)] \cong (1/2)D(p_0||p_w)$ in the neighborhood $L(w) - L_0 = 0$, the renormalizable condition ensures that the fluctuation of the random Hamiltonian is bounded by the average one. This is the intuitive reason why the universal law holds.

Secondly, we study scale invariantness of renormalizablity. Let $f_1(x,w)$ and $f_2(x,w)$ be log likelihood ratio functions of two different statistical problems. If they are renormalizable and satisfy the relations

$$\begin{aligned} E_X[f_1(X,w)] &= E_X[f_2(X,w)], \\ E_X[f_1(X,w)f_1(X,w')] &= E_X[f_2(X,w)f_2(X,w')], \end{aligned}$$

then they have the same birational invariants (λ, ν). In other words, the learning curves are determined only by the average and covariance of the log density ratio function. It might seem that $E_X[f(X,w)]^2 \propto E_X[f(X,w)^2]$, but such a relation does not hold even in a trivial case. In a realizable and regular case, $a \in \mathbb{R}^1$,

$$p(x|a) = \frac{1}{(2\pi)^{1/2}} \exp(-\frac{1}{2}(x-a)^2)$$

and $q(x) = p(x|0)$, then $f(x,a) = a^2/2 - ax$, resulting that $E_X[f(X,a)] = a^2/2$ and $E_X[f(X,a)^2] \cong a^2 + a^4/4$. Therefore, in the neighborhood of $a = 0$, both $E_X[f(X,a)]$ and $E_X[f(X,a)^2]$ are in proportion to a^2. The renormalizable condition in this case is invariant under a scaling transform $f(X,w) \to sf(X,w)$ for an arbitrary constant $s > 0$. The renormalizable condition of this paper is a generalized concept of such invariantness.

6.3. *Bayes Observables as Random Variables*

In statistical learning theory, Bayes observables are random variables. In this paper, we mainly studied the expectation values of them. Note that the generating function $F_n(\alpha)$ does not have sufficient information about randomness of Bayes observables. If a true distribution is regular or realizable, then stochastic properties of Bayes observables were clarified [10, 11]. It is a future study to clarify the stochastic behavior of Bayes observables as random variables.

7. Conclusion

In this paper, we defined the renormalizable condition of a learning system, and proved that, in the renormalizable case, the universal law holds. Also we showed that, in nonrenormalizable case, the universal law does not hold in general. It is the future study to clarify the more general universal learning theory, which contains both renormalizable and nonrenormalizable statistical problems.

Acknowledgment. This research was partially supported by the Ministry of Education, Science, Sports and Culture in Japan, Grant-in-Aid for Scientific Research 18079007.

References

[1] Watanabe S 1995 *Proc. of Int. Symp. on NOLTA* 207-210
[2] Watanabe S 2001 *Neur. Netw.* **14**(8) 1049-60
[3] Yamazaki K and Watanabe S 2003 *Neur. Netw.* **16**(7) 1029-38
[4] Aoyagi M and S.Watanabe S 2005 *Neur. Netw.* **18**(7) 924-33
[5] Rusakov D and Geiger D 2005 *J. of Mach. Lear. Res.* **6** 1-35
[6] Hartigan J A 1985 Proc. *Barkeley Conference in Honor of J. Neyman and J. Kiefer* **2** 807-10
[7] Watanabe S 2001 *Neur. Comput.* **13**(4) 899-933
[8] Drton M, Sturmfes B and Sullivant S 2009 *Lectures on Algebraic Statistics* (Berlin: Birkhäuser Verlag)
[9] Watanabe S 2009 *Algebraic geometry and statistical learning theory* (Cambridge: Cambridge University Press)
[10] Watanabe S 2010 *Neur. Netw.* **23**(1) 20-34
[11] Watanabe S 2010 *IEICE Trans* **E93-A**(3) 617-26
[12] Watanabe S 2010 *Adv. Stud. Pure Math.* **57** 473-92
[13] Schwarz G 1978 *Ann. Stat.* **6**(2) 461-4
[14] Levin E, Tishby N and Solla S A 1990 *Proc. of IEEE,* **78**(10) 1568-74
[15] Amari S 1993 *Neur. Netw.* **6**(2) 161-6
[16] Akaike H 1974 *IEEE Trans. on Aut. Cont.* **10** 716-23
[17] Hironaka H 1964 *Ann. of Math.* **79** 109-326
[18] van der Vaart A W and Wellner J A 1996 *Weak Convergence and Empirical Processes* (New York: Springer)
[19] Atiyah M F 1970 *Comm. Pure and Appl. Math.* **13** 145-50
[20] Bernstein I N 1972 *Func. Anal. Appl.* **6** 26-40
[21] Gelfand I M and Shilov G E 1964 *Generalized Functions* (San Diego: Academic Press)
[22] Kashiwara M 1976 *Invent. Math.* **38** 33-53
[23] Kollór J, Mori S, Clemens C H and Corti A 1998 *Birational geometry of algebraic varieties* (Cambridge: Cambridge University Press)
[24] Mustata M 2002 *J. Amer. Math. Soc.* **15** 599-615
[25] Oaku T 1997 *J. Pure Appl. Alg.* **117** 495-518
[26] Saito M 2007 *arXiv:0707.2308v1*

On the robustness of random Boolean formulae

Alexander Mozeika[1], **David Saad**[1] **and Jack Raymond**[2]

[1] The Non-linearity and Complexity Research Group, Aston University, Birmingham B4 7ET, UK

[2] Department of Physics, The Hong Kong University of Science and Technology, Clear Water Bay, Hong Kong, China

E-mail: {a.s.mozeika,d.saad}@aston.ac.uk

Abstract. Random Boolean formulae, generated by a growth process of noisy logical gates are analyzed using the generating functional methodology of statistical physics. We study the type of functions generated for different input distributions, their robustness for a given level of gate error and its dependence on the formulae depth and complexity and the gates used. Bounds on their performance, derived in the information theory literature for specific gates, are straightforwardly retrieved, generalized and identified as the corresponding typical-case phase transitions. Results for error-rates, function-depth and sensitivity of the generated functions are obtained for various gate-type and noise models.

1. Introduction

Boolean functions are typically represented by formulae based on simple Boolean operations and implemented using basic logical gates. They form the basis of most computing devices and are the subject of extensive and varied research efforts in the theoretical computer science and information theory communities. The main sources of errors in classical computing circuits based on semiconductor technology are heat generation, cosmic rays and production defects [1], the impact of which is exacerbated by the ever-increasing drive towards miniaturization and more complex devices of large scale [1].

von Neumann [2] was the first to address the problem of computation with noisy gates and its limitations; and suggested simple gate-constructions, based only on noisy gates to limit the resulting function error. He showed that a bounded function error can be implemented in such cases for a sufficiently small gate-error. His motivation for investigating the problem had arisen from the study of biologically-motivated circuits and an attempt to explain the robustness of neural activities represented by a circuit (or formula) composed of ϵ-noisy Boolean gates.

Before progressing any further, a few formal definitions are required: (i) A *circuit* may be regarded as a directed acyclic graph in which the nodes of in-degree zero are either Boolean constants or references to arguments, the nodes of in-degree $k \geq 1$ are logical gates of k arguments and the nodes of out-degree zero correspond to the circuit outputs. (ii) A *formula* is a single-output circuit where the output of each gate is used as an input to at most one gate. (iii) The ϵ-noisy gate is designed to compute a Boolean function $\alpha : \{-1,1\}^k \to \{-1,1\}$, but for each input $\boldsymbol{S} \in \{-1,1\}^k$ there is an error probability ϵ such that $\alpha(\boldsymbol{S}) \to -\alpha(\boldsymbol{S})$. To simplify the analysis, error-probability is taken to be independent for each gate in the circuit. Clearly, a noisy circuit ($\epsilon > 0$) cannot perform any given computation in a deterministic manner: for any circuit-input there is a non-vanishing probability that the circuit will produce the wrong output. (iv) The maximum of this error probability δ over all circuit-inputs determines *reliability* of the circuit.

von Neumans's study was later revisited by Pippenger [3] using information theory methods. He showed that if a noisy k-ary formula is used to compute a Boolean function f with the error probability $\delta < 1/2$, then (i) there is an upper bound for the gate-error $\epsilon(k)$ which is strictly less than $1/2$ and (ii) there is a lower bound for the formula-depth $\hat{d}(k, \epsilon, \delta) \geq d$, where d is the depth of a noiseless formula computing f; the *depth* of a formula being the number of gates on the longest path from an input node to the output node. In comparison to its noiseless counterpart, a noisy formula that computes reliably has greater depth due to the presence of restitution-gates, implying longer computation times [3]. These results have been refined and extended for circuits [4], for k-ary Boolean formulae [5, 6] and different gates [7, 8].

The aim of this study is to investigate the properties of noisy circuits via methods of statistical physics, identify physical properties that characterize existing bounds and convergence rates, and provide insight and quantitative results beyond what is accessible via the information theory methodology. Key aspect of our analysis is the study of *typical* rather than *worst* case bounds.

The paper is organized as follows: In section 2 we discuss the generation of typical Boolean functions and the model used for generating them followed by a description of the model used in section 3 and main thrust of the derivation in section 4. The results obtained are summarized in section 5 followed by a conclusion and discussion section.

2. Generating random functions

As the aim of this work is to study the properties of typical cases, one should first identify a mechanism for randomly generating *typical* functions. This is not a simple task as most of the commonly known paradigms in the theoretical computer science literature aim to prove that canonical representations can describe any given arbitrary function using elementary gates or processes, but when applied at random they tend generate trivial functions showing weak dependence on the input variables.

A good example is the use of covering or bi-decomposition to represent arbitrary functions [9] resulting in a disjunctive normal form (DNF). The DNF or its dual CNF (conjunctive normal form) is a depth-2 formula with AND and OR gates used as internal nodes and with the input Boolean variables and their negations distributed on the leaves. However, *random* DNF (CNF) formulae offer very low sensitivity [10] to the input values and highly simple and uncharacteristic Boolean functions.

To generate formulae that represent all Boolean functions with uniform probability, using *randomly generated* circuits, we employ a variant of the growth process suggested by Savický [11]. Firstly, one defines an initial distribution over a set of simple Boolean functions. Secondly, and in further steps, the formulae chosen from the distributions defined in previous steps are combined by Boolean gates. One such process, described by Savický [11], uses only a single Boolean gate α and is defined by the recursion on the set of formulae A_ℓ:

$$\begin{aligned} A_0 &= \{1, -1, S_1, \ldots, S_n, -S_1, \ldots, -S_n,\} \\ A_{\ell+1} &= \{\alpha(\phi_1, \ldots, \phi_k); \phi_j \in A_\ell \text{ for } j = 1, 2, \ldots, k\}. \end{aligned} \quad (1)$$

Savický showed, under a very broad conditions on α, that the probability of computing a Boolean function by a formula $\phi \in A_\ell$ tends to the uniform distribution over all Boolean functions of n variables when $\ell \to \infty$ [11]. Furthermore, depending on the initial conditions A_0 and the gate α the process converges to a *single Boolean function* or to the *uniform distribution* over some class of Boolean functions [12].

In this framework, all Boolean functions of n variables are represented with equal statistical weight when $\ell \to \infty$, but the number of gates in formulae grows exponentially with the formula

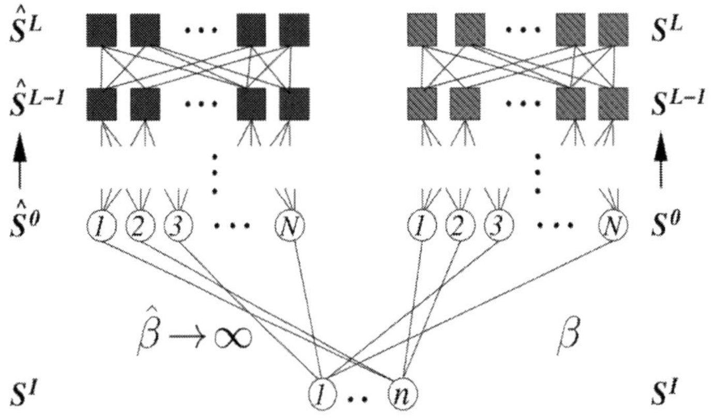

Figure 1. The model of two coupled systems with identical topology and different inverse temperatures β and $\hat{\beta}\to\infty$. Gates are indicated by squares, S^I and input nodes by circles. Blue indicates noiseless gates, red noisy gates.

depth ℓ. To contain the number of gates, we propose a layered variant of the Savický growth process. The first step in our process is to sample randomly and uniformly exactly N entries of an input vector $\hat{\boldsymbol{S}}^0 = (S^0_1, \dots, S^0_N)$. In the second, and all subsequent steps for $\ell = 1, \dots, L-1$, we construct a vector $\hat{\boldsymbol{S}}^{\ell+1} = (S^{\ell+1}_1, \dots, S^{\ell+1}_N)$ where the i-th entry $S^{\ell+1}_i$ is an output of the gate $\alpha(S^\ell_{i_1}, \dots, S^\ell_{i_k})$ with k input-indices sampled uniformly from the set of all possible (unordered) indices $\{i_1, \dots, i_k\}$. The result of the process is the layered $N\times(L+1)$ Boolean circuit shown in figure 1 (left construction, in blue). For large N, the variable S^ℓ_i in our model corresponds to the output of a random k-ary formula of depth ℓ, which computes a Boolean function $\{-1,1\}^N \to \{-1,1\}$. We expect that in the limit $N\to\infty$, with $\ell \in O(N^0)$, the statistical properties of the formulae generated in our process and in the Savický's growth process are equivalent; this is supported by the results reported later. The layered representation allows us to explore the typical behavior of noisy random Boolean formulae using methods of statistical physics.

While the vector $\hat{\boldsymbol{S}}^0$ represents randomly sampled single entries, one would also like to study cases where entries are statistically dependent and are sampled from a smaller set. To cater for a possible higher level of correlation, the 0-layer boundary conditions are generated by selecting randomly $\hat{\boldsymbol{S}}^0$ entries from members of the finite set $S^I = \{S^I_1, \dots, S^I_n\}$. This allows to investigate the properties of the functions generated and their dependence on properties of the set S^I.

3. Model

The noisy computation model considered here is a feed-forward layered $N\times(L+1)$ Boolean circuit. The layers in the circuit are numbered from 0 (input) to L (output). Each layer $\ell \in \{1, \dots, L\}$ in the circuit is composed of exactly N ϵ-noisy, k-ary Boolean gates. Noise at gate α^ℓ_i on site (i,ℓ) operates independently and in a stochastic manner according to the microscopic law

$$P(S^\ell_i|S^{\ell-1}_{i_1}, \dots, S^{\ell-1}_{i_k}) = \frac{\mathrm{e}^{\beta S^\ell_i \alpha^\ell_i(S^{\ell-1}_{i_1}, \dots, S^{\ell-1}_{i_k})}}{2\cosh[\beta\alpha^\ell_i(S^{\ell-1}_{i_1}, \dots, S^{\ell-1}_{i_k})]}\,, \tag{2}$$

where $\beta = 1/T$ is the inverse temperature, related to the gate noise ϵ via $\tanh\beta = 1-2\epsilon$. The gate-output S^ℓ_i is completely random when $\beta\to 0$ ($\epsilon = 1/2$) and completely deterministic when

$\beta \to \infty$ ($\epsilon = 0$). Our model is acyclic by definition, so given the state of gates at layer ℓ, gates at layer $\ell+1$ operate independently of each other. The latter suggests that the probability of the microscopic state $\boldsymbol{S}^0, \ldots, \boldsymbol{S}^L$, where $\boldsymbol{S}^\ell \in \{-1,1\}^N$, is just a product of (2) over all sites and layers in the circuit. Furthermore, to investigate the properties of noisy formulas we consider two copies of the same topology, shown in figure 1, but with different temperatures $\beta < \infty$ (noisy) and $\hat{\beta} \to \infty$ (noiseless), comparing the two will enable us to study the effect of noise on the resulting functions. Following similar arguments to those of the single circuit case, the probability of microscopic states in the two systems are given by

$$P[\{\boldsymbol{S}^\ell\};\{\hat{\boldsymbol{S}}^\ell\}] = P(\boldsymbol{S}^0,\hat{\boldsymbol{S}}^0|\boldsymbol{S}^I)\prod_{\ell=1}^{L} P(\boldsymbol{S}^\ell|\boldsymbol{S}^{\ell-1})P(\hat{\boldsymbol{S}}^\ell|\hat{\boldsymbol{S}}^{\ell-1}) \tag{3}$$

where

$$P(\boldsymbol{S}^\ell|\boldsymbol{S}^{\ell-1}) = \prod_{i=1}^{N} \frac{e^{\beta S_i^\ell \sum_{j_1,\ldots,j_k}^N A^{\ell,i}_{j_1,\ldots,j_k}\alpha_i^\ell(S^{\ell-1}_{j_1},\ldots,S^{\ell-1}_{j_k})}}{2\cosh[\beta\sum_{j_1,\ldots,j_k}^N A^{\ell,i}_{j_1,\ldots,j_k}\alpha_i^\ell(S^{\ell-1}_{j_1},\ldots,S^{\ell-1}_{j_k})]}. \tag{4}$$

The set of connectivity tensors $\{A^{\ell,i}_{i_1,\ldots,i_k}\}$, where $A^{\ell,i}_{i_1,\ldots,i_k} \in \{0,1\}$, denotes connections in the circuit. The conditional probability $P(\hat{\boldsymbol{S}}^\ell|\hat{\boldsymbol{S}}^{\ell-1})$ is the same as in (4) but with $\beta \to \hat{\beta}$.

The sources of disorder in our model are the random connections, random boundary conditions and random gates. The former two arise in the layered growth process described in the last two paragraphs of section 2. The basic step in this growth process is the addition of a new gate with probability $P(A^{\ell,i}_{j_1,\ldots,j_k}) = \frac{1}{N^k}\delta_{A^{\ell,i}_{j_1,\ldots,j_k};1} + (1-\frac{1}{N^k})\delta_{A^{\ell,i}_{j_1,\ldots,j_k};0}$ of being connected to any labeled set of k gate-outputs on the previous layer $\ell-1$. This procedure is carried out independently for all gates in the circuit giving rise to the probability distribution

$$P(\{A^{\ell,i}_{i_1,\ldots,i_k}\}) = \frac{1}{Z_A}\prod_{\ell,i=1}^{L,N}\left[\delta\left[1;\sum_{j_1,\ldots,j_k}^{N} A^{\ell,i}_{j_1,\ldots,j_k}\right]\prod_{i_1,\ldots,i_k}^{N}\left[\frac{1}{N^k}\delta_{A^{\ell,i}_{i_1,\ldots,i_k};1} + (1-\frac{1}{N^k})\delta_{A^{\ell,i}_{i_1,\ldots,i_k};0}\right]\right] \tag{5}$$

where Z_A is a normalization constant. The Kronecker delta function inside the definition (5) enforces the constraint $\sum_{j_1,\ldots,j_k}^N A^{\ell,i}_{j_1,\ldots,j_k} = 1$, i.e. the gate on site (i,ℓ) is mapped to exactly one element from the set of all possible output-indices $\{i_1,\ldots,i_k\}$ from the previous layer. Other sparse connectivity profiles can be easily accommodated into our framework by incorporating additional constraints into the definition (5) via the appropriate delta functions.

Random boundary conditions in the growth process are generated by selecting members of the input set S^I, where $|S^I| \in O(N^0)$, with uniform probability, and assigning them to the input layer 0. The boundary condition is identical for the two systems yielding the distribution

$$P(\boldsymbol{S}^0,\hat{\boldsymbol{S}}^0|\boldsymbol{S}^I) = \prod_{i=1}^{N}\delta_{S_i^0;S^I_{n_i}}\,\delta_{\hat{S}_i^0;S_i^0} \tag{6}$$

where $\{n_i\}$ are independent random indices pointing to the members of input set S^I with probability $P(n_i) = \frac{1}{|S^I|}$. Further correlations can be introduced by defining the probability function $P(S^I)$.

In addition to the topological disorder, induced by the growth process, we assume that the gate α_i^ℓ added at each step of the process can be sampled randomly and independently from the set G of k-ary Boolean gates. Under this assumption the distribution over gates takes the form

$$P(\{\alpha_i^\ell\}) = \prod_{\ell,i=1}^{L,N} P(\alpha_i^\ell) \tag{7}$$

where $P(\alpha_i^\ell) = \sum_{\alpha\in G} p_\alpha \delta_{\alpha;\alpha_i^\ell}$ with $\sum_{\alpha\in G} p_\alpha = 1$ and $p_\alpha \geq 0$.

4. Analysis

Several methods are at our disposal to study this model. For instance, the replica and cavity methods have been successfully employed to study of similar problems from theoretical computer science and information theory [13] ranging from classical combinatorial optimization problems (graph coloring, K-SAT, reconstruction on trees and graph-isomorphism to name but a few) to source and channel coding. However, given the directed nature of the formulae studied we prefer to use the generating functional approach where formula layers take the role of time.

To compute the probability distribution (3) directly for a circuit of finite but significant size is difficult. However, the structure of (3) is similar to the one that describes evolution of the disordered Ising spin system [14]. This similarity becomes more apparent if one regards the layers in our model as discrete time-steps of parallel dynamics. We use the generating functional method of statistical mechanics [15]; the generating functional for the current model is given by

$$\Gamma[\boldsymbol{\psi};\hat{\boldsymbol{\psi}}] \;=\; \left\langle \mathrm{e}^{-\mathrm{i}\sum_{\ell,i}\{\psi_i^\ell S_i^\ell + \hat{\psi}_i^\ell \hat{S}_i^\ell\}} \right\rangle \tag{8}$$

where the shorthand $\langle \ldots \rangle$ denotes the average over the joint probability (3). The generating functional (8) can be regarded as a characteristic function of (3) from which moments of the distribution can be obtained by taking partial derivatives with respect to the generating fields $\{\psi_i^\ell, \hat{\psi}_j^{\ell'}\}$, for example $\langle S_i^\ell \hat{S}_j^{\ell'} \rangle = -\lim_{\boldsymbol{\psi},\hat{\boldsymbol{\psi}}\to\mathbf{0}} \frac{\partial^2}{\partial\psi_i^\ell \partial\hat{\psi}_j^{\ell'}} \Gamma[\boldsymbol{\psi};\hat{\boldsymbol{\psi}}]$. Following prescription of [15], we assume that for $N\to\infty$ the system is self-averaging and compute $\overline{\Gamma[\boldsymbol{\psi};\hat{\boldsymbol{\psi}}]}$, where $\overline{[\cdots]}$ denotes an average over the disorder. The disorder-averaged generating function (8) gives rise to the following macroscopic observables

$$\begin{aligned} m(\ell) &= \frac{1}{N}\sum_{i=1}^{N} \overline{\langle S_i^\ell \rangle} = \lim_{\boldsymbol{\psi},\hat{\boldsymbol{\psi}}\to\mathbf{0}} \frac{\mathrm{i}}{N}\sum_{i=1}^{N} \frac{\partial}{\partial\psi_i^\ell}\overline{\Gamma[\boldsymbol{\psi};\hat{\boldsymbol{\psi}}]} \\ C(\ell) &= \frac{1}{N}\sum_{i=1}^{N} \overline{\langle S_i^\ell \hat{S}_i^\ell \rangle} = -\lim_{\boldsymbol{\psi},\hat{\boldsymbol{\psi}}\to\mathbf{0}} \frac{1}{N}\sum_{i=1}^{N} \frac{\partial^2}{\partial\psi_i^\ell \partial\hat{\psi}_i^\ell}\overline{\Gamma[\boldsymbol{\psi};\hat{\boldsymbol{\psi}}]} \end{aligned} \tag{9}$$

where $m(\ell)$ is the average activity (magnetization) on layer ℓ and $C(\ell)$ is the overlap between two systems. Averaging over the disorder in (8) leads to the saddle-point integral

$$\overline{\Gamma} \;=\; \int \{\mathrm{d}\boldsymbol{P}\,\mathrm{d}\hat{\boldsymbol{P}}\,\mathrm{d}\boldsymbol{\Omega}\,\mathrm{d}\hat{\boldsymbol{\Omega}}\}\mathrm{e}^{N\Psi[\boldsymbol{P},\hat{\boldsymbol{P}};\boldsymbol{\Omega},\hat{\boldsymbol{\Omega}}]}\,, \tag{10}$$

where Ψ is the macroscopic saddle-point surface

$$\begin{aligned}\Psi[\ldots] &= \mathrm{i}\sum_{\ell=0}^{L-1}\sum_{S,\hat{S}}\hat{P}^{\ell}(S,\hat{S})P^{\ell}(S,\hat{S})+\mathrm{i}\sum_{\ell=0}^{L-1}\int \mathrm{d}x\,\mathrm{d}\hat{x}\,\mathrm{d}\omega\,\hat{\Omega}^{\ell}(x,\hat{x},\omega)\Omega^{\ell}(x,\hat{x},\omega) \\ &+ \sum_{\ell=0}^{L-1}\sum_{\{S_j,\hat{S}_j\}}\prod_{j=1}^{k}\left[P^{\ell}(S_j,\hat{S}_j)\right]\int \mathrm{d}x\,\mathrm{d}\hat{x}\,\mathrm{d}\omega\,\Omega^{\ell}(x,\hat{x},\omega)\left\langle \mathrm{e}^{-\mathrm{i}\{x\alpha(\{S_j\})+\hat{x}\alpha(\{\hat{S}_j\})+\omega\}}\right\rangle_{\alpha} \\ &+ \sum_{n}P(n)\log\int\{\mathrm{d}\boldsymbol{H}\,\mathrm{d}\boldsymbol{x}\,\mathrm{d}\hat{\boldsymbol{H}}\,\mathrm{d}\hat{\boldsymbol{x}}\}\int \mathrm{D}\boldsymbol{\omega}\sum_{\boldsymbol{S},\hat{\boldsymbol{S}}}M_n[\boldsymbol{H},\boldsymbol{x};\hat{\boldsymbol{H}},\hat{\boldsymbol{x}};\boldsymbol{\omega};\boldsymbol{S},\hat{\boldsymbol{S}}]\,,\end{aligned} \tag{11}$$

$\langle\cdot\rangle_{\alpha}$ represents an average over gate distribution and M is an effective single-site measure

$$\begin{aligned}M_n[\ldots] &= \delta_{S^0;S_n^I}\delta_{\hat{S}^0;S^0}\prod_{\ell=0}^{L-1}\Big[\mathrm{e}^{\mathrm{i}x^{\ell}H^{\ell}+\mathrm{i}\hat{x}^{\ell}\hat{H}^{\ell}+\beta S^{\ell+1}H^{\ell}+\hat{\beta}\hat{S}^{\ell+1}\hat{H}^{\ell}} \\ &\times \mathrm{e}^{-\log 2\cosh(\beta H^{\ell})-\log 2\cosh(\hat{\beta}\hat{H}^{\ell})-\mathrm{i}\hat{P}^{\ell}(S^{\ell},\hat{S}^{\ell})-\mathrm{i}\hat{\Omega}^{\ell}(x^{\ell},\hat{x}^{\ell},\omega^{\ell+1})+\mathrm{i}\omega^{\ell+1}}\Big].\end{aligned} \tag{12}$$

The generating fields $\boldsymbol{\psi},\hat{\boldsymbol{\psi}}$ have been removed from the above as they are no longer needed. For $N\to\infty$ the path-integral (10) is dominated by the extremum of the functional $\Psi[\ldots]$ of (11). Functional variation of (11) with respect to the order parameters $\{P,\hat{P},\Omega,\hat{\Omega}\}$ gives rise to four saddle-point equations

$$P^{\ell}(S,\hat{S}) = \sum_{n}P(n)\left\langle\delta_{S^{\ell};S}\delta_{\hat{S}^{\ell};\hat{S}}\right\rangle_{M_n} \tag{13}$$

$$\begin{aligned}\hat{P}^{\ell}(S,\hat{S}) &= \mathrm{i}\sum_{i=1}^{k}\sum_{\{S_j,\hat{S}_j\}}\delta_{S_i;S}\delta_{\hat{S}_i;\hat{S}}\prod_{j\neq i}^{k}\left[P(S_j,\hat{S}_j)\right] \\ &\times\int \mathrm{d}x\,\mathrm{d}\hat{x}\,\mathrm{d}\omega\,\Omega^{\ell}(x,\hat{x},\omega)\left\langle \mathrm{e}^{-\mathrm{i}\{x\alpha(\{S_j\})+\hat{x}\alpha(\{\hat{S}_j\})+\omega\}}\right\rangle_{\alpha}\end{aligned} \tag{14}$$

$$\Omega^{\ell}(x,\hat{x},\omega) = \sum_{n}P(n)\left\langle\delta(x-x^{\ell})\delta(\hat{x}-\hat{x}^{\ell})\delta(\omega-\omega^{\ell+1})\right\rangle_{M_n} \tag{15}$$

$$\hat{\Omega}^{\ell}(x,\hat{x},\omega) = \mathrm{i}\sum_{\{S_j,\hat{S}_j\}}\prod_{j=1}^{k}\left[P^{\ell}(S_j,\hat{S}_j)\right]\left\langle \mathrm{e}^{-\mathrm{i}\{x\alpha(\{S_j\})+\hat{x}\alpha(\{\hat{S}_j\})+\omega\}}\right\rangle_{\alpha} \tag{16}$$

where $\langle\cdots\rangle_{M_n}$ is the average over the probability distribution (12). The saddle-point equations (13)–(16) can be simplified significantly and it turns out that in order to solve this problem we only need to compute the order parameter (13). The physical meaning of this order parameter is given by $P^{\ell}(S,\hat{S}) = \lim_{N\to\infty}\frac{1}{N}\sum_{i=1}^{N}\overline{\langle\delta_{S_i^{\ell};S}\delta_{\hat{S}_i^{\ell};\hat{S}}\rangle|_{S^I}}$, i.e. the disorder-averaged joint probability of sites in the two systems. The single-site effective measure (12) also benefits from the simplification; in particular, if we integrate out the continuous variables in (12) we are led to the expression

$$\begin{aligned}M_n[S^L,\hat{S}^L,\ldots,S^0,\hat{S}^0] &= \delta_{S^0;S_n^I}\delta_{\hat{S}^0;S^0}\prod_{\ell=0}^{L-1}\Bigg\{\sum_{\{S_j,\hat{S}_j\}}\prod_{j=1}^{k}\left[P^{\ell}(S_j,\hat{S}_j)\right] \\ &\times\left\langle\frac{\mathrm{e}^{\beta S^{\ell+1}\alpha(\{S_j\})}}{2\cosh\beta[\alpha(\{S_j\})]}\frac{\mathrm{e}^{\hat{\beta}\hat{S}^{\ell+1}\alpha(\{\hat{S}_j\})}}{2\cosh\hat{\beta}[\alpha(\{\hat{S}_j\})]}\right\rangle_{\alpha}\Bigg\}.\end{aligned} \tag{17}$$

Using (17) the macroscopic observables (9) can be easily computed from the joint probability distribution (13), resulting in the set of equations

$$m(\ell+1) = \sum_{\{S_j\}} \prod_{j=1}^{k} \left[\frac{1}{2}\{1 + S_j m(\ell)\}\right] \langle \tanh[\beta\alpha(S_1, \dots, S_k)]\rangle_\alpha \tag{18}$$

$$\begin{aligned} C(\ell+1) = & \sum_{\{S_j, \hat{S}_j\}} \prod_{j=1}^{k} \left[\frac{1}{4}\{1 + S_j m(\ell) + \hat{S}_j \hat{m}(\ell) + S_j \hat{S}_j C(\ell)\}\right] \\ & \times \left\langle \tanh[\beta\alpha(S_1, \dots, S_k)] \tanh[\hat{\beta}\alpha(\hat{S}_1, \dots, \hat{S}_k)] \right\rangle_\alpha , \end{aligned} \tag{19}$$

where the magnetization $\hat{m}(\ell)$ is computed by a similar equation to (18) but with $\beta \to \hat{\beta}$. The initial conditions for the above system of equations are given by $m(0) = \hat{m}(0) = \frac{1}{|S^I|}\sum_{S\in S^I} S$, $C(0) = 1$. Equations (18) and (19) describe the coupled evolution of the magnetization and overlap from layer to layer. While in general sparsely connected systems one may expect a richer structure of cross and multi-layer correlations, the current case is characterized by different connectivities between layers and tree like formulae, as their depth is much smaller than the layer size, where these macroscopic observables are sufficient for describing the system's properties.

The connectivity profile considered in our model leads to a simple mean-field theory, where the macroscopic behaviors of the two copies of the same system is completely determined by the set of observables $\{m(\ell), \hat{m}(\ell), C(\ell)\}$; which relate to the order parameter (13) via $P^\ell(S, \hat{S}) = \frac{1}{4}(1 + Sm(\ell) + \hat{S}\hat{m}(\ell) + S\hat{S}C(\ell))$, while the single system behavior is described by $\{m(\ell)\}$. Furthermore, since $\langle \prod_j S^\ell_{i_j}\rangle \to \prod_j \langle S^\ell_{i_j}\rangle$ for finite j, when $N \to \infty$ the spins in layer ℓ are uncorrelated. The reason for this behavior is that in our model, the site (i, ℓ) is a root of a full k-ary tree growing from the input layer $\ell = 0$, which in turn points to the input set S^I. The loops in the circuit are rare, so that trees can be regarded as random Boolean formulas and when presented with the input, operate independently of each other. The output of a typical formula at layer ℓ is determined by the probability $P^\ell(S)$.

The overlap order parameter of (9) is related to the normalized Hamming distance $D(\ell)$ between the states $\boldsymbol{S}^\ell$ and $\hat{\boldsymbol{S}}^\ell$ via the identity $D(\ell) = \frac{1}{2}(1 - C(\ell)) = \frac{1}{2}(1 - \frac{1}{N}\sum_{i=1}^{N} \overline{\langle S^\ell_i \hat{S}^\ell_i\rangle})$. This allows one to define the order parameter $\Delta(\ell) = \lim_{\beta,\hat{\beta}\to\infty} \frac{1}{2}(1 - C(\ell))$, used to probe sensitivity of the circuit with respect to its input, an indication to the complexity of the functions represented by the given circuit. The Hamming distance $D(\ell)$ is also related to the probability $P(S^\ell_i \neq \hat{S}^\ell_i)$ and facilitates the estimate of the error probability $\delta(\ell)$ on the ℓ-th layer of a noisy circuit. More specifically, we define this error probability $\delta(\ell) = \max_{S^I} \lim_{\hat{\beta}\to\infty} \frac{1}{2}(1 - C(\ell))$, comparing the maximal error between the noisy and noiseless version of the same circuit with respect to all possible inputs. Obviously, in the absence of noise ($\beta \to \infty$) one trivially obtains $\delta(\ell) = 0$ for all ℓ.

5. Results

While (18) and (19) are general for any gate or distribution of gates, we will focus here on a particular Boolean gate, a majority gate with k inputs (MAJ-k), k being odd. The reasons for choosing this gate are twofold: it was proved to be optimal for noisy computation in formulas [5, 6], and a formula constructed from majority gates can in principle compute any Boolean function [11]. A convenient representation of the MAJ-k gate is given by the identity

$\mathrm{MAJ}(S_1,\ldots,S_k) = \mathrm{sgn}[\sum_{j=1}^{k} S_j]$. Inserting this identity into (18) and (19) one obtains

$$m(\ell+1) = (1-2\epsilon)\sum_{n=0}^{k}\binom{k}{n}\left[\frac{1+m(\ell)}{2}\right]^{n}\left[\frac{1-m(\ell)}{2}\right]^{k-n}\mathrm{sgn}\,[2n-k] \quad (20)$$

$$C(\ell+1) = (1-2\epsilon)\sum_{k_1+..+k_4=k}\frac{k!}{k_1!\times..\times k_4!}\,P^{k_1}(-1,-1)\,P^{k_2}(1,-1) \quad (21)$$
$$\times\, P^{k_3}(-1,1)\,P^{k_4}(1,1)\,\mathrm{sgn}\,[k_1-k_2+k_3-k_4]\,\mathrm{sgn}\,[k_1+k_2-k_3-k_4]$$

where $P(S,\hat{S}) = \frac{1}{4}(1 + Sm(\ell) + \hat{S}\hat{m}(\ell) + S\hat{S}C(\ell))$. In the derivation of the above we have used the identity $\tanh\beta = 1-2\epsilon$ which relates the gate-error ϵ to the inverse temperature β. Also, we have taken the limit $\hat{\beta}\to\infty$ ($\hat{\epsilon}=0$) which allows us to compare noisy circuit to its noiseless counterpart later on. Equations (20) and (21) form the basis for the results presented in the remainder of the paper.

5.1. *Critical behavior in MAJ-k based formulae*

Equation (20) describes the evolution of the magnetization from layer to layer. The point $m(\infty)=0$ is always a stationary solution of this equation. Expanding (20) around this stationary solution gives us the value of noise $\epsilon^*(k) = 1/2 - 2^{k-2}/k\binom{k-1}{(k-1)/2}$, shown in figure 2 (a), below which the solution $m(\infty)=0$ becomes unstable and two stable solutions $|m(\infty)|\neq 0$ emerge. This result is identical to those reported in [5, 6], and reduces to $\epsilon^*(3) = 1/6$ for $k=3$; the non-vanishing magnetization values then become [16] $m(\infty) = \pm\sqrt{\frac{1-6\epsilon}{1-2\epsilon}}$.

For $\epsilon > \epsilon^*(k)$ the magnetization $m(\ell)$ decays to 0 when $\ell\to\infty$, while for $\epsilon < \epsilon^*(k)$ $\lim_{\ell\to\infty} m(\ell) = \pm m(\infty)$ where the positive and negative stationary solutions correspond to the positive and negative initial magnetizations $m(0) = \frac{1}{|S^I|}\sum_{S\in S^I} S$ respectively. The function error δ grows respectively. Thus $\epsilon^*(k)$ constitutes the critical noise level that separates the unordered phase of the system from ordered one (see the inset of figure 2 (a) for the case $k=3$). A single input bit of information will be preserved by the circuit for arbitrary many layers only when $\epsilon < \epsilon^*(k)$. The probability of an error $P^\ell(-S) = \frac{1}{2}(1-Sm(\ell))$ is a measure of how well this one bit of the information is preserved after passing through ℓ layers. A complicated computational task may require significant number of layers, hence only relatively simple operations can be performed by the circuit reliably when $\epsilon > \epsilon^*(k)$.

5.2. *Boolean functions generated*

The analysis of (20) also reveals the type of Boolean functions generated. Stationary solutions for the noiseless system ($\epsilon=0$) result in $m(\infty)=\pm1$ for the initial conditions $m(0)>0$ and $m(0)<0$, respectively. For $m(0)=0$ the resulting asymptotic magnetization is $m(\infty)=0$ and each site is associated with an output of the formula which computes some Boolean function. The average formula on layer ℓ operates according to the probability $P^\ell(S) = \frac{1}{2}(1-Sm(\ell))$. This behavior of the magnetization suggests that when $\epsilon=0$ the average formula on layer ℓ converges to the formula that computes the random Boolean functions $F=1$ and $F=-1$ for the initial conditions $m(0)>0$ and $m(0)<0$, respectively, and $F=\pm1$ with equal probability if $m(0)=0$; where $m(0) = \frac{1}{|S^I|}\sum_{S\in S^I} S$.

Depending on initial conditions the formulas converge to a single Boolean function or to the uniform distribution over some set of functions. For example, if we take $S^I = \{-1, -S_1^I, -S_2^I\}$ then the formulas converge to the NAND function. Taking $S^I = \{-S_1^I, -S_2^I\}$, on the other hand, gives a uniform distribution over the inverse functions $-S_1^I$ and $-S_2^I$; this follows from the majority property of the gate. In general, when $m(0)=0$, is difficult to say if the formulas compute all Boolean functions in the set generated or only the subset of these functions. However, our result is consistent with other studies [11, 12] with a similar growth process that generates uniformly all random Boolean formulas. In particular, for $S^I = \{-1, 1, S_1^I, \ldots, S_n^I\}$ the formulas converge to the MAJ-n function when n is odd and to the uniform distribution over so-called slice functions when n is even [12]. The same happens when the constants $\{-1, 1\}$ are removed from the set S^I [12]. The functions generated can be written in a more compact form as $F = \mathrm{sgn}[\sum_{i=1}^{n} \theta_i S_i^I + \theta_0]$, where $\theta_i \in \mathbb{Z}$, using the definition $\mathrm{sgn}[0] = \pm 1$ with probability 1/2. If $\sum_{i=1}^{n} \theta_i S_i^I + \theta_0 \neq 0$ for $\forall\ (S_1^I, \ldots, S_n^I) \in \{-1, 1\}^n$ then all formulas in the circuit converge to a single linear threshold function, which can represent any linearly-separable Boolean function.

5.3. *Sensitivity*

To measure the evolving overlap between the noiseless and noisy systems one employs (21); the initial conditions are the same for both systems with $m(0) = \hat{m}(0)$ and $C(0) = 1$. For the noiseless system $\epsilon = 0$ the stationary unstable solution of this equation is given by $C(\infty) = 1$. Perturbations to the initial state $C(0) = 1$ lead to the stable stationary state $C(\infty) = 0$. This implies that the circuit is very input-sensitive when $m(0) = 0$. In particular, the Hamming distance $\Delta(\ell) = \frac{1}{2}(1 - C(\ell))$ increases for small perturbations, amplified by the circuit, to $\Delta(0)$ (see figure 2 (b)). This implies that when $\epsilon > 0$ the circuit also amplifies the noise-perturbation and the error $\delta(\ell)$ is growing; nevertheless, it can be controlled for a number of layers for sufficiently small noise level ϵ (see figure 2 (c)).

5.4. *Convergence rates*

In general, we find that the number of layers needed for the magnetization to converge in the noiseless case scales as $O(\log n)$ for $m(0) = 1/n$ where $n \in \mathbb{N}$. This rate of convergence is consistent with the results of [12] for the Savický growth process [11]. Considering a particular function, namely the representation of noiseless MAJ-k circuit computing MAJ-n function, one can obtain the number of layers L required for the various k values as presented in figure 2 (d). This result holds for all linear threshold functions. The figure also shows the upper $k2^k \log(n)$ bound obtained in [12] for $k = 3$.

As we increase ϵ towards the critical value ϵ^* the rate of convergence decreases. Very close to the phase boundary ϵ^* the differences $m(\ell+1) - m(\ell)$ are very small and the difference equation in the case $k = 3$ can be approximated by the differential equation $\frac{\mathrm{d}}{\mathrm{d}\ell} m(\ell) = -m(\ell) + \frac{1}{2}(1-2\epsilon)[3m(\ell) - m^3(\ell)]$, where ℓ is continuous now. The solution to this equation is given by $m^2(\ell) = \left\{ \left[\frac{1}{m^2(0)} - \frac{1-2\epsilon}{1-6\epsilon} \right] \mathrm{e}^{-(1-6\epsilon)\,\ell} + \frac{1-2\epsilon}{1-6\epsilon} \right\}^{-1}$. This solution is only accurate for $\epsilon = 1/6 \pm \Delta\epsilon$, where $0 < \Delta\epsilon \ll 1$, where it gives us the asymptotic form $|m(\ell) - m(\infty)| \approx \mathrm{e}^{-\gamma(3)\Delta\epsilon\ell}$ with $\gamma(3) = 3$ when $\epsilon = 1/6 + \Delta\epsilon$ and $\gamma(3) = 6$ when $\epsilon = 1/6 - \Delta\epsilon$. Decay rates for $k > 3$ are expected to be qualitatively similar ($|m(\ell) - m(\infty)| \approx \mathrm{e}^{-\gamma(k)\Delta\epsilon\ell}$) but to obtain them analytically directly is difficult. For $k > 3$ one may use a different method that relies on the fact

$$\mathrm{const} \times \left| \left\{ \frac{\mathrm{d}}{\mathrm{d}m} F_\epsilon^k[m] \right\}_{m=m(\infty)} \right|^{\ell} \leq |m(\ell) - m(\infty)|, \tag{22}$$

where $F_\epsilon^k[m]$ is the right hand side of (20), i.e. the distance $|m(\ell) - m(\infty)|$ for an arbitrary point $m(\ell)$ is always greater than the distance when $m(\ell) = m(\infty) + \Delta m(\ell)$. The inequality (22) can lead to an approximation by choosing an appropriate constant. Computing (22) for the MAJ-k circuit with $\epsilon = \epsilon^*(k) \pm \Delta\epsilon$ one obtains

$$\left\{ \left[1 - \frac{2\Delta\epsilon}{1 - 2\epsilon^*(k)} \right] (1 - m^2(\infty))^{(k-1)/2} \right\}^{\ell} \leq |m(\ell) - m(\infty)|. \tag{23}$$

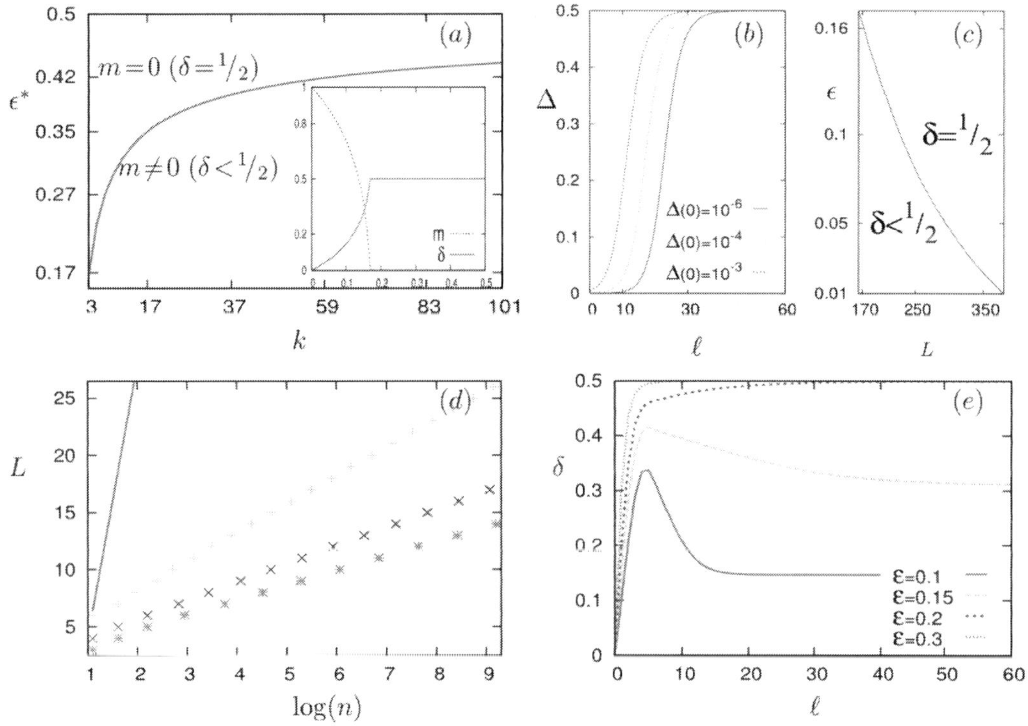

Figure 2. (a)The critical noise level ϵ^* as a function of k. Inset - Properties of MAJ-3 gate-based formulae: Magnetization m and output error δ as a function of gate noise ϵ. (b) Sensitivity of $\Delta(\ell)$ to input mismatch $\Delta(0)$ for $m(0)=0$. (c) Phase diagram for gate noise ϵ at layer L. (d) Number of layers in the noiseless MAJ-k circuit computing MAJ-n function. Theoretic data-points are represented by the symbols $+$ $(k=3)$, $\times$ $(k=5)$ and $*$ $(k=7)$. Slopes of the respective straight lines (not shown on this figure) fitted to the data are 2.473, 1.595 and 1.282. The straight line corresponds to the bound [12] $k2^k\log(n)$, plotted here for $k=3$ only. (e) For the MAJ-7 *function*, we plot the evolution, in layers (ℓ), of the output error δ.

Since $\left[1-\frac{2\Delta\epsilon}{1-2\epsilon^*(k)}\right](1-m^2(\infty))^{(k-1)/2} \leq \left[1-\frac{2\Delta\epsilon}{1-2\epsilon^*(k)}\right]$, the rate of convergence in the paramagnetic region ($\epsilon > \epsilon^*(k)$) is slower than in the ferromagnetic one ($\epsilon < \epsilon^*(k)$). This is due to the amplification of thermal fluctuations that are only suppressed as $\ell\to\infty$. For large k the critical noise value $\epsilon^*(k)$ can be approximated by $\epsilon^*(k) \approx \frac{1}{2}\left(1-\frac{\sqrt{\pi}}{\sqrt{2k}}\right)$. Inserting this value into (23) for $m(\infty)=0$ gives rise to the asymptotic form

$$|m(\ell)-m(\infty)| \approx \mathrm{e}^{-2\frac{\sqrt{2k}}{\sqrt{\pi}}\Delta\epsilon\ell}, \qquad (24)$$

where increasing k speeds up convergence in both paramagnetic and ferromagnetic regimes.

5.5. Dynamics

The function error $\delta(\ell)$, calculated for $k=3$ majority gates and shown in figure 2 (e), for different ϵ values, exhibits two distinct stages in the dynamics. It is calculated using (20) and (21) for a formula that represents the MAJ-7 function. Initially, the error increases until it reaches its maximum value at $\ell=5$, before the MAJ-7 function is computed exactly at $\ell=8$ for $\epsilon=0$; the location of this maximum is independent of ϵ. This suggests that gate-inputs at layers $\ell\leq 5$ are non-uniform, contributing to noise-amplification, but become more uniform later leading to noise-suppression and decreasing error. As we approach ϵ^* the number of layers needed for the error to reach stationarity increases; in the region $\epsilon = 1/6 \pm \Delta\epsilon$ it can be estimated from the asymptotic form derived for $m(\ell)$. The dynamic behavior of the error changes to monotonically increasing at $\epsilon^0 = \frac{1}{2}(1 - m^2(0))/(3 - m^2(0))$ above which noise cannot be reduced by additional layers. For $\epsilon \gg 1/6$ the error evolution becomes strictly monotonic and relaxes exponentially fast to its stationary value 1/2.

6. Discussion

By employing the generating functional methodology of statistical physics we analyzed the robustness of random functions composed of logical gates to gate-noise. To generate random function via a growth process we employed a variant of the framework devised by Savický [11], which results in Boolean functions that are randomly sampled from the uniform distribution over the function space. The layered variant used facilitates the study of the random functions generated by the growth process as well as their robustness to gate noise.

The results obtained for typical functions retrieve the ones obtained in the theoretical computer science literature and identify them as phase transitions in the corresponding physical system. In addition, it facilitates the derivation of the function properties at any depth and their convergence rate towards the asymptotic values. In addition we obtain typical convergence rate in both the noiseless case and close to the critical noise level; the results obtained for the noiseless case are in agreement with the bounds presented in the literature; the convergence rate results for the noisy case are completely new.

The main advantages of the statistical physics methodology presented here over existing methods are the derivation of typical case results, the flexibility in accommodating different gate types or distribution of gates and the accessibility to results at various depths. However, in order to find out exactly which functions are computed when the formulas generated by the (noiseless) growth process compute more than one function, or in a more general noisy setting, one needs a more sophisticated version of the mean-field theory presented here [17].

While in this paper we concentrated on the majority gates, we have looked at other gate types [16], hard noise (fabrication errors) and other properties of noisy circuits [18]. We believe that much can be explored about the properties of noisy circuits using the methodology developed here, for instance, the type of functions generated depending on the gate types and the level of gate noise. Work in these areas is underway [17].

Acknowledgments

Support by the Leverhulme trust (grant F/00 250/H) is gratefully acknowledged.

References

[1] Borkar S 2005 *IEEE Micro* **25**(6) 10–6
[2] Von Neumann J 1956 *Probabilistic Logics and the Synthesis of Reliable Organisms from Unreliable Components* (Princeton, NJ: Princeton University Press) pp 43–98 Automata Studies
[3] Pippenger N 1988 *IEEE Trans. Inf. Theory* **34** 194–7
[4] Feder T 1989 *IEEE Trans. Inf. Theory* **35** 569–71
[5] Hajek B and Weller T 1991 *IEEE Trans. Inf. Theory* **37** 388–91
[6] Evans W and Schulman L 2003 *IEEE Trans. Inf. Theory* **49** 3094–8
[7] Evans W and Pippenger N 1998 *IEEE Trans. Inf. Theory* **44** 1299–305
[8] Unger F 2008 *IEEE Trans. Inf. Theory* **54** 3693–8
[9] Steinbach B and Lang C 2003 *Artif. Intell. Rev.* **20**(3) 319–60
[10] Boppana R B 1997 *Inform. Process. Lett.* **63** 257–61
[11] Savický P 1990 *Discrete Math.* **83** 95–103
[12] Brodsky A and Pippenger N 2005 *Random Struct. Algor.* **27** 490–519
[13] Mézard M and Montanari A 2009 *Information, Physics, and Computation* (Oxford: Oxford University Press)

[14] Hatchett J P L, Wemmenhove B, Castillo I P, Nikoletopoulos T, Skantzos N S and Coolen A C C 2004 *J. Phys. A: Math. Gen.* **37** 6201–20
[15] De Dominicis C 1978 *Phys. Rev.* B **18** 4913–9
[16] Mozeika A, Saad D and Raymond J 2009 *Phys. Rev. Lett.* **103** 248701
[17] Mozeika A and Saad D 2010 In preparation.
[18] Mozeika A, Saad D and Raymond J 2010 Noisy Random Boolean Formulae - a Statistical Physics Perspective (*Preprint* arXiv:1003.3806v1 [cond-mat.dis-nn])

International Workshop on Statistical-Mechanical Informatics 2010 IOP Publishing
Journal of Physics: Conference Series **233** (2010) 012016 doi:10.1088/1742-6596/233/1/012016

Quantum channel estimation and asymptotic bound

Masahito Hayashi

Graduate School of Information Sciences, Tohoku University, Sendai, 980-8579, Japan
Centre for Quantum Technologies, National University of Singapore, 3 Science Drive 2, Singapore 117542

E-mail: `hayashi@math.is.tohoku.ac.jp`

Abstract. Quantum channel estimation in the one-parameter case is treated with the asymptotic setting. This problem sometimes has square speedup. A sufficient condition for this speedup is obtained. In this case, the existing results in Cramér-Rao approach seem to contradict the existing results in covariant approach. In order to resolve this problem, we introduce two kinds of asymptotic bounds.

1. Introduction

In quantum information technology, it is usual to use quantum channel for sending quantum state. Since a quantum channel has noise, it is important to identify quantum channel. In this paper, we consider theoretical optimal performance of quantum channel estimation when we can apply the same unknown channel several times. In order to treat this problem, we employ quantum state estimation theory. In our setting, we can optimize our input state and our measurement[1, 2, 3, 4, 5, 6, 7, 8, 9, 10, 11, 12]. We also assume that use of entanglement with reference system is available in the measurement process.

The main topic is the following. In the state estimation, when the number n of prepared states goes to infinity, the mean square error behaves as the order $O(\frac{1}{n})$ as in the estimation of probability distribution. However, in the estimation of quantum channel, two different cases were reported concerning asymptotic behavior of mean square error. As the first case, in the estimations of depolarizing channels and Pauli channels, the optimal mean square error behaves as $O(\frac{1}{n})$[1, 3]. As the second case, in the estimation of unitary, the optimal mean square error behaves as $O(\frac{1}{n^2})$[4, 12, 6, 7, 5, 9, 8]. In order to clarify this point, we consider what case has the order $O(\frac{1}{n})$ concerning the mean square error. As the first result, we derive a necessary condition for the order $O(\frac{1}{n})$ by using Cramér-Rao approach for quantum state estimation. While Fujiwara and Imai[2] derived this kind of condition, our condition is much simpler than their condition. That is, it is easier to check our condition than to check their condition.

Next, we consider the meaning of Cramér-Rao approach when the order is $O(\frac{1}{n^2})$. In this case, two different results concerning estimation of unitary were reported. One is based on the Cramér-Rao approach[4, 12]. The other is based on the group covariant approach[6, 7, 5, 9, 8]. In particular, Imai and Hayashi [9] treated this problem in the case of phase estimation. They seem to contradict with each other. In this paper, in order to resolve this problem, we introduce another asymptotic bound for general channel estimation. Using this bound and the Cramér-Rao bound, we consider the relation between these results and find a relation with superefficiency.

In estimation of probability distribution, if we assume a weaker condition for our estimator, there exists an estimator that surpasses the Cramér-Rao bound only in measure zero points. Such an estimator is called a superefficient estimator[23]. As the second result, in the estimation of phase action, we find that the bound by Cramér-Rao approach can be attained only by a quantum channel version of a superefficient estimator that works at specific points and the bound by group covariant approach can be attained by a suitable estimator that works at all points. Cramér-Rao approach is based on the asymptotically locally unbiased condition. So, we can conclude that the asymptotically locally unbiased condition is too weak for deriving a bound that can be attained in all points. Indeed, a similar phenomena happens in quantum state estimation when we use the large deviation criterion[15].

This paper is organized as follows. Some of obtained results are based on quantum state estimation with Cramér-Rao approach. Section 2 is devoted to a review of Cramér-Rao approach in quantum state estimation. In this section, the symmetric logarithmic derivative (SLD) Fisher information and the right logarithmic derivative (RLD) Fisher information are explained. However, the bound based on Cramér-Rao approach is obtained only by the locally unbiased condition. So, it is needed to discuss its relation with the estimator that works globally. In section 3, we treat SLD Fisher information and RLD Fisher information in the quantum channel estimation, and discuss the increasing order of SLD Fisher information. Section 4 is devoted to the global attainability of the Cramér-Rao bound in the channel estimation and a quantum channel version of superefficiency in the phase estimation.

2. Cramér-Rao bound in quantum state estimation

In quantum state estimation, we estimate the true state through the quantum measurement under the assumption that the true state of the given quantum system $\mathcal{H}$ belongs to a certain parametric state family $\{\rho_\theta|\theta \in \Theta \subset \mathbb{R}^d\}$. Usually, we assume that n quantum systems are prepared in the state ρ_θ. Hence, the total system is described by the tensor product space $\mathcal{H}^{\otimes n}$, and the state of the total system is given by $\rho_\theta^{\otimes n}$.

In this case, when we choose a suitable measurement, the mean square error decreases in proportion to n^{-1} as in the estimation of probability distribution. So, we focus on the first order coefficient of the mean square error concerning n^{-1}. In the most general setting, when a positive operator valued measure M^n on the total system $\mathcal{H}^{\otimes n}$ takes values in the parameter space $\Theta \subset \mathbb{R}^d$, it is allowed as an estimator.

For simplicity, in the following, we consider the case when the number d of parameters is one. So, the mean square error is given as

$$\mathrm{MSE}_\theta(M^n) := \int (\hat{\theta} - \theta)^2 \operatorname{Tr} \rho_\theta^{\otimes n} M^n(d\hat{\theta}).$$

The optimal first order coefficient of this formulation is described by C_θ and is called Quantum Cramér-Rao bound.

In the quantum case, there are several quantum extensions of Fisher information. The largest one is the right logarithmic derivative (RLD) Fisher information J_θ^R, and the smallest one is symmetric logarithmic derivative (SLD) Fisher information J_θ^S. For these definitions, we define the RLD L_θ^R and the SLD L_θ^S as the operators satisfying

$$\frac{d\rho_\theta}{d\theta} = \rho_\theta L_\theta^R, \quad \frac{d\rho_\theta}{d\theta} = \frac{1}{2}\left(L_\theta^S \rho_\theta + \rho_\theta L_\theta^S\right).$$

Then, the RLD and SLD Fisher informations are given by [16, 17, 18]

$$J_\theta^R := \operatorname{Tr} \rho_\theta L_\theta^R (L_\theta^R)^\dagger, \quad J_\theta^S := \operatorname{Tr} \rho_\theta (L_\theta^S)^2.$$

When the range of ρ_θ contains the range of $(\frac{d\rho_\theta}{d\theta})^2$, the RLD Fisher information has another expression:

$$J_\theta^R = \mathrm{Tr}(\frac{d\rho_\theta}{d\theta})^2 \rho_\theta^{-1}. \tag{1}$$

When the state family $\{\rho_\theta\}$ is given by $\rho_\theta := e^{\theta X}|u\rangle\langle u|e^{-\theta X}$, the condition (1) does not hold, where $X^\dagger = -X$. In this case, the SLD Fisher information is calculated as follows[14].

$$4(\langle u|(iX)^2|u\rangle - (\langle u|iX|u\rangle)^2). \tag{2}$$

Now, we introduce the unbiased condition by

$$\int \hat{\theta}\, \mathrm{Tr}\, \rho_\theta^{\otimes n} M^n(d\hat{\theta}) = \theta, \quad \forall \theta \in \Theta.$$

However, this condition is sometimes too restrictive in the asymptotic setting. So, we consider the Taylor expansion at a point θ_0 and focus on the first order. Then, we obtain the locally unbiased condition at θ_0:

$$\int \hat{\theta}\, \mathrm{Tr}\, \rho_{\theta_0}^{\otimes n} M^n(d\hat{\theta}) = \theta_0, \quad \frac{d}{d\theta} \int \hat{\theta}\, \mathrm{Tr}\, \rho_\theta^{\otimes n} M^n(d\hat{\theta})|_{\theta=\theta_0} = 1$$

Under the locally unbiased condition at θ_0, an application of Schwarz inequality similar to the classical case yields the quantum Cramér-Rao inequalities for both quantum Fisher information.

$$\mathrm{MSE}_\theta(M^n) \geq \frac{1}{n}(J_\theta^R)^{-1} \tag{3}$$

$$\mathrm{MSE}_\theta(M^n) \geq \frac{1}{n}(J_\theta^S)^{-1}. \tag{4}$$

Since J_θ^R is greater than J_θ^S, the inequality (4) is more informative than the inequality (3). When the estimator M^n is the spectral decomposition of the operator $\theta_0 I + \frac{1}{nJ_\theta^S}(L_\theta^{S,(1)} + \cdots + L_\theta^{S,(n)})$, the equality in (4) holds, where $X^{(j)}$ is given as $I^{\otimes j-1} \otimes X \otimes I^{\otimes n-j}$. Then, we obtain the following inequality

$$J_\theta^R \geq J_\theta^S.$$

In fact, in the asymptotic setting, a suitable estimator usually satisfies the asymptotic locally unbiased condition:

$$\lim_{n\to\infty} \int \hat{\theta}\, \mathrm{Tr}\, \rho_{\theta_0} M^n(d\hat{\theta}) = \theta_0, \quad \lim_{n\to\infty} \frac{d}{d\theta} \int \hat{\theta}\, \mathrm{Tr}\, \rho_\theta M^n(d\hat{\theta})\Big|_{\theta=\theta_0} = 1$$

for all points θ_0. Under the above condition, using (4), we obtain the inequality

$$\liminf_{n\to\infty} n\, \mathrm{MSE}_\theta(M^n) \geq (J_\theta^S)^{-1}. \tag{5}$$

Further, by using the two-step method, the bound $(J_\theta^S)^{-1}$ can be universally attained for any true parameter θ [19, 20, 21]. So, defining the Cramér-Rao bound:

$$C_\theta := \inf_{\{M^n\}} \left\{ \liminf_{n\to\infty} n\, \mathrm{MSE}_\theta(M^n) \,\middle|\, \begin{array}{l} \{M^n\} \text{ satisfies the asymptotic} \\ \text{locally unbiased condition.} \end{array} \right\},$$

we obtain

$$C_\theta = (J_\theta^S)^{-1}.$$

3. Maximum SLD Fisher information in quantum channel estimation

In this section, we discuss applicability of the Cramér-Rao approach to estimation of channel. In the quantum system, the channel is given by a trace preserving completely positive (TP-CP) map Λ from the set of densities on the input system $\mathcal{H} := \mathbb{C}^d$ to the set of densities on the output system $\mathcal{K} := \mathbb{C}^{d'}$. By using dd' linear maps F_i from $\mathcal{S}(\mathcal{H})$ to $\mathcal{S}(\mathcal{K})$, any TP-CP map Λ can be described by $\Lambda(\rho) = \sum_{i=1}^{dd'} F_i \rho F_i^\dagger$. Hence, our task is to estimate the true TP-CP map under the assumption that the true TP-CP map belongs to a certain family of TP-CP maps $\{\Lambda_\theta\}$.

In order to characterize a TP-CP map Λ_θ, we formulate the notation concerning states on the tensor product system $\mathcal{H} \otimes \mathcal{R}$, where $\mathcal{R}$ is a system of the same dimensionality as $\mathcal{H}$ and is called the reference system. Using a linear map A from $\mathcal{R}$ to $\mathcal{H}$, we define an element of $\mathcal{H} \otimes \mathcal{R}$ as follows.

$$|A\rangle\rangle := \sum_{j,k} A_{j,k} |j\rangle_H \otimes |k\rangle_R, \tag{6}$$

where $\{|j\rangle_H\}_{j=1,\dots,d}$ and $\{|k\rangle_R\}_{k=1,\dots,d}$ are complete orthonormal systems (CONSs) of $\mathcal{H}$ and $\mathcal{R}$. Hence, the relation

$$B \otimes C|A\rangle\rangle = |BAC^T\rangle\rangle \tag{7}$$

holds. This notation is applied to the cases of $\mathcal{K} \otimes \mathcal{H}$ and $\mathcal{K} \otimes \mathcal{R}$.

Now, we focus on the matrix $\rho[\Lambda_\theta] := (\Lambda_\theta \otimes \mathrm{id})(|I\rangle\rangle\langle\langle I|)$. Then, when the input state is the maximally entangled state $|\frac{1}{\sqrt{d}} I\rangle\rangle = \sum_{j=1}^d \frac{1}{\sqrt{d}} |j\rangle \otimes |j\rangle$, the output state is $\frac{1}{d}\rho[\Lambda_\theta] = (\Lambda_\theta \otimes \mathrm{id})(|\frac{1}{\sqrt{d}} I\rangle\rangle\langle\langle \frac{1}{\sqrt{d}} I|)$. When the matrix $\overline{A}A^T$ is a density matrix on $\mathcal{H}$, $|A\rangle\rangle\langle\langle A|$ is a pure state on the product system $\mathcal{H} \otimes \mathcal{R}$. Thus, the output state is given as

$$(\Lambda_\theta \otimes \mathrm{id})(|A\rangle\rangle\langle\langle A|) = (I \otimes A^T)(\Lambda_\theta \otimes \mathrm{id})(|I\rangle\rangle\langle\langle I|)(I \otimes \overline{A}) = (I \otimes A^T)\rho[\Lambda_\theta](I \otimes \overline{A}). \tag{8}$$

In the one-parameter case, we express the derivative $\frac{d\rho(\Lambda_\theta)}{d\theta}$ by $D[\Lambda_\theta]$.

When the input state is the product state $|v\rangle\langle v| \otimes |u\rangle\langle u|$, the output state is $(I \otimes u \cdot v^T)\rho[\Lambda_\theta](I \otimes \overline{v} \cdot u^\dagger)$. Since

$$\mathrm{Tr}_{\mathcal{K}}(I \otimes u \cdot v^T)\rho[\Lambda_\theta](I \otimes \overline{v} \cdot u^\dagger) = |u\rangle\langle u|, \tag{9}$$

we have

$$(I \otimes u \cdot v^T)\rho[\Lambda_\theta](I \otimes \overline{v} \cdot u^\dagger) = (\mathrm{Tr}_{\mathcal{R}}(I \otimes u \cdot v^T)\rho[\Lambda_\theta](I \otimes \overline{v} \cdot u^\dagger)) \otimes |u\rangle\langle u|. \tag{10}$$

This relation implies that the system $\mathcal{R}$ has no information concerning Λ_θ. In this case, it is enough to treat the state $\mathrm{Tr}_{\mathcal{R}}(I \otimes u \cdot v^T)\rho[\Lambda_\theta](I \otimes \overline{v} \cdot u^\dagger)$ for estimation of Λ_θ. Since this state is independent of $|u\rangle$, it is abbreviated by $\langle \overline{v}|\rho[\Lambda_\theta]|\overline{v}\rangle$. Thus,

$$\langle \overline{v}|(\mathrm{Tr}_{\mathcal{K}}\, \rho[\Lambda_\theta])|\overline{v}\rangle = \mathrm{Tr}_{\mathcal{K}} \langle \overline{v}|\rho[\Lambda_\theta]|\overline{v}\rangle 1, \tag{11}$$

which implies that $\langle \overline{v}|(\mathrm{Tr}_{\mathcal{K}}\, \rho[\Lambda_\theta])|\overline{v}\rangle = 1$. Thus, we obtain

$$\mathrm{Tr}_{\mathcal{K}}\, \rho[\Lambda_\theta] = I. \tag{12}$$

Taking the derivative in (12), we obtain

$$\mathrm{Tr}_{\mathcal{K}}\, D[\Lambda_\theta] = 0. \tag{13}$$

Now, we back to our estimation problem. In this problem, our choice is given by a pair of the input state ρ and quantum measurement M. When we fix the input state, our estimation problem can be reduced to the state estimation with the state family $\{\Lambda_\theta(\rho)|\theta \in \Theta\}$. In the one-parameter case, we focus on the suprema

$$J^R[\Lambda_\theta] := \sup_\rho J^R[\Lambda_\theta, \rho], \quad J^S[\Lambda_\theta] := \sup_\rho J^S[\Lambda_\theta, \rho], \tag{14}$$

where $J^S[\Lambda_\theta, \rho]$ and $J^R[\Lambda_\theta, \rho]$ are the SLD and RLD Fisher informations when the input state is ρ. In particular, it is important to calculate the supremum $J^S[\Lambda_\theta]$ which is smaller than $J^R[\Lambda_\theta]$.

When n applications of the unknown channel Λ_θ are available, the input state ρ_n and the measurement M^n are given as a state on $(\mathcal{H}\otimes\mathcal{R})^{\otimes n}$ and a POVM on $(\mathcal{K}\otimes\mathcal{R})^{\otimes n}$. For a sequence of estimators $\{(\rho_n, M^n)\}$, we consider the asymptotic locally unbiased condition:

$$\lim_{n\to\infty} \int \hat{\theta} \operatorname{Tr} \Lambda_{\theta_0}[\rho_n] M^n(d\hat{\theta}) = \theta_0, \quad \lim_{n\to\infty} \frac{d}{d\theta} \int \hat{\theta} \operatorname{Tr} \Lambda_{\theta_0}[\rho_n] M^n(d\hat{\theta}) \Big|_{\theta=\theta_0} = 1$$

for all points θ_0, and denotes the mean square error of (ρ_n, M^n) by $\mathrm{MSE}_\theta(\rho_n, M^n)$. Assume that $J^S[\Lambda_\theta^{\otimes n}]$ behaves as $O(n^\alpha)$ when n goes to infinity. When $\{(\rho_n, M^n)\}$ satisfies the asymptotic locally unbiased condition, the inequality (5) yields that

$$\limsup_{n\to\infty} n^\alpha \, \mathrm{MSE}_\theta(\rho_n, M^n) \geq \limsup_{n\to\infty} \frac{n^\alpha}{J^S[\Lambda_\theta^{\otimes n}]}.$$

We define the Cramér-Rao bound:

$$\tilde{C}_\alpha[\Lambda_\theta] := \inf_{\{\rho_n, M^n\}} \left\{ \limsup_{n\to\infty} n^\alpha \, \mathrm{MSE}_\theta(\rho_n, M^n) \,\middle|\, \begin{array}{l} \{(\rho_n, M^n)\} \text{ satisfies the asymptotic} \\ \text{locally unbiased condition.} \end{array} \right\}.$$

Thus, we obtain

$$\tilde{C}_\alpha[\Lambda_\theta] = \limsup_{n\to\infty} \frac{n^\alpha}{J^S[\Lambda_\theta^{\otimes n}]}. \tag{15}$$

In order to treat the above values, we consider the following condition:

(C) The range of $\rho[\Lambda_\theta]$ contains the range of $D[\Lambda_\theta]^2$.

Assume that the condition (C) does not hold. When the input state is the maximally entangled state $|\frac{1}{\sqrt{d}} I\rangle\rangle$, the RLD Fisher information diverges. So, $J^R[\Lambda_\theta]$ is infinity.

Theorem 1 *When the condition (C) holds,*

$$J^R[\Lambda_\theta] = \| \operatorname{Tr}_{\mathcal{K}} D[\Lambda_\theta] \rho[\Lambda_\theta]^{-1} D[\Lambda_\theta] \|. \tag{16}$$

Proof: Assume that the input state is given by $|A\rangle\rangle\langle\langle A|$ and A is an invertible matrix. Then, the range of $(I \otimes A^T)\rho[\Lambda_\theta](I \otimes \overline{A})$ contains the range of $((I \otimes A^T)D[\Lambda_\theta](I \otimes \overline{A}))^2$.

Using the formula (1), we obtain

$$\begin{aligned} & J^R[\Lambda_\theta, |A\rangle\rangle\langle\langle A|] = \operatorname{Tr}((I \otimes A^T)D[\Lambda_\theta](I \otimes \overline{A}))^2((I \otimes A^T)\rho[\Lambda_\theta](I \otimes \overline{A}))^{-1} \\ & = \operatorname{Tr}(I \otimes A^T)D[\Lambda_\theta](I \otimes \overline{A})(I \otimes A^T)D[\Lambda_\theta](I \otimes \overline{A})(I \otimes \overline{A})^{-1}\rho[\Lambda_\theta]^{-1}(I \otimes A^T)^{-1} \\ & = \operatorname{Tr}(I \otimes \overline{A}A^T)D[\Lambda_\theta]\rho[\Lambda_\theta]^{-1}D[\Lambda_\theta] \\ & = \operatorname{Tr} \overline{A}A^T(\operatorname{Tr}_{\mathcal{K}} D[\Lambda_\theta]\rho[\Lambda_\theta]^{-1}D[\Lambda_\theta]), \end{aligned}$$

where $((I \otimes A^T)\rho[\Lambda_\theta](I \otimes \overline{A}))^{-1}$ is the inverse of $(I \otimes A^T)\rho[\Lambda_\theta](I \otimes \overline{A})$ on its range. So, the supremum of $\mathrm{Tr}\,\overline{A}A^T(\mathrm{Tr}_{\mathcal{K}}\, D[\Lambda_\theta]\rho[\Lambda_\theta]^{-1}D[\Lambda_\theta])$ with the condition $\mathrm{rank}\,\overline{A}A^T = \dim\mathcal{H}$ equals $\|\,\mathrm{Tr}_{\mathcal{K}}\, D[\Lambda_\theta]\rho[\Lambda_\theta]^{-1}D[\Lambda_\theta]\|$.

For a non-invertible matrix A, the inequality $J^R[\Lambda_\theta, |A\rangle\rangle\langle\langle A|] \le \|\,\mathrm{Tr}_{\mathcal{K}}\, D[\Lambda_\theta]\rho[\Lambda_\theta]^{-1}D[\Lambda_\theta]\|$ can be shown by modifying the above discussion, which will be given in [25]. ∎

Theorem 2 *When the condition (C) holds,*

$$J^R[\Lambda_\theta^{\otimes n}] = nJ^R[\Lambda_\theta]. \tag{17}$$

Since $nJ^S[\Lambda_\theta] \le J^S[\Lambda_\theta^{\otimes n}] \le J^R[\Lambda_\theta^{\otimes n}]$, $J^S[\Lambda_\theta^{\otimes n}]$ increases in order n under the assumption of Theorem 1, i.e., $J^S[\Lambda_\theta^{\otimes n}] = O(n)$. When the rank of $\rho[\Lambda_\theta]$ is the maximum, i.e., dd', this condition holds and $J^S[\Lambda_\theta^{\otimes n}] = O(n)$. Even if the rank of $\rho[\Lambda_\theta]$ is not the maximum, we have an example satisfying the condition (C) as follows.

A channel Λ is called a phase damping channel when the output system $\mathcal{K}$ equals the input system $\mathcal{H}$ and there exist complex numbers $d_{k,l}$ such that

$$\Lambda(\rho) = \sum_{k,l} d_{k,l}\rho_{k,l}|k\rangle\langle l|,$$

where $\rho = \sum_{k,l}\rho_{k,l}|k\rangle\langle l|$. In this case, the state $\rho[\Lambda]$ is written as the following form

$$\rho[\Lambda] = \sum_{k,l} d_{k,l}|k\rangle|k\rangle\langle l|\langle l|.$$

That is, its range is included by the space spanned by $\{|k\rangle_K|k\rangle_R\}$. When a channel family $\{\Lambda_\theta\}$ consists of phase damping channels and the range of $\rho[\Lambda_\theta]$ is the space spanned by $\{|k\rangle_K|k\rangle_R\}$, the condition (C) holds.

Proof: For any matrix X, we define matrixes:

$$[X]_k := \underbrace{\rho[\Lambda_\theta]\otimes\cdots\otimes\rho[\Lambda_\theta]}_{k-1}\otimes X\otimes\underbrace{\rho[\Lambda_\theta]\otimes\cdots\otimes\rho[\Lambda_\theta]}_{n-k} \tag{18}$$

$$(X)_k := \underbrace{I\otimes\cdots\otimes I}_{k-1}\otimes X\otimes\underbrace{I\otimes\cdots\otimes I}_{n-k} \tag{19}$$

$$[X]_{k,j} := \underbrace{\rho[\Lambda_\theta]\otimes\cdots\otimes\rho[\Lambda_\theta]}_{k-1}\otimes X\otimes\underbrace{\rho[\Lambda_\theta]\otimes\cdots\otimes\rho[\Lambda_\theta]}_{j-k-1}\otimes X\otimes\underbrace{\rho[\Lambda_\theta]\otimes\cdots\otimes\rho[\Lambda_\theta]}_{n-j}. \tag{20}$$

Here, (20) is the definition in the case of $j > k$. In the case of $j < k$, $[X]_{k,j}$ is defined to be $[X]_{j,k}$. Then,

$$D[\Lambda_\theta^{\otimes n}] = \sum_{k=1}^{n}[D[\Lambda_\theta]]_k \tag{21}$$

$$\|\sum_{k=1}^{n}(X)_k\| = n\|X\|. \tag{22}$$

Using (12), we have

$$\mathrm{Tr}_{\mathcal{K}^{\otimes n}}[X]_k = (\mathrm{Tr}_{\mathcal{K}}\, X)_k. \tag{23}$$

For $k \neq j$, using (12) and (13), we have

$$\mathrm{Tr}_{\mathcal{K}^{\otimes n}}[D[\Lambda_\theta]]_k(\rho[\Lambda_\theta]^{\otimes n})^{-1}[D[\Lambda_\theta]]_j = \mathrm{Tr}_{\mathcal{K}^{\otimes n}}[D[\Lambda_\theta]]_{k,j} = 0. \tag{24}$$

For any k, using (23), we have

$$\begin{aligned}&\mathrm{Tr}_{\mathcal{K}^{\otimes n}}[D[\Lambda_\theta]]_k(\rho[\Lambda_\theta]^{\otimes n})^{-1}[D[\Lambda_\theta]]_k = \mathrm{Tr}_{\mathcal{K}^{\otimes n}}[D[\Lambda_\theta]\rho[\Lambda_\theta]^{-1}D[\Lambda_\theta]]_k \\ =&(\mathrm{Tr}_{\mathcal{K}}\, D[\Lambda_\theta]\rho[\Lambda_\theta]^{-1}D[\Lambda_\theta]])_k\end{aligned} \tag{25}$$

Thus, combining (21), (22), (24), and (25), we obtain

$$\begin{aligned}&J^R[\Lambda_\theta^{\otimes n}] = \| \mathrm{Tr}_{\mathcal{K}^{\otimes n}} \sum_{k,j}[D[\Lambda_\theta]]_k(\rho[\Lambda_\theta]^{\otimes n})^{-1}[D[\Lambda_\theta]]_j\| \\ =&\| \mathrm{Tr}_{\mathcal{K}^{\otimes n}} \sum_{k=1}^{n}[D[\Lambda_\theta]]_k(\rho[\Lambda_\theta]^{\otimes n})^{-1}[D[\Lambda_\theta]]_k + \sum_{k\neq j}[D[\Lambda_\theta]]_{k,j}\| \\ =&\| \mathrm{Tr}_{\mathcal{K}^{\otimes n}} \sum_{k=1}^{n}(\mathrm{Tr}_{\mathcal{K}}\, D[\Lambda_\theta]\rho[\Lambda_\theta]^{-1}D[\Lambda_\theta]])_k\| = n\| \mathrm{Tr}_{\mathcal{K}}\, D[\Lambda_\theta]\rho[\Lambda_\theta]^{-1}D[\Lambda_\theta]\|.\end{aligned}$$

■

Next, we consider the case where the condition (C) does not hold. As the simplest example, we consider the one-parameter unitary case, i.e., the case when $\Lambda_\theta(\rho) = e^{\theta X}\rho e^{-\theta X}$ and $X^\dagger = -X$. Using (2), we obtain

$$J^S[\Lambda_\theta] = (\lambda_{\max}(iX) - \lambda_{\min}(iX))^2, \tag{26}$$

where $\lambda_{\max}(iX)$ and $\lambda_{\min}(iX)$ are the maximum and minimum of eigenvalues of iX. So, we obtain

$$J^S[\Lambda_\theta^{\otimes n}] = n^2(\lambda_{\max}(iX) - \lambda_{\min}(iX))^2. \tag{27}$$

4. Global attainability in channel estimation

The discussion in the previous section treats the bound only with the locally unbiased condition. In order to check the operational meaning, we have to consider the global attainability for this bound as in the case of state estimation. Here, remember that this bound can be attained by using the two-step method in the case of state estimation. For this purpose, we introduce another quantity. In the following definition, we focus on an ϵ-neighborhood $U_{\theta,\epsilon}$ of θ. Define

$$C_\alpha[\Lambda_{\theta_0}, \{(\rho_n, M^n)\}] := \lim_{\epsilon\to 0}\limsup_{n\to\infty} n^\alpha \sup_{\theta\in U_{\theta_0,\epsilon}} \mathrm{MSE}_\theta(\rho_n, M^n).$$

$$C_\alpha[\Lambda_{\theta_0}] := \inf_{\{(\rho_n,M^n)\}} C_\alpha[\Lambda_{\theta_0}, \{(\rho_n, M^n)\}].$$

Concerning this quantity, we have the following two propositions, whose proofs will be given in [25].

Proposition 3 *When $\tilde{C}_\alpha[\Lambda_{\theta_0}]$ is continuous,*

$$C_\alpha[\Lambda_{\theta_0}] \geq \tilde{C}_\alpha[\Lambda_{\theta_0}]. \tag{28}$$

Proposition 4 *Assume that $R := \sup_{\theta\in\Theta}|\theta| < \infty$. When the order parameter α equals 1 and $\tilde{C}_\alpha[\Lambda_{\theta_0}]$ is continuous,*

$$C_1[\Lambda_{\theta_0}] = \tilde{C}_1[\Lambda_{\theta_0}].$$

By using the two-step method[22], the bound $C_\alpha[\Lambda_{\theta_0}]$ can be attained at all points θ as follows.

Proposition 5 *Assume that* $R := \sup_{\theta\in\Theta}|\theta| < \infty$ *and* $C_1[\Lambda_{\theta_0}]$ *is continuous. For any* $\delta > 0$, *there exists a sequence of estimators* $\{(\rho_n, M^n)\}$ *such that*

$$C_\alpha[\Lambda_\theta, \{(\rho_n, M^n)\}] \le C_\alpha[\Lambda_\theta] + \delta \tag{29}$$

for all points θ.

A proof of this proposition will be given in [25].

As a typical example, we consider the case of $\Lambda_\theta(\rho) = e^{\theta X}\rho e^{-\theta X}$ with $X = \begin{pmatrix} i/2 & 0 \\ 0 & -i/2 \end{pmatrix}$. In this case, we replace mean square error by $\min_k(\hat{\theta} + 2\pi k - \theta)^2$. The minimum average error behaves as $\frac{\pi^2}{n^2}$ [8, 9]. That is, the leading decreasing order is $O(1/n^2)$ and the leading decreasing coefficient is π^2 when we apply the optimal estimator. Since the bound $C_2[\Lambda_\theta]$ can be globally attained, $C_2[\Lambda_\theta] = \pi^2$. Since $J^S[\Lambda_\theta^{\otimes n}] = n^2$, $\tilde{C}_2[\Lambda_\theta] = 1$. Hence, the bound $\tilde{C}_2[\Lambda_\theta]$ can be attained only in a specific point as follows.

Proposition 6 *Assume that* $R := \sup_{\theta\in\Theta}|\theta| < \infty$ *and* $C_1[\Lambda_{\theta_0}]$ *is continuous. For any* $\delta > 0$ *and any* $\theta_0 \in \Theta$, *there exists a sequence of estimators* $\{(\rho_{n,\theta_0}, M^n_{\theta_0})\}$ *satisfying the asymptotically locally unbiased condition and the relations:*

$$\limsup_{n\to\infty} n^\alpha \,\mathrm{MSE}_\theta(\rho_{n,\theta_0}, M^n_{\theta_0}) \le C_\alpha[\Lambda_\theta] + \delta, \quad \forall\theta \ne \theta_0 \tag{30}$$

$$\limsup_{n\to\infty} n^\alpha \,\mathrm{MSE}_{\theta_0}(\rho_{n,\theta_0}, M^n_{\theta_0}) \le \tilde{C}_\alpha[\Lambda_{\theta_0}] + \delta. \tag{31}$$

A proof of this proposition will be given in [25].

In estimation of probability distribution, there exists a superefficient estimator that has smaller error at a discrete set than Cramér-Rao bound[23]. Since such a superefficient estimator cannot be useful, it is thought to be better to choose a condition for our estimators in order to remove superefficient estimators in statistics. In this classical case, if we assume the asymptotic locally unbiased condition, we have no superefficient estimator. Proposition 6 means that even if the asymptotic locally unbiased condition is assumed, there exists an estimator that behaves in the similar way to a superefficient estimator in the case of unitary estimation. So, we call such an estimator a q-channel-superefficient estimator. That is, a sequence of estimators $\{(\rho_n, M^n)\}$ is called *q-channel-superefficient* at θ with the order $\frac{1}{n^\alpha}$ when $\lim_{n\to\infty} n^\alpha \,\mathrm{MSE}_\theta(\rho_n, M^n) < C_\alpha[\Lambda_\theta]$. Hence, in order to remove the q-channel-superefficiency problem, it is better to adopt the bound $C_\alpha[\Lambda_\theta]$ as the criterion instead of $\tilde{C}_\alpha[\Lambda_\theta]$.

Finally, we consider the relation with the adaptive method proposed by Nagaoka[13]. In this method, we apply our POVM to each single system $\mathcal{H}$, and we decide k-th POVM based on the knowledge of previous $k-1$ outcomes. In this case, Fujiwara [24] analyzed the asymptotic behaviour of the MSE of this estimator. Now, we consider the case of nm applications of the unknown channel Λ_θ. In this case, we divide nm applications into n groups consisting of m applications. When we apply the adaptive method mentioned in Fujiwara[24] to these groups, the MSE of this estimator behaves as $\frac{1}{nJ^S[\Lambda_\theta^{\otimes m}]}$, which is close to $\frac{C_\alpha[\Lambda_\theta]}{nm^\alpha}$. So, when $\alpha > 1$, this method cannot realize the optimal order $O(\frac{1}{(nm)^\alpha})$.

5. Discussion

We have treated quantum channel estimation as a general framework containing quantum state estimation. We have introduced two kinds of asymptotic bound in quantum channel estimation.

One is given from the limit of maximum of SLD Fisher information $J^S[\Lambda_\theta^{\otimes n}]$. The other is given by more operational quantities. The relation between both bounds is closely related to the asymptotic behavior of $J^S[\Lambda_\theta^{\otimes n}]$. We have shown that both bounds coincide when $J^S[\Lambda_\theta^{\otimes n}]$ behaves as $O(n)$. The case of state estimation can be regarded as a special case of the above case. That is, the conventional state estimation has no difference between both quantities. However, in the case of unitary estimation, $J^S[\Lambda_\theta^{\otimes n}]$ behaves as $O(n^2)$, and there exists a difference between both bounds.

So, the asymptotic behavior of $J^S[\Lambda_\theta^{\otimes n}]$ is quite important. In order to clarify this behavior, we have derived the condition (C) as a sufficient condition for $J^S[\Lambda_\theta^{\otimes n}] = O(n)$. Indeed, Fujiwara and Imai [2] also obtained another sufficient condition. Since the relation with their condition is not clear, it is an open problem. Our condition (C) trivially contains the case when the state $\rho[\Lambda_\theta]$ is a full rank state on the tensor product system while it is not so easy to derive the above full rank condition from Fujiwara and Imai's condition. Further, we have also obtained another example for $J^S[\Lambda_\theta^{\otimes n}] = O(n^2)$[25] under the condition (C). This example is a larger class than the unitary model. So, we can expect that $J^S[\Lambda_\theta^{\otimes n}]$ behaves as $O(n^2)$ if the condition (C) does not hold. This is a challenging open problem.

Acknowledgment

The author was partially supported by a Grant-in-Aid for Scientific Research in the Priority Area 'Deepening and Expansion of Statistical Mechanical Informatics (DEX-SMI)', No. 18079014 and a MEXT Grant-in-Aid for Young Scientists (A) No. 20686026. The Center for Quantum Technologies is funded by the Singapore Ministry of Education and the National Research Foundation as part of the Research Centres of Excellence programme.

References

[1] Fujiwara A 2001 *Phys. Rev.* A **63** 042304
[2] Fujiwara A and Imai H 2008 *J. Phys. A: Math. Theor.* **41** 255304
[3] Fujiwara A and Imai H 2003 *J. Phys. A: Math. Gen.* **36** 8093–103
[4] Fujiwara A 2002 *Phys. Rev.* A **65** 012316
[5] Hayashi M 2006 *Phys. Lett. A* **354** 183–9
[6] Chiribella G, D'Ariano G M, Perinotti P and Sacchi M F 2004 *Phys. Rev. Lett.* **93** 180503
[7] Bagan E, Baig M and Munoz-Tapia R 2004 *Phys. Rev.* A **70** 030301
[8] Buzek V, Derka R and Massar S 1999 *Phys. Rev. Lett.* **82** 2207–10
[9] Imai H and Hayashi M 2009 Fourier Analytic Approach to Phase Estimation in Quantum Systems *New J. Phys.* **11** 043034
[10] Kahn J 2007 *Phys. Rev.* A **75** 022326
[11] Hotta M, Karasawa T and Ozawa M 2006 *J. Phys. A: Math. Gen.* **39** 14465–70
[12] Imai H and Fujiwara F 2007 *J. Phys. A: Math. Theor.* **40** 4391–400
[13] Nagaoka H 1989 *Proc. 12th Symp. on Inform. Theory and its Appl.* p 577; Nagaoka H 2005 *Asymptotic Theory of Quantum Statistical Inference* ed M Hayashi (Singapore: World Scientific) p 125 (reprinted)
[14] Fujiwara A and Nagaoka H 1995 *Phys. Lett. A* **201** 119–24
[15] Hayashi M 2002 *J. Phys. A: Math. Gen.* **35** 7689–727
[16] Helstrom C W 1976 *Quantum Detection and Estimation Theory* (New York: Academic Press)
[17] Holevo A S 1982 *Probabilistic and Statistical Aspects of Quantum Theory* (Amsterdam: North-Holland); (Originally published in Russian 1980)
[18] Hayashi M 2006 *Quantum Information: An Introduction* (Berlin: Springer)
[19] Hayashi M and Matsumoto K 1998 *RIMS koukyuroku* No 1055 (Kyoto: Kyoto University) pp 96–110 (In Japanese); Hayashi M and Matsumoto K 2005 *Asymptotic Theory of Quantum Statistical Inference* ed M Hayashi (Singapore: World Scientific) pp 162–9 (reprinted, English translation)
[20] Gill R and Massar S 2000 *Phys. Rev.* A **61** 042312
[21] Hayashi M and Matsumoto K 2008 *J. Math. Phys.* **49** 102101
[22] Hayashi M 2009 *Selected Papers on Probability and Statistics* (American Mathematical Society Translations Series 2 Vol. 277) ed American Mathematical Society Staff (American Mathematical Society) p 95 (*Preprint quant-ph/0608198*)

[23] LeCam L 1986 *Asymptotic Methods in Statistical Decision Theory* (New York: Springer)
[24] Fujiwara A 2006 *J. Phys. A: Math. Gen.* **39** 12489–504
[25] Hayashi M 2010 Two kinds of asymptotic bounds in quantum channel estimation *Preprint* arXiv:1003.4575v1 [quant-ph]

International Workshop on Statistical-Mechanical Informatics 2010 IOP Publishing
Journal of Physics: Conference Series **233** (2010) 012017 doi:10.1088/1742-6596/233/1/012017

Error detection by binary sparse matrices

Tadashi Wadayama
Nagoya Institute of Technology, Gokiso, Showa-ku, Naogya, Aichi, Japan, 466-8555
E-mail: wadayama@nitech.ac.jp

Abstract. Error detection is a crucial process in an information transmission protocol like TCP for attaining reliable communications over unreliable channels. The most important performance measure for an error detection scheme is its undetected error probability, which is the probability corresponding to the event such that an erroneous received word passes the detection test. The paper reviews the author's results on an analysis of the undetected error probability of ensembles of $m \times n$ binary matrices. The ensemble of binary sparse matrices called the *Bernoulli ensemble* whose members are considered as matrices generated from i.i.d. Bernoulli source is mainly discussed here. The error exponent of the average undetected error probability and closed form expressions for the variance of the undetected error probability will be presented.

1. Introduction

An error detection scheme is a crucial part of feedback error correction schemes such as ARQ (Automatic Repeat reQuest), which are required for attaining reliable communications over unreliable channels. The most important performance measure for an error detection scheme is its undetected error probability, which is the probability corresponding to the event such that an erroneous received word passes the detection test. A parity check matrix used in an error detection process determines its undetected error probability. Detailed knowledge on the error detection performance would be useful for assessing the performance of a feedback error correction scheme.

Random coding is an extremely powerful technique to show the existence of a code satisfying certain properties. It has been used for proving the direct part (achievability) of many types of coding theorems. Recently, the idea of random coding has also come to be regarded as important from a practical point of view. An LDPC (Low-density parity-check) code can be constructed by choosing a parity check matrix from an ensemble of sparse matrices. Thus, there is a growing interest in randomly generated codes.

One of the main difficulties associated with the use of randomly generated codes is the difficulty in evaluating the properties or performance of such codes. For example, it is difficult to evaluate minimum distance, weight distribution, ML decoding performance, etc. for these codes. To overcome this problem, we can take a *probabilistic approach*. In such an approach, we consider an ensemble of parity check matrices: i.e., probability is assigned to each matrix in the ensemble. A property of a matrix (e.g., minimum distance, weight distributions) can then be regarded as a random variable. It is natural to consider statistics of the random variable, such as mean, variance, higher moments and covariance. In some cases, we can show that a property is strongly concentrated around its expectation. Such a concentration result justifies the use of the probabilistic approach.

The evaluation of the error detection probability of a given code (or given parity check matrix) is a classical problem in coding theory [2, 3] and some results on this topic have been derived from the view point of a probabilistic approach. For example, for a linear code ensemble the inequality $P_U < 2^{-m}$ has long been known, where P_U is the undetected error probability and m is the number of rows of a parity check matrix. Since the undetected error probability can be expressed as a linear combination of the weight distribution of a code, there is a natural connection between the expectation of the weight distribution and the expectation of the undetected error probability.

The paper reviews the author's results on an analysis of the undetected error probability of ensembles of $m \times n$ binary matrices according to [6, 7]. The ensemble of binary sparse matrices called the *Bernoulli ensemble* whose members are considered as matrices generated from i.i.d. Bernoulli source is mainly discussed here.

2. Average undetected error probability

This section discusses the error exponents related to the undetected error probability. The error exponent shows asymptotic behaviors of an error detection system.

2.1. Notation

Let H be an $m \times n$ $(m, n \geq 1)$ binary parity check matrix. The set $C(H)$ is the binary linear code of length n defined by H, namely, $C(H) \stackrel{\triangle}{=} \{\boldsymbol{x} \in \mathbb{F}_2^n : H\boldsymbol{x}^t = 0^m\}$ where $\mathbb{F}_2$ is the Galois field with two elements $\{0, 1\}$ (the addition over $\mathbb{F}_2$ is denoted by $\oplus$). The notation 0^m denotes the zero vector of length m. In this paper, a boldface letter, such as $\boldsymbol{x}$ for example, denotes a binary row vector.

Throughout the paper, a binary symmetric channel (BSC) with crossover probability ϵ $(0 < \epsilon < 1/2)$ is assumed. In this paper, we use the conventional scenario for error detection: a transmitter sends a codeword $\boldsymbol{x} \in C(H)$ to a receiver via a BSC with crossover probability ϵ. The receiver obtains a received word $\boldsymbol{y} = \boldsymbol{x} \oplus \boldsymbol{e}$, where $\boldsymbol{e}$ denotes an error vector. The receiver firstly computes the syndrome $\boldsymbol{s}^t = H\boldsymbol{y}^t$ and then checks whether $\boldsymbol{s} = 0^m$ holds or not.

An undetected error event occurs when $H\boldsymbol{e}^t = 0^m$ and $\boldsymbol{e} \neq 0^m$. This means that the error vector $\boldsymbol{e} \in C(\boldsymbol{e} \neq 0^n)$ causes an undetected error event. Thus, the undetected error probability $P_U(H)$ can be expressed as

$$P_U(H) = \sum_{\boldsymbol{e} \in C(H), \boldsymbol{e} \neq 0^m} \epsilon^{w(\boldsymbol{e})}(1-\epsilon)^{n-w(\boldsymbol{e})} \tag{1}$$

where $w(\boldsymbol{x})$ denotes the Hamming weight of vector $\boldsymbol{x}$. The above equation can be rewritten as

$$P_U(H) = \sum_{w=1}^{n} A_w(H)\epsilon^w(1-\epsilon)^{n-w}, \tag{2}$$

where $A_w(H)$ is defined by

$$A_w(H) \stackrel{\triangle}{=} \sum_{\boldsymbol{x} \in Z^{(n,w)}} I[H\boldsymbol{x}^t = 0^m]. \tag{3}$$

The set $\{A_w(H)\}_{w=0}^n$ is usually called the *weight distribution* of $C(H)$. The notation $Z^{(n,w)}$ denotes the set of n-tuples with weight w. The notation $I[condition]$ is the indicator function such that $I[condition] = 1$ if *condition* is true; otherwise, it evaluates to 0.

Suppose that $\mathcal{G}$ is a set of binary $m \times n$ matrices $(m, n \geq 1)$. A probability $P(H)$ is associated with each matrix H in $\mathcal{G}$. Thus, $\mathcal{G}$ can be considered as an *ensemble* of binary matrices. Let $f(H)$ be a real-valued function which depends on $H \in \mathcal{G}$. The expectation of $f(H)$ with respect to the ensemble $\mathcal{G}$ is defined by

$$E_{\mathcal{G}}[f(H)] \stackrel{\triangle}{=} \sum_{H \in \mathcal{G}} P(H) f(H). \tag{4}$$

The average weight distribution of a given ensemble $\mathcal{G}$ is given by $E_{\mathcal{G}}[A_w(H)]$. This quantity is very useful for analyzing the performance of binary linear codes, including analysis of the undetected error probability.

2.2. Bernoulli ensemble

In this paper, we will focus on a parameterized ensemble $\mathcal{B}_{m,n,k}$ which is called the *Bernoulli ensemble* because the Bernoulli ensemble is amenable to ensemble analysis. The Bernoulli ensemble $\mathcal{B}_{m,n,k}$ contains all the binary $m \times n$ matrices $(m, n \geq 1)$, whose elements are regarded as i.i.d. binary random variables such that an element takes the value 1 with probability $p \stackrel{\triangle}{=} k/n$. The parameter k $(0 < k \leq n/2)$ is a positive real number which represents the average number of ones for each row. In other words, a matrix $H \in \mathcal{B}_{m,n,k}$ can be considered as an output from the Bernoulli source such that symbol 1 occurs with probability p.

From the above definition, it is clear that a matrix $H \in \mathcal{B}_{m,n,k}$ is associated with the probability

$$P(H) = p^{\bar{w}(H)}(1-p)^{mn-\bar{w}(H)}, \tag{5}$$

where $\bar{w}(H)$ is the number of ones in H (i.e., Hamming weight of H). The average weight distribution of the Bernoulli ensemble is given by

$$E_{\mathcal{B}_{m,n,k}}[A_w(H)] = \left(\frac{1+z^w}{2}\right)^m \binom{n}{w} \tag{6}$$

for $w \in [0, n]$ where $z \stackrel{\triangle}{=} 1-2p$. The notation $[a, b]$ denotes the set of consecutive integers from a to b. The average weight distribution of this ensemble was first discussed by Litsyn and Shevelev [4].

If k is a constant (i.e., not a function of n), this ensemble can be considered as an ensemble of sparse matrices. In the spacial case where $k = n/2$, equal probability $1/2^{mn}$ is assigned to every matrix in the Bernoulli ensemble. As a simplified notation, we will denote $\mathcal{R}_{m,n} \stackrel{\triangle}{=} \mathcal{B}_{m,n,n/2}$, where $\mathcal{R}_{m,n}$ is called the *random ensemble*.

2.3. Average undetected error probability of an ensemble

For a given $m \times n$ matrix H, the evaluation of the undetected error probability $P_U(H)$ is in general computationally difficult because we need to know the weight distribution of $C(H)$ for such evaluation. On the other hand, in some cases, we can evaluate the average of $P_U(H)$ for a given ensemble. Such an average probability is useful for the estimation of the undetected error probability of a matrix which belongs to the ensemble.

Taking the ensemble average of the undetected error probability over a given ensemble $\mathcal{G}$, we have

$$\begin{aligned} E_{\mathcal{G}}[P_U(H)] &= E_{\mathcal{G}}\left[\sum_{w=1}^{n} A_w(H)\epsilon^w(1-\epsilon)^{n-w}\right] \\ &= \sum_{w=1}^{n} E_{\mathcal{G}}[A_w(H)]\epsilon^w(1-\epsilon)^{n-w}. \end{aligned} \tag{7}$$

In the above equations, H can be regarded as a random variable. From this equation, it is evident that the average of $P_U(H)$ can be evaluated if we know the average weight distribution of the ensemble. For example, in the case of the random ensemble $\mathcal{R}_{m,n}$, the average undetected error probability has a simple closed form:

$$E_{\mathcal{R}_{m,n}}[P_U(H)] = 2^{-m}(1 - (1-\epsilon)^n). \tag{8}$$

2.4. Error exponent of undetected error probability

For a given sequence of $(1-R)n \times n$ matrix ensembles ($n = 1, 2, 3, \dots,$), the average undetected error probability is usually an exponentially decreasing function of n, where R is a real number satisfying $0 < R < 1$ (called the *design rate*). Thus, the exponent of the undetected error probability is of prime importance in understanding the asymptotic behavior of the undetected error probability.

2.4.1. Definition of error exponent Let $\{\mathcal{G}_n\}_{n>0}$ be a series of ensembles such that $\mathcal{G}_n$ consists of $(1-R)n \times n$ binary matrices. In order to see the asymptotic behavior of the undetected error probability of this sequence of ensembles, it is reasonable to define the error exponent of undetected error probability in the following way:

Definition 1 *The asymptotic error exponent of the average undetected error probability for a series of ensembles $\{\mathcal{G}_n\}_{n>0}$ is defined by*

$$T_{\mathcal{G}_n} \triangleq \lim_{n\to\infty} \frac{1}{n} \log_2 E_{\mathcal{G}_n}[P_U] \tag{9}$$

if the limit exists. □

Henceforth we will not explicitly express the dependence of P_U on H, writing instead P_U to denote $P_U(H)$ in all cases where there is no fear of confusion.

The following example describes the exponent of the random ensemble.

Example 1 *Consider the series of the random ensembles $\{\mathcal{R}_{n,(1-R)n}\}_{n>0}$. It is easy to evaluate $T_{\mathcal{R}_{(1-R)n,n}}$:*

$$\begin{aligned} T_{\mathcal{R}_{(1-R)n,n}} &= \lim_{n\to\infty} \frac{1}{n} \log_2 E_{\mathcal{R}_{(1-R)n,n}}[P_U] \\ &= \lim_{n\to\infty} \frac{1}{n} \log_2 2^{-(1-R)n}(1-(1-\epsilon)^n) \\ &= -(1-R). \end{aligned} \tag{10}$$

This equality implies that the average undetected error probability of the sequence of random ensembles behaves like

$$E_{\mathcal{R}_{(1-R)n,n}}[P_U] \simeq 2^{-n(1-R)} \tag{11}$$

if n is sufficiently large. Note that the exponent $-(1-R)$ is independent of the crossover probability ϵ. □

2.4.2. Error exponent and asymptotic growth rate The *asymptotic growth rate* of the average weight distribution (for simplicity henceforth abbreviated as the asymptotic growth rate), which is the basis of the derivation of the error exponent, is defined as follows.

Definition 2 *Suppose that a series of ensembles* $\{\mathcal{G}_n\}_{n>0}$ *is given. If*

$$\lim_{n\to\infty} \frac{1}{n} \log_2 E_{\mathcal{G}_n}[A_{\ell n}]$$

exists for $0 \le \ell \le 1$*, then we define the* asymptotic growth rate $f(\ell)$ *by*

$$f(\ell) \triangleq \lim_{n\to\infty} \frac{1}{n} \log_2 E_{\mathcal{G}_n}[A_{\ell n}]. \tag{12}$$

The parameter ℓ *is called the* normalized weight. □

From this definition, it is clear that

$$E_{\mathcal{G}_n}[A_{\ell n}] = 2^{n(f(\ell)+o(1))}, \tag{13}$$

where the notation $o(1)$ denotes terms which converge to 0 in the limit as n goes to infinity. The asymptotic growth rate of some ensembles of binary matrices can be found in [1, 4, 5].

The next theorem [6, 7] gives the error exponent of the undetected error probability for a series of ensembles $\{\mathcal{G}_n\}_{n>0}$.

Theorem 1 *The error exponent of* $\{\mathcal{G}_n\}_{n>0}$ *is given by*

$$T_{\mathcal{G}_n} = \sup_{0<\ell\le 1} [f(\ell) + \ell \log_2 \epsilon + (1-\ell)\log_2(1-\epsilon)], \tag{14}$$

where $f(\ell)$ *is the asymptotic growth rate of* $\{\mathcal{G}_n\}_{n>0}$*.*

□

The next example discusses the case of the random ensemble.

Example 2 *Let us again consider the series of the random ensembles given by* $\{\mathcal{R}_{(1-R)n,n}\}_{n>0}$*. These ensembles have the asymptotic growth rate* $f(\ell) = h(\ell) - (1-R)$*, where the function* $h(x)$ *is the binary entropy function defined by*

$$h(x) \triangleq -x\log_2 x - (1-x)\log_2(1-x). \tag{15}$$

In this case, by using Theorem 1, we have

$$T_{\mathcal{R}_{(1-R)n,n}} = \sup_{0<\ell\le 1} [h(\ell) - (1-R) + \ell\log_2 \epsilon + (1-\ell)\log_2(1-\epsilon)]. \tag{16}$$

Let

$$D_{\ell,\epsilon} \triangleq \ell \log_2 \left(\frac{\ell}{\epsilon}\right) + (1-\ell)\log_2\left(\frac{1-\ell}{1-\epsilon}\right). \tag{17}$$

By using $D_{\ell,\epsilon}$*, we can rewrite (16) as*

$$T_{\mathcal{R}_{(1-R)n,n}} = \sup_{0<\ell\le 1} [-(1-R) - D_{\ell,\epsilon}]. \tag{18}$$

Since $D_{\ell,\epsilon}$ *can be considered as the Kullback-Leibler divergence between two probability distributions* $(\epsilon, 1-\epsilon)$ *and* $(\ell, 1-\ell)$*,* $D_{\ell,\epsilon}$ *is always non-negative and* $D_{\ell,\epsilon} = 0$ *holds if and only if* $\ell = \epsilon$*. Thus, we obtain*

$$\sup_{0<\ell\le 1} [-(1-R) - D_{\ell,\epsilon}] = -(1-R), \tag{19}$$

which is identical to the exponent obtained in expression (10).

Let $g_\epsilon^{(rnd)}(\ell) \triangleq h(\ell) - (1-R) + \ell\log_2\epsilon + (1-\ell)\log_2(1-\epsilon)$*. Figure 1 displays the behavior of* $g_\epsilon^{(rnd)}(\ell)$ *when* $R = 0.5$*. This figure confirms the result that the maximum* ($\sup_{0<\ell\le 1} g_\epsilon^{(rnd)}(\ell) = -0.5$) *is attained at* $\ell = \epsilon$*.* □

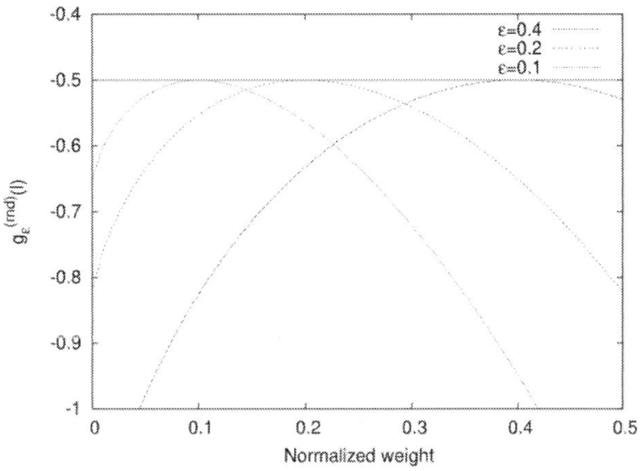

Figure 1. The curves of $g_\epsilon(\ell)$ for random ensembles with $R = 0.5$. The curves of $g_\epsilon^{(rnd)}(\ell)$ corresponding to the parameters $\epsilon = 0.1, 0.2, 0.4$ from left to right are presented. As a reference, line of $-(1-R) = -0.5$ is also included in the figure.

2.5. *Error exponent of the Bernoulli ensemble with constant k*

The asymptotic growth rate of the Bernoulli ensemble $\mathcal{B}_{m,n,k}$ with a constant k and design rate R is given by

$$f(\ell) = h(\ell) + (1-R)\log_2\left(\frac{1+e^{-2k\ell}}{2}\right). \tag{20}$$

This formula is presented in [4]. The error exponent of this ensemble shows a different behavior from that for random ensembles.

Example 3 *Consider the Bernoulli ensemble with parameters $R = 0.5$ and $k = 20$. Let*

$$g_\epsilon^{(Ber)}(\ell) \triangleq h(\ell) + (1-R)\log_2\left(\frac{1+e^{-2k\ell}}{2}\right) + \ell\log_2\epsilon + (1-\ell)\log_2(1-\epsilon). \tag{21}$$

The function $g_\epsilon^{(Ber)}(\ell)$ corresponds to $f(\ell) + \ell\log_2\epsilon + (1-\ell)\log_2(1-\epsilon)$ in (14). □

Figure 2 includes the curves of $g_\epsilon^{(Ber)}(\ell)$ where $\epsilon = 0.1, 0.2, 0.4$. In contrast to $g_\epsilon^{(rnd)}(\ell)$ of a random ensemble, we can see that $g_\epsilon^{(Ber)}(\ell)$ is not a concave function. The shape of the curve of $g_\epsilon^{(Ber)}(\ell)$ depends on the crossover probability ϵ. For large ϵ, $g_\epsilon(\ell)$ takes its largest value around $\ell = \epsilon$. On the other hand, for small ϵ, $g_\epsilon^{(Ber)}(\ell)$ has the supremum at $\epsilon = 0$.

Figure 3 presents the error exponent of Bernoulli ensembles with parameters $R = 0.3, 0.5, 0.7, 0.9$ and $k = 20$. As an example, consider the exponent for $R = 0.5$. In the regime where ϵ is smaller than (around) 0.3, the error exponent is a monotonically decreasing function of ϵ.

The examples suggest that a Bernoulli ensemble has less powerful error detection performance than that of a dense ensemble (such as the random ensemble) in terms of the error exponent. However, if the crossover probability is sufficiently large, the difference in exponent of Bernoulli and random ensembles is negligible. For example, the exponent of the Bernoulli ensemble in figure 3 is almost equal to that of the random ensemble when ϵ is larger than (around) 0.3.

The above properties of the error exponents of the Bernoulli ensembles can be explained with reference to their average weight distributions (or asymptotic growth rate). Figure 4 displays the asymptotic growth rates of a random ensemble and a Bernoulli ensemble.

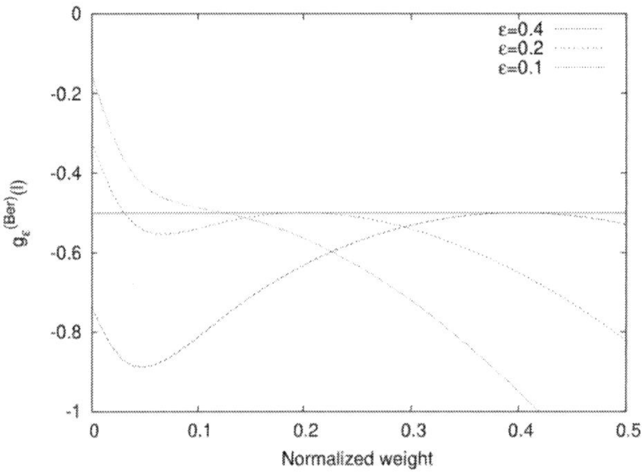

Figure 2. The curves of $g_\epsilon^{(Ber)}(\ell)$ for Bernoulli ensembles. The curves of $g_\epsilon^{(Ber)}(\ell)$ corresponding to the parameters $\epsilon = 0.1, 0.2, 0.4$ are presented. The parameters $R = 0.5, k = 20$ are assumed. As a reference, line of $-(1-R) = -0.5$ is also included in the figure.

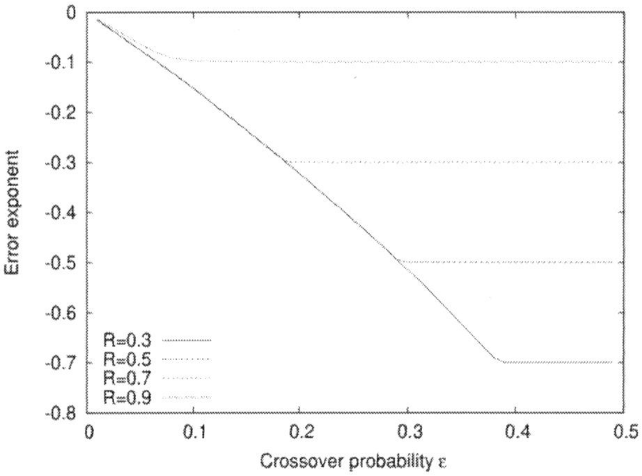

Figure 3. Error exponent of Bernoulli ensemble. The curves of $T_{\mathcal{B}_{m,n,k}}$ corresponding to the parameters $R = 0.3, 0.5, 0.7, 0.9$ and $k = 20$ are presented.

The weight of typical error vectors is very close to ϵn when n is sufficiently large. For a large value of ϵ, such as $\epsilon = 0.4$, the average weight distribution around $w = 0.4n$, namely $E_{\mathcal{G}}[A_{0.4n}]$, dominates the undetected error probability. In such a range, the difference in the average weight distributions corresponding to the random and the Bernoulli ensembles is small. On the other hand, if the crossover probability is small, weight distributions of low weight become the most influential parameter. The difference in the average weight distributions of small weight results in a difference in the error exponent.

Note that the time complexity of the error detection operation (multiplication of received vector and a parity check matrix) is $O(n^2)$-time for a typical instance of a random ensemble, and is $O(n)$-time for a typical instance of a Bernoulli ensemble with constant k. A sparse matrix offers almost the same error detection performance of a dense matrix with linear time complexity if ϵ is sufficiently large.

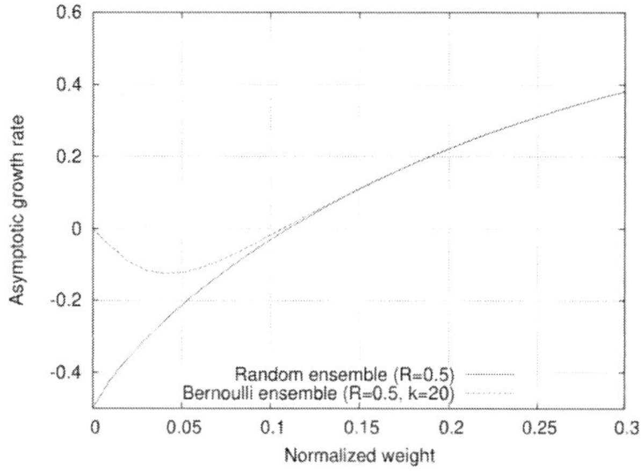

Figure 4. Asymptotic growth rate of a random ensemble and a Bernoulli ensemble.

3. Variance of undetected error probability

In the previous section, we have seen that the average weight distribution plays an important role in the derivation of average undetected error probability. Similarly, we need to examine the *covariance of weight distribution* in order to analyze the variance of undetected error probability.

3.1. Covariance formula

The covariance between two real-valued functions $f(\cdot), g(\cdot)$ defined on an ensemble $\mathcal{G}$ is given by

$$\mathrm{Cov}_{\mathcal{G}}[f,g] \triangleq E_{\mathcal{G}}[fg] - E_{\mathcal{G}}[f]E_{\mathcal{G}}[g]. \tag{22}$$

The next theorem forms the basis of the derivation of the variance of the undetected error probability for the Bernoulli ensemble. The covariance of the weight distribution for the Bernoulli ensemble is given in the following theorem [6, 7].

Theorem 2 *The covariance of the weight distribution for the Bernoulli ensemble $\mathcal{B}_{m,n,k}$ is given by*

$$\begin{aligned}
&\mathrm{Cov}_{\mathcal{B}_{m,n,k}}(A_{w_1}, A_{w_2}) \\
&\triangleq \left(\frac{1+z^{w_1}}{2}\right)^m \left(\frac{1+z^{w_2}}{2}\right)^m \\
&\times \sum_{v=\max\{0,w_1+w_2-n\}}^{w_1} \binom{n}{w_1}\binom{w_1}{v}\binom{n-w_1}{w_2-v}\left(\left(1+\frac{z^{w_1+w_2-2v}-z^{w_1+w_2}}{(1+z^{w_1})(1+z^{w_2})}\right)^m - 1\right)
\end{aligned} \tag{23}$$

for $1 \le w_1 \le w_2 \le n$ *and*

$$\mathrm{Cov}_{\mathcal{B}_{m,n,k}}(A_{w_1}, A_{w_2}) = \mathrm{Cov}_{\mathcal{B}_{m,n,k}}(A_{w_2}, A_{w_1}) \tag{24}$$

for $1 \le w_2 < w_1 \le n$, *where* $z = 1 - 2p$ *and* $p = k/n$. □

3.2. Variance of undetected error probability

The variance of the undetected error probability is a straightforward consequence of Theorem 2.

Corollary 1 *The variance of the undetected error probability of the Bernoulli ensemble* $\sigma^2_{\mathcal{B}_{m,n,k}}$ *is given by*

$$\sigma^2_{\mathcal{B}_{m,n,k}} = \sum_{w_1=1}^{n} \sum_{w_2=1}^{n} \mathrm{Cov}_{\mathcal{B}_{m,n,k}}(A_{w_1}, A_{w_2}) \epsilon^{w_1+w_2}(1-\epsilon)^{2n-w_1-w_2}. \quad (25)$$

□

In the case of $k = n/2$ (i.e. the case of a random ensemble), we can derive a closed form expression for the variance.

Corollary 2 *For the random ensemble* $\mathcal{R}_{m,n}$, *the variance of the undetected error probability* P_U *is given by* $\sigma^2_{\mathcal{R}_{m,n}} = (1-2^{-m})2^{-m}\left((\epsilon^2+(1-\epsilon)^2)^n - (1-\epsilon)^{2n}\right)$. □

The next numerical example facilitates an understanding of how the average and the variance of P_U behave.

Example 4 *We consider the random ensemble with* $m = 20, n = 40$, *and the Bernoulli ensemble with* $m = 20, n = 40, k = 5$. *Figure 5 shows the variance of* P_U *for the above two ensembles. The two curves have a similar shape, but the variance of the Bernoulli ensemble is always larger than that of the random ensemble.* □

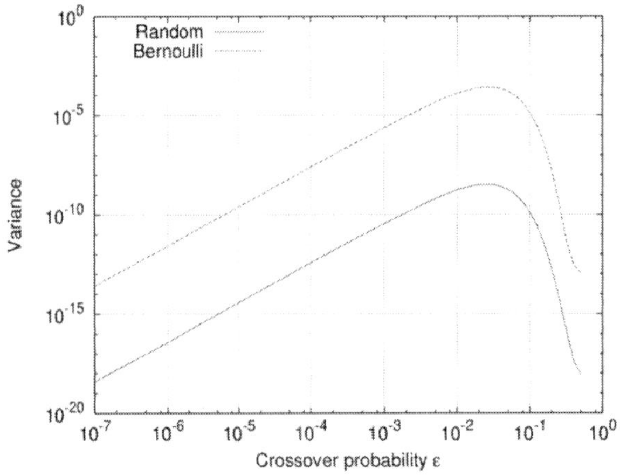

Figure 5. Variance of undetected error probability (Random ensemble: $m = 20, n = 40$, Bernoulli ensemble: $m = 20, n = 40, k = 5$).

3.3. Concentration to average: random ensemble

The variance derived in Corollary 2 can be used to prove the following concentration result for the random ensemble.

Corollary 3 *The ratio of* P_U *and* $E_{\mathcal{R}_{m,n}}[P_U]$ *converges to 1 in probability, namely,*

$$\frac{P_U}{E_{\mathcal{R}_{m,n}}[P_U]} \to 1 \quad \textit{in probability} \quad (26)$$

as n goes to infinity if ϵ $(0 < \epsilon < 1/2)$ *satisfies* $1 - R + \log_2(\epsilon^2 + (1-\epsilon)^2) < 0$. □

Let ϵ^* be the root of the equation

$$1 - R + \log_2(\epsilon^{*2} + (1 - \epsilon^*)^2) = 0. \tag{27}$$

Table 1 presents some values of ϵ^* for values of R from 0.1 to 0.9. When $\epsilon > \epsilon^*$, we have $1 - R + \log_2(\epsilon^{*2} + (1 - \epsilon^*)^2) < 0$. In such a region, it is evident that P_U concentrates around its average value in the limit as n tends to infinity.

Table 1. Roots of $1 - R + \log_2(\epsilon^{*2} + (1 - \epsilon^*)^2) = 0$

R	ϵ^*
0.1	0.366 047
0.2	0.307 193
0.3	0.259 613
0.4	0.217 375
0.5	0.178 203
0.6	0.140 933
0.7	0.104 872
0.8	0.069 564
0.9	0.034 687

4. Concluding Summary

The paper gives a succinct review of highlights of [6, 7]. The ensemble of binary sparse matrices called the Bernoulli ensemble whose members are considered as matrices generated from i.i.d. Bernoulli source is mainly discussed here. The error exponent of the average undetected error probability and closed form expressions for the variance of the undetected error probability are main contributions of this work.

Acknowledgment

This work was supported by the Ministry of Education, Science, Sports and Culture, Japan, Grant-in-Aid for Scientific Research on Priority Areas (Deepening and Expansion of Statistical-Mechanical Informatics) 18079009 and SRC (Storage Research Consortium). The author wishes to express his sincere thanks to anonymous reviewers whose comments help to improve the quality of the paper.

References

[1] Burshtein D and Miller G 2004 *IEEE Trans. Inform. Theory* **50** 1115–31
[2] Kløve T 2007 *Codes for Error Detection* (Singapore: World Scientific)
[3] Kløve T and Korzhik V I 1995 *Error Detecting Codes: General Theory and Their Application in Feedback Communication Systems* (Massachusetts: Kluwer Academic Publishers)
[4] Litsyn S and Shevelev V 2002 *IEEE Trans. Inform. Theory* **48** 887–908
[5] Litsyn S and Shevelev V 2003 *IEEE Trans. Inform. Theory* **49** 3140–59
[6] Wadayama T 2008 *Proc. IEEE Int. Symp. Inform. Theory* pp 1045–9
[7] Wadayama T 2010 *IEEE Trans. Inform. Theory* to appear

International Workshop on Statistical-Mechanical Informatics 2010 IOP Publishing
Journal of Physics: Conference Series **233** (2010) 012018 doi:10.1088/1742-6596/233/1/012018

Properties of a certain stochastic dynamical system, channel polarization, and polar codes

Toshiyuki Tanaka

Graduate School of Informatics, Kyoto University, Yoshida Hon-machi, Sakyo-ku, Kyoto-shi, Kyoto 606-8501, Japan

E-mail: tt@i.kyoto-u.ac.jp

Abstract. A new family of codes, called polar codes, has recently been proposed by Arıkan. Polar codes are of theoretical importance because they are provably capacity achieving with low-complexity encoding and decoding. We first discuss basic properties of a certain stochastic dynamical system, on the basis of which properties of channel polarization and polar codes are reviewed, with emphasis on our recent results.

1. Introduction

The concept "More is different" is relevant not only in statistical physics but also in various other research fields, such as information theory, where one can find several examples in which properties of a system drastically change as the system size becomes large. As an example, channel coding theorem states that there exist error-correcting codes such that their error probabilities are arbitrarily small and yet their coding rates are arbitrarily close to the channel capacity. Although channel coding theorem is one of the most important theorems in information theory, it is also widely understood that channel coding theorem is a theorem of existence, so that one requires different arguments as to how one can construct practical codes with high performance. The problem of constructing codes which have practically low computational complexity in encoding and decoding, as well as good performance in view of channel coding theorem, remains a problem of particular importance in information theory and coding theory.

Proposal of turbo codes in 1990s, as well as subsequent rediscovery of low-density parity-check (LDPC) codes, has ignited extensive research activities on the combination of codes with sparse graphical representation and probability-based iterative decoding algorithms. As a consequence, by now we have practical means of how to construct codes with low computational complexity and high performance. However, rigorous results justifying the high performance of LDPC codes are mostly limited to the case with erasure channels, and we have not arrived at a thorough understanding of LDPC codes for other channels.

Arıkan [1] recently proposed polar codes as a class of error-correcting codes which is completely different from LDPC codes. Polar codes are theoretically very interesting because it saturates symmetric channel capacity (maximum mutual information when input distribution is assumed uniform, which is equal to the channel capacity for a symmetric channel) in the large codelength limit, whereas computational complexity of encoding and decoding is polynomial in the codelength. In this paper, we review recent developments regarding polar codes, as well as

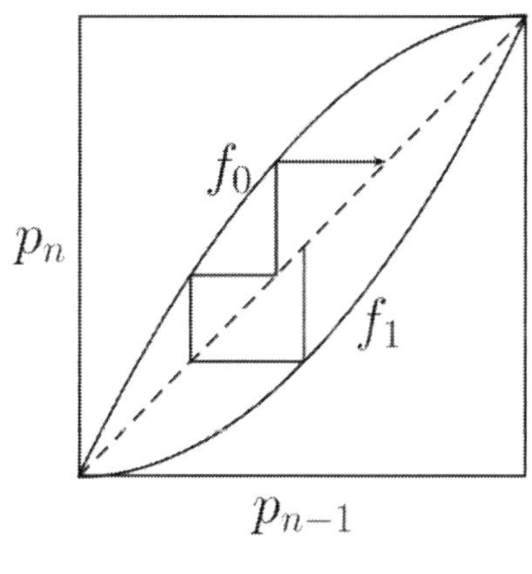

Figure 1. Discrete-time stochastic system.

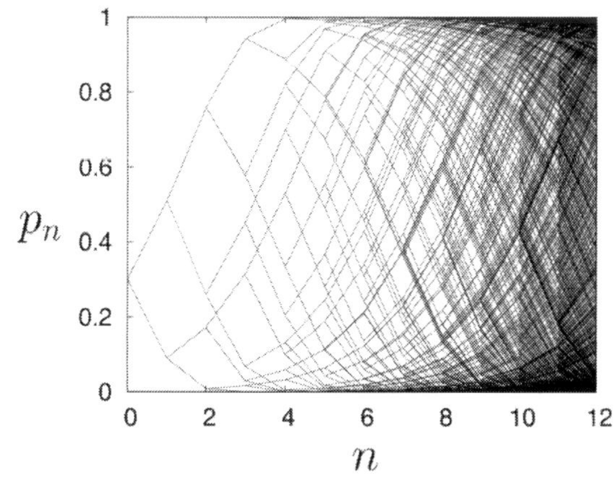

Figure 2. Time evolution of the discrete-time stochastic system.

channel polarization which forms mathematical foundation of polar codes, putting emphasis on our own results.

2. Properties of a stochastic dynamical system

In this section we define a stochastic dynamical system and argue its basic properties. The argument in this section will be utilized in understanding fundamentals of channel polarization and polar codes.

Let f_0 and f_1 be functions on the interval [0, 1] defined as

$$f_0(x) = 1 - (1 - x)^2,$$
$$f_1(x) = x^2. \tag{1}$$

Let B_1, B_2, ... be a sequence of independent and identically-distributed (i.i.d.) Bernoulli random variables, each of which takes 0 or 1 with probability 1/2. On the basis of them, we define a stochastic dynamical system as follows.

$$p_0 = p \in (0,\, 1)$$
$$p_n = f_{B_n}(p_{n-1}) = \begin{cases} f_0(p_{n-1}), & \text{if } B_n = 0 \\ f_1(p_{n-1}), & \text{if } B_n = 1 \end{cases} \tag{2}$$

In other words, the stochastic dynamical system is defined by the stochastic updating rule in which one throws a fair coin at every discrete time, and updates the state, denoted by p_n indexed by discrete time n, with either of the functions f_0 and f_1 according to the outcome of the coin flip. There are 2^n possible values of p_n with equal probability, and a realization of p_n depends on a realization of $\{B_1,\, B_2,\, \ldots,\, B_n\}$, so that $\{p_0,\, p_1,\, \ldots,\, \}$ is a stochastic process. Figure 1 shows a sample process that can be obtained from this dynamical process. Figure 2 shows all possible trajectories from the initial value $p = 0.3$.

For the stochastic dynamical system (2), it is not difficult to show the following proposition [1] to hold.

Proposition 1 *The limit* $\lim_{n\to\infty} p_n = p_\infty$ *exists, and*

$$P(p_\infty) = \begin{cases} 1-p & (p_\infty = 0) \\ p & (p_\infty = 1) \end{cases} \tag{3}$$

holds.

We show an outline of the proof, essentially due to Arıkan and Telatar [2], in the following. First, from the identity $[f_0(x) + f_1(x)]/2 = x$, one can prove $\mathbb{E}(p_n|p_{n-1}) = p_{n-1}$. It means that the stochastic process $\{p_0,\, p_1,\, \ldots,\, \}$ defined by the stochastic dynamical system (2) is a martingale. It is obvious that this martingale is bounded, so that one can apply the martingale convergence theorem to prove that p_n converges in the limit $n \to \infty$ almost surely to a random variable p_∞. It is also obvious that $\mathbb{E}(p_\infty) = p$ holds.

Next, in order to study properties of the limiting random variable p_∞, let us define a stochastic process $\{q_0,\, q_1,\, \ldots\}$ by

$$q_n := p_n(1 - p_n). \tag{4}$$

For this stochastic process, one has

$$q_n = q_{n-1} \times \begin{cases} (2 - p_{n-1})(1 - p_{n-1}) & \text{if } B_n = 0, \\ p_{n-1}(1 + p_{n-1}) & \text{if } B_n = 1. \end{cases} \tag{5}$$

On the other hand, from Schwartz inequality, one has, for $^\forall x \in [0,\, 1]$,

$$\left[\sqrt{(2-x)(1-x)} + \sqrt{x(1+x)}\right]^2 \le [(2-x) + (1+x)] \cdot [(1-x) + x] = 3, \tag{6}$$

which proves

$$\mathbb{E}\big(q_n^{1/2} \mid q_{n-1}\big) \le \sqrt{\frac{3}{4}} q_{n-1}^{1/2}. \tag{7}$$

Recursive application of the above inequality gives

$$\mathbb{E}(q_n^{1/2}) \le \frac{1}{2}\left(\frac{3}{4}\right)^{n/2}. \tag{8}$$

Since q_n is non-negative, (8) implies that q_n takes values close to 0 with probability close to 1. More precisely, application of Markov inequality yields, for any $\rho > 0$,

$$P\left(q_n \ge \frac{\rho^n}{4}\right) \le \left(\frac{3}{4\rho}\right)^{n/2} \tag{9}$$

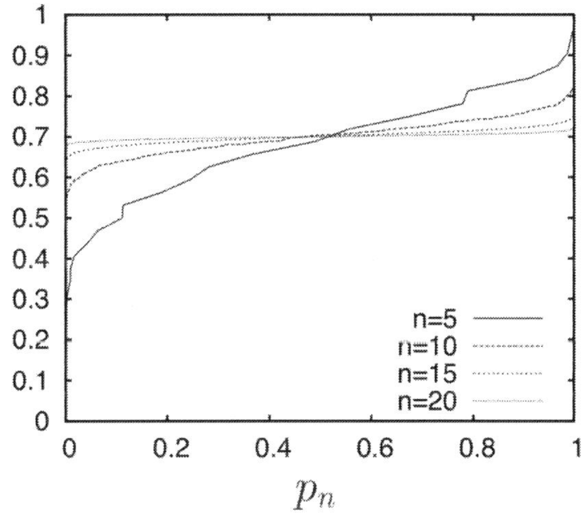

Figure 3. Cumulative distributions of p_n.

holds. Let $\rho \in (3/4, 1)$, then the probability of the event $q_n < \rho^n/4$ approaches 1 exponentially as n tends to large. That q_n approaches 0 corresponds to p_n approaching 0 or 1. More explicitly, one has

$$P(p_n < \rho^n) + P(p_n > 1 - \rho^n) \geq P\left(q_n < \frac{\rho^n}{4}\right) \geq 1 - \left(\frac{3}{4\rho}\right)^{n/2}, \tag{10}$$

proving that p_∞ takes 0 or 1 with probability 1. This observation, along with the fact $\mathbb{E}(p_\infty) = p$, proves Proposition 1.

Figure 3 shows cumulative distributions of p_n with the initial condition $p = 0.3$, for the cases with $n = 5, 10, 15, 20$. It can be observed that the distributions approach the limiting distribution as stated in Proposition 1.

3. Channel polarization and polar codes

3.1. Successive cancellation decoding

We first construct a code with codelength 2 and code rate 1, which forms a basis for polar codes. The code is a linear code on GF(2) (Galois field of order 2), whose generator matrix is

$$F = \begin{pmatrix} 1 & 0 \\ 1 & 1 \end{pmatrix}. \tag{11}$$

In other words, letting $(u_0, u_1) \in \{0, 1\}^2$ be a message (for later convenience, we adopt the convention to start the index of a vector from 0 rather than from 1), encoding is represented as an operation of F from right

$$(u_0, u_1)F = (u_0, u_1)\begin{pmatrix} 1 & 0 \\ 1 & 1 \end{pmatrix} = (u_0 + u_1, u_1), \tag{12}$$

producing a codeword $(x_0, x_1) = (u_0 + u_1, u_1)$. Here, $u_0 + u_1$ means the addition in GF(2).

Assume that every symbol of the codeword is to be transmitted over a binary erasure channel (BEC) with erasure probability p. Let (y_0, y_1) be the received word, then y_0 is equal to x_0 with probability $(1 - p)$, and an erasure symbol with probability p. Similarly, y_1 is equal to x_1 with probability $(1 - p)$, and an erasure symbol with probability p. In addition, y_0 and y_1 are independent given (x_0, x_1).

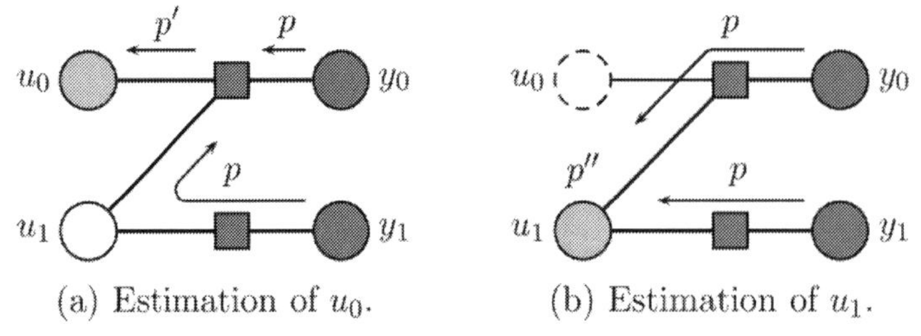

Figure 4. Successive cancellation decoding (codelength: 2).

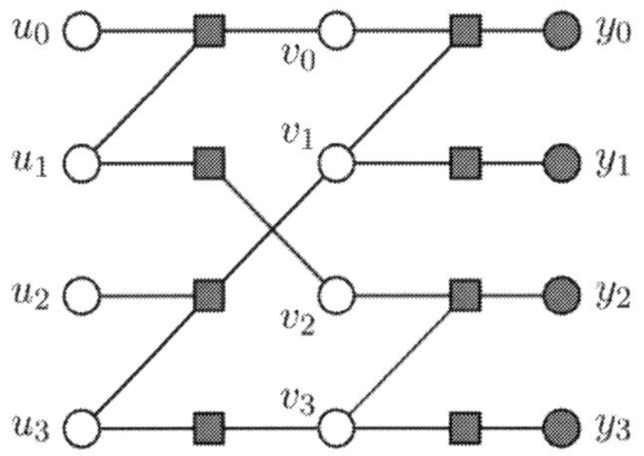

Figure 5. Factor graph representation of decoding of code of length 4.

The problem of decoding here is to estimate (u_0, u_1) on the basis of the received word (y_0, y_1). Although there would be several ways to do the decoding, we consider in the following the decoding in which u_0 and u_1 are estimated individually and sequentially. We first consider the estimation of u_0. The estimation of u_0 is not possible if either y_0 or y_1 is erased: If y_0 is unknown, information about u_0 is no longer available. Even if $y_0 = x_0 = u_0 + u_1$ is available, estimation of u_0 is not possible if $y_1 = u_1$ is unknown. The probability p' of failure in estimating u_0 is, therefore, equal to the probability that either y_0 or y_1 (or both) is erased, and is given by $f_0(p) = 1 - (1-p)^2$. Figure 4 (a) depicts a factor-graph representation of the problem of estimating u_0.

We next consider the problem of estimating u_1, assuming that we have the correct estimation result for u_0. One can estimate u_1 correctly if $y_1 = u_1$ is not erased. Or even though y_1 is erased, one can still estimate u_1 correctly if $y_0 = u_0 + u_1$ is not erased, by making use of the knowledge of u_0 which is assumed to be correctly estimated. Thus, the probability p'' that one cannot correctly estimate u_1 on the assumption that u_0 has been correctly estimated is equal to the probability that y_0 and y_1 are both erased, and therefore given by $f_1(p) = p^2$. The problem of estimating u_1 is represented by a factor graph as in figure 4 (b). The approach of decoding discussed so far, in which the components u_0 and u_1 are estimated one-by-one sequentially is called successive cancellation decoding [1].

On the basis of the above-defined code, we next construct a code with codelength 4 and code rate 1. It is a linear code over GF(2), as before, and is constructed so that the problem of decoding a codeword (u_0, u_1, u_2, u_3) from a received word (y_0, y_1, y_2, y_3) has a factor-graph representation as shown in figure 5.

Let us consider successive cancellation decoding of this code applied to a binary erasure channel with erasure probability p. Remembering the above argument on successive cancellation decoding of the code with codelength 2, in order to correctly estimate u_0, one needs correct estimation results for the intermediate variables v_0 and v_2 in figure 5. Since the probabilities

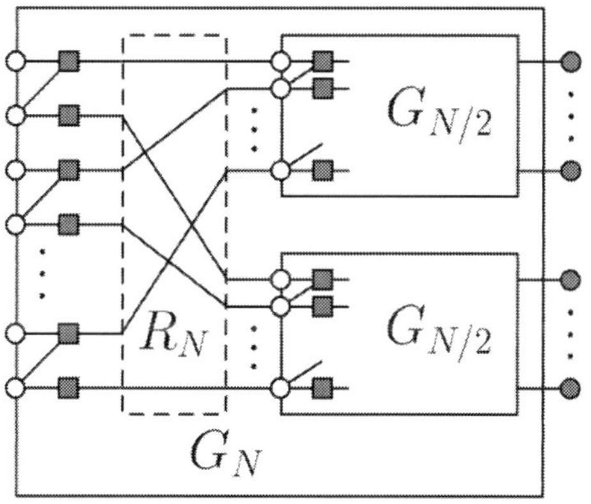

Figure 6. Recursive construction of generator matrices.

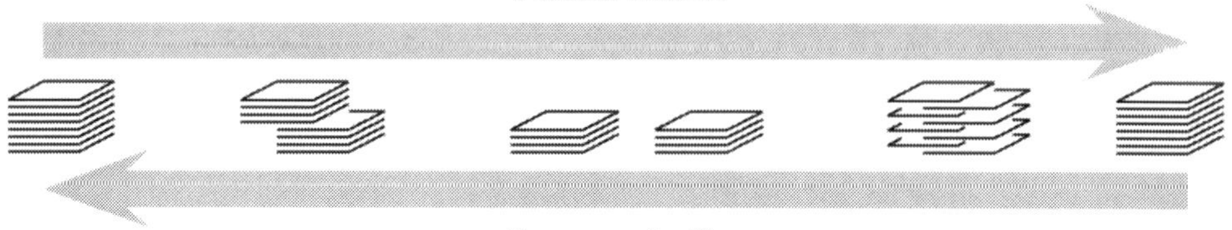

Figure 7. Reverse shuffle.

that v_0 and v_2 are not correctly estimated are both equal to $p' = f_0(p)$, the probability that u_0 is not correctly estimated is equal to $1-(1-p')^2 = f_0(f_0(p))$. Next, we consider the probability that one cannot correctly estimate u_1 under the assumption that one knows the correct estimation result for u_0. The probability is equal to the probability that neither v_0 nor v_2 is correctly estimated, and is thus given by $(p')^2 = f_1(f_0(p))$. Similarly, the probability that u_2 cannot be correctly estimated provided that the correct estimates of u_0 and u_1 are known is equal to $f_0(f_1(p))$, and the probability that u_3 cannot be correctly estimated provided that the correct estimates of u_0, u_1, and u_2 are known is given by $f_1(f_1(p))$.

In a similar way, we can recursively construct codes whose codelengths are integer powers of 2 (see figure 6): Let $G_2 = F$, and construct the generator matrices recursively as

$$G_N = (I_{N/2} \otimes F) R_N (I_2 \otimes G_{N/2}). \tag{13}$$

The matrix R_N is called a reverse-shuffle operator which, when applied to a vector from right, induces a permutation among elements of the vector that corresponds to the inverse of the so-called perfect shuffle of a deck of playing cards (see figure 7). For example,

$$R_8 = \begin{pmatrix} 1 & 0 & 0 & 0 & 0 & 0 & 0 & 0 \\ 0 & 0 & 0 & 0 & 1 & 0 & 0 & 0 \\ 0 & 1 & 0 & 0 & 0 & 0 & 0 & 0 \\ 0 & 0 & 0 & 0 & 0 & 1 & 0 & 0 \\ 0 & 0 & 1 & 0 & 0 & 0 & 0 & 0 \\ 0 & 0 & 0 & 0 & 0 & 0 & 1 & 0 \\ 0 & 0 & 0 & 1 & 0 & 0 & 0 & 0 \\ 0 & 0 & 0 & 0 & 0 & 0 & 0 & 1 \end{pmatrix}. \tag{14}$$

For the successive cancellation decoding of the code of codelength $N = 2^n$ and a binary erasure channel with erasure probability p, the probability of not correctly estimating u_0 is given by

$$\underbrace{f_0 \circ f_0 \circ \cdots \circ f_0}_{n \text{ operations}}(p), \tag{15}$$

and, assuming that we know the correct estimation results for $u_0, \ldots, u_{i-1}$, the probability that one fails to estimate u_i is given by

$$f_{b_1} \circ f_{b_2} \circ \cdots \circ f_{b_n}(p), \tag{16}$$

where $b_n b_{n-1} \ldots b_2 b_1$ denotes binary expansion of i.

3.2. Polar codes

The codes defined in the previous subsection has code rate 1, so that they cannot be used as error-correcting codes as they are. One can construct codes with code rate less than 1 from the above rate-1 codes, by simply clamping some components of $\boldsymbol{u} = (u_0, u_1, \ldots, u_{N-1})$ to 0, which is called shortening of codes. Let $\mathcal{I}$ denote the set of indices of unclamped components of $\boldsymbol{u}$. A component u_i with $i \notin \mathcal{I}$ is called a frozen bit. A polar code is constructed by appropriately choosing $\mathcal{I}$ to make error probability of successive cancellation decoding small.

One could consider a strategy that defines $\mathcal{I}$ as a set of indices of the rows of G_N containing the largest numbers of 1s. The resulting codes are equivalent to the Reed-Muller codes. Polar codes are constructed on the basis of a different strategy to define $\mathcal{I}$, as described below.

In performing successive cancellation decoding, one does not have to decode the frozen bits $\{u_i;\ i \notin \mathcal{I}\}$ since one knows that these values have been clamped to 0. Let $\hat{\boldsymbol{u}}$ be the decoding result of successive cancellation decoding. Let $\mathcal{E}_N$ be the event that successive cancellation decoding fails in a polar code with codelength N, and let $\mathcal{B}_{i,N}$ for $i \in \mathcal{I}$ be the event that successive cancellation decoding fails for the first time at the ith component. Then, $\{\mathcal{B}_{i,N};\ i \in \mathcal{I}\}$ are mutually exclusive, and

$$\mathcal{E}_N = \bigcup_{i \in \mathcal{I}} \mathcal{B}_{i,N} \tag{17}$$

holds. Let $\mathcal{A}_{i,N}$ be the event that successive cancellation decoding fails for u_i provided that the decoder knows the correct decoding results for u_0^{i-1}, where and hereafter, we adopt the notational convention of expressing a subvector as $u_i^j = (u_i, u_{i+1}, \ldots, u_{j-1}, u_j)$, etc. Noting that $\mathcal{B}_{i,N} \subset \mathcal{A}_{i,N}$, one obtains an upper bound of the error probability of successive cancellation decoding as

$$P(\mathcal{E}_N) = \sum_{i \in \mathcal{I}} P(\mathcal{B}_{i,N}) \le \sum_{i \in \mathcal{I}} P(\mathcal{A}_{i,N}). \tag{18}$$

Arıkan [1] proposed constructing a polar code by choosing $\mathcal{I}$ to minimize the upper bound (18).

Since the resulting code is a linear code, computational complexity of encoding is at most $O(N^2)$: In fact, one can exploit the recursive structure of the generator matrix to reduce the complexity of encoding to $O(N \log N)$. On the other hand, computational complexity of successive cancellation decoding is $O(N)$ per component of $\boldsymbol{u}$, but the total complexity can be reduced to $O(N \log N)$ since intermediate results for decoding of one component can be utilized in decoding of other components. Computational complexity of polar codes is therefore $O(N \log N)$ for encoding as well as for successive cancellation decoding.

3.3. Channel polarization

Knowledge of the probabilities $\{P(\mathcal{A}_{i,N})\}$ is essential in order to construct polar codes on the basis of the method given in the previous subsection. It is also important when one wants to know performance of polar codes.

As already seen in section 3.1, the problem of evaluating the values of the probabilities $\{P(\mathcal{A}_{i,N})\}$ has been completely solved: Let $b_n b_{n-1} \dots b_2 b_1$ be the binary expansion of i, then

$$P(\mathcal{A}_{i,N}) = f_{b_1} \circ f_{b_2} \circ \cdots \circ f_{b_n}(p) \tag{19}$$

holds. Moreover, on the basis of the results reviewed in section 2, one immediately has, as $N = 2^n$ tends to be large, that $\{P(\mathcal{A}_{i,N})\}$ will take values close to 0 or 1 with high probability, and that fractions of indices i for which $P(\mathcal{A}_{i,N})$ takes a value close to 0 and 1 are $(1-p)$ and p, respectively. Arıkan called such a phenomenon channel polarization. From the above-mentioned facts, the number of indices i such that $P(\mathcal{A}_{i,N})$ take values close to 0 is approximately $N(1-p)$ as N gets large, one can collect those indices to define $\mathcal{I}$ to construct a polar code with code rate less than $(1-p)$ and a vanishing error probability. In order to demonstrate that the above scenario is justified, one has to prove that the error probability $P(\mathcal{E}_N)$ is $o(1)$ by showing that the fraction of indices i for which $P(\mathcal{A}_{i,N}) = o(1/N)$ hold can be made arbitrarily close to $(1-p)$.

Intuitively, if p_m takes a value close to 0, then it is very unlikely that p_n takes a large value for any $n \gg m$. In order to see this observation in a more concrete setting, we define a stochastic process $\{\tilde{p}_0, \tilde{p}_1, \dots\}$ as

$$\tilde{p}_n = p_n \quad (n \le m) \tag{20}$$

$$\tilde{p}_n = \begin{cases} 2\tilde{p}_{n-1}, & \text{if } B_n = 0, \\ \tilde{p}_{n-1}^2, & \text{if } B_n = 1. \end{cases} \quad (n > m) \tag{21}$$

Obviously $\tilde{p}_n \ge p_n$ holds because of the relation $2x \ge f_0(x)$. Taking logarithm of both sides of (21) yields

$$\log_2 \tilde{p}_n = \begin{cases} \log_2 \tilde{p}_{n-1} + 1, & \text{if } B_n = 0, \\ 2\log_2 \tilde{p}_{n-1}, & \text{if } B_n = 1. \end{cases} \quad (n > m) \tag{22}$$

One observes that the addition by 1 when $B_n = 0$ has a relatively negligible effect if $\tilde{p}_m$ is sufficiently close to 0, because $\log_2 \tilde{p}_{n-1}$ should take large negative values. Therefore, from the law of large numbers, one typically has

$$\tilde{p}_n \approx p_m^{2^{(n-m)/2}}. \tag{23}$$

More precisely, the probability that

$$p_n = o\left(2^{-2^{\beta n}}\right) = o\left(2^{-N^\beta}\right) \tag{24}$$

holds for any $\beta < 1/2$ tends to $(1-p)$ as $n \to \infty$, which implies that block error probability $P(\mathcal{E}_N)$ satisfies

$$P(\mathcal{E}_N) = o\left(2^{-2^{\beta n}}\right) \tag{25}$$

whenever the code rate R is less than $(1-p)$. This result is due to Arıkan and Telatar [2]. We have further refined the above result by applying the central limit theorem, to obtain the following result depending on the code rate R [3, 4]:

$$P(\mathcal{E}_N) = o\left(2^{-2^{(n+t\sqrt{n})/2}}\right) \quad \text{for any } t \text{ satisfying } Q(t) > \frac{R}{1-p} \tag{26}$$

Here $Q(t) = \int_t^\infty e^{-u^2/2} du/\sqrt{2\pi}$ is the conventional Q-function.

It should be noted that $(1-p)$ is nothing but the channel capacity of a binary erasure channel with erasure probability p. The argument so far has therefore proved that polar codes can achieve the channel capacity of binary erasure channels asymptotically in the large codelength limit, and that the error probability of successive cancellation decoding decays like $o\left(2^{-2^{[n+Q^{-1}(R/(1-p))\sqrt{n}]/2}}\right)$ provided that the code rate R is less than the channel capacity $(1-p)$.

3.4. Polarization for general binary-input memoryless channel

We have so far discussed channel polarization and polar codes assuming binary erasure channels. Even if one assumes a general binary-input memoryless channel, one can construct polar codes which asymptotically saturates the symmetric capacity on the basis of channel polarization.

A binary-input memoryless channel is characterized by the conditional probability $W(y|x)$ of output y given input x, where $x \in \{0, 1\}$. We first discuss the combination of the code with the generator matrix F defined in (11), which has codelength 2 and code rate 1, and successive cancellation decoding, as discussed in section 3.1. Assuming that u_0 and u_1 take 0 or 1 independently with probability 1/2, one can regard that the combination of the code and successive cancellation decoding generates, from the channel W, a pair of two channels defined by

$$\begin{aligned} W_2^0(y_0, y_1|u_0) &= \sum_{u_1=0,1} \frac{1}{2} W(y_0|u_0+u_1) W(y_1|u_1), \\ W_2^1(y_0, y_1, u_0|u_1) &= \frac{1}{2} W(y_0|u_0+u_1) W(y_1|u_1). \end{aligned} \tag{27}$$

The factor 1/2 on the right-hand sides of the above equations corresponds to the prior probabilities of u_1 and u_0, respectively. In successive cancellation decoding, one first obtains an estimation result $\hat{u}_0$ from (y_0, y_1) via maximum-likelihood decoding of the channel $W_2^0(y_0, y_1|u_0)$, as

$$\hat{u}_0 = \arg\max_{u_0=0,1} W_2^0(y_0, y_1|u_0). \tag{28}$$

Assuming that the estimation result be correct, one next obtains an estimate $\hat{u}_1$ from (y_0, y_1) and u_0 via maximum-likelihood decoding of the channel $W_2^1(y_0, y_1, u_0|u_1)$, as

$$\hat{u}_1 = \arg\max_{u_1=0,1} W_2^1(y_0, y_1, u_0|u_1). \tag{29}$$

One can extend the above transformation of channels in accordance with the recursive construction of codes of codelength $N = 2^n$ described in section 3.1, and define the recursive process of generating pairs of channels (W_N^{2i}, W_N^{2i+1}) from $W_{N/2}^i$ starting with the given channel $W(y|x) = W_1^0(y|x)$, as

$$\begin{aligned} &W_N^{2i}(y_0^{N-1}, u_0^{2i-1}|u_{2i}) \\ &= \sum_{u_{2i+1}=0,1} \frac{1}{2} W_{N/2}^i(y_0^{N/2-1}, u_{0,e}^{2i-1} + u_{0,o}^{2i-1}|u_{2i}+u_{2i+1}) W_{N/2}^i(y_{N/2}^{N-1}, u_{0,o}^{2i-1}|u_{2i+1}), \end{aligned} \tag{30}$$

$$\begin{aligned} &W_N^{2i+1}(y_0^{N-1}, u_0^{2i}|u_{2i+1}) \\ &= \frac{1}{2} W_{N/2}^i(y_0^{N/2-1}, u_{0,e}^{2i-1} + u_{0,o}^{2i-1}|u_{2i}+u_{2i+1}) W_{N/2}^i(y_{N/2}^{N-1}, u_{0,o}^{2i-1}|u_{2i+1}), \end{aligned} \tag{31}$$

where $u_{0,e}^{2i-1}$ and $u_{0,o}^{2i-1}$ denote the vectors formed by extracting the elements of the vector u_0^{2i-1} with even and odd indices, respectively. It should be noted that the equalities $u_{0,e}^{2i-1} + u_{0,o}^{2i-1} =$

$\left(u_0^{N-1}(I_{N/2}\otimes F)R_N\right)_0^{i-1}$ and $u_{0,o}^{2i-1} = \left(u_0^{N-1}(I_{N/2}\otimes F)R_N\right)_{N/2}^{N/2+i-1}$ hold, which clearly indicate the relation between the above recursive procedure of channel generation and the recursive structure (13) of the generator matrices $\{G_N\}$. Successive cancellation decoding is defined as a method of estimating u_i sequentially via the maximum-likelihood estimation for the channel W_N^i, as

$$\hat{u}_i = \arg\max_{u_i=0,\,1} W_N^i(y_0^{N-1},\, u_0^{i-1}|u_i). \tag{32}$$

The factor graph regarding estimation of u_i has a tree structure for each i, so that the estimation can be performed efficiently via belief propagation without iteration.

As in the case with binary erasure channels, one can construct polar codes if one knows the probabilities $P(\mathcal{A}_{i,N})$ of incorrectly estimating u_i via successive cancellation decoding, conditional on the event that one knows the correct estimation results for u_0^{i-1}. Moreover, one can argue theoretical performance of polar codes if one can evaluate how many indices i there are for which $P(\mathcal{A}_{i,N})$ take values close to 0. We have discussed, in the argument of channel polarization in section 3.3, properties of the stochastic process $\{p_n\}$ corresponding to the error probabilities $P(\mathcal{A}_{i,N})$, where we had to confirm the following three things in order for the combination of polar codes and successive cancellation decoding to achieve channel capacity: First, to show that p_n converges to a random variable p_∞ as $n\to\infty$. Second, to show that p_∞ takes either 0 or 1. And third, to show that the probability that p_n takes values close to 0 asymptotically approaches the channel capacity. It is not an easy task, however, to confirm the above three points when one considers the error probabilities $P(\mathcal{A}_{i,N})$ directly. Arıkan [1] considered symmetric channel capacity

$$I(W) = \sum_{x=0,\,1}\sum_y \frac{1}{2}W(y|x)\log\frac{W(y|x)}{\frac{1}{2}\left[W(y|0)+W(y|1)\right]}, \tag{33}$$

which is defined for a general binary-input memoryless channel $W(y|x)$, as well as Bhattacharyya parameter

$$Z(W) = \sum_y \sqrt{W(y|0)W(y|1)}. \tag{34}$$

The Bhattacharyya parameter is known to give an upper bound of maximum-likelihood decoding for the channel W. If the channel is a binary erasure channel, then the symmetric capacity and the Bhattacharyya parameter have simple relationship with the erasure probability p as $I(W)=1-p$ and $Z(W)=p$. For a general binary-input memoryless channel, it is known that the relations

$$I(W)+Z(W)\ge 1 \tag{35}$$

$$I(W)^2+Z(W)^2\le 1 \tag{36}$$

hold between $I(W)$ and $Z(W)$, and the equality in (35) holds if and only if W is a binary erasure channel.

On the basis of the series of i.i.d. Bernoulli random variables $B_1, B_2, \ldots$, each with success probability 1/2, we define a random variable $i\in\{0, 1, \ldots, N-1\}$ for $N=2^n$ whose binary expansion is $B_nB_{n-1}\cdots B_2B_1$. The stochastic processes $\{I_n\}$ and $\{Z_n\}$ are defined therefrom, as

$$\begin{aligned} I_n &= I(W_N^i),\\ Z_n &= Z(W_N^i). \end{aligned} \tag{37}$$

In the argument of channel polarization for a binary erasure channel, it was sufficient to consider properties of the stochastic process $\{p_n\}$. However, in the argument of channel polarization for an arbitrary binary-input memoryless channel, one has to consider two stochastic processes $\{I_n\}$ and $\{Z_n\}$, as described in the following. The stochastic process $\{I_n\}$ is a martingale because the condition

$$I(W_N^{2i}) + I(W_N^{2i+1}) = 2I(W_{N/2}^i) \tag{38}$$

holds. On the other hand, $\{Z_n\}$ is in general a supermartingale because the inequality

$$Z(W_N^{2i}) + Z(W_N^{2i+1}) \leq 2Z(W_{N/2}^i) \tag{39}$$

holds. Arıkan [1] showed that the stochastic process $\{Z_n\}$ converges almost surely to a random variable Z_∞ which takes 0 or 1, on the basis of an argument similar to that discussed in section 2. Using the fact that the bounded martingale $\{I_n\}$ also converges almost surely to a random variable I_∞, as well as the relations (35) and (36), he then showed that I_∞ also takes 0 or 1 with probability

$$P(I_\infty) = \begin{cases} I(W) & (I_\infty = 0) \\ 1 - I(W) & (I_\infty = 1) \end{cases} \tag{40}$$

The above discussion assures that the number of indices i for which $P(\mathcal{A}_{i,N})$ take values close to 0 is approximately $NI(W)$ as N becomes sufficiently large. In order to show that polar codes constructed by collecting those indices to form $\mathcal{I}$ achieve the symmetric capacity asymptotically, one has to prove that the probability of the event $Z_n = o(1/N)$ approaches $I(W)$. This statement was proved by Arıkan [1]. A proof can be obtained, for example, by applying an argument similar to that given in section 3.3 to the stochastic process $\{Z_n\}$. We have obtained the following result [3, 4], refining the result of Arıkan and Telatar [2] to include dependence on the code rate.

Theorem 1 *Let $I(W)$ be the symmetric capacity of a binary-input memoryless channel W. Then the error probability of a polar code with codelength $N = 2^n$ and code rate $R < I(W)$ via successive cancellation decoding satisfies*

$$P(\mathcal{E}_N) = o\left(2^{-2^{(n+t\sqrt{n})/2}}\right) \tag{41}$$

for an arbitrary t satisfying $t < Q^{-1}(R/I(W))$.

A detailed derivation will be given elsewhere [4]. Hassani and Urbanke [5] independently obtained the same result. It should also be worth noting that Korada et al. [6] have shown that (40) can be directly derived without recourse to discussion of the stochastic process $\{Z_n\}$ for a binary-input memoryless symmetric channel.

3.5. Construction of codes for general binary-input memoryless channel

It is not an easy task to evaluate error probabilities $\{P(\mathcal{A}_{i,N})\}$ efficiently for a general binary-input memoryless channel. Although Arıkan [1] proposed a method that is based on approximate evaluation of Bhattacharyya parameters, it is no longer guaranteed for resulting codes to have the capacity-achieving property. For a general binary-input memoryless symmetric channel, we have proposed use of density evolution to evaluate $\{P(\mathcal{A}_{i,N})\}$, whose computational complexity, in terms of the number of convolution of densities, scales as $O(N)$ [7]. The use of density evolution is justified because decoding of a component is regarded as belief propagation without cycles. We also showed that the proposed method produces codes that are better in error probability than those obtained by Arıkan's approximate method [8].

4. Extensions

4.1. Construction using $\ell \times \ell$ matrix

Although polar codes are provably capacity-achieving, performance of polar codes with finite codelength is empirically not so remarkable, if compared with that of LDPC codes with the same codelength. Rather poor performance of polar codes may be ascribed to the code construction and/or the decoding, so that there are several attempts to improve performance of polar codes by modifying the code construction method and/or the decoding algorithm.

One can consider an extension in which an $\ell \times \ell$ matrix, with $\ell \geq 3$, is used in place of the 2×2 matrix F (11) to construct polar codes. For the case with a binary-input memoryless symmetric channel, Korada et al. [6] obtained the following results. First, they showed that a necessary and sufficient condition for an $\ell \times \ell$ matrix F to induce channel polarization via recursive construction similar to that discussed in section 3.1 is that one cannot make the matrix F upper-triangular by any permutation of columns of F. Second, defining the "error exponent" of a matrix F to be the supremum of β such that the error probability of successive cancellation decoding for codes of codelength ℓ^n is $o\left(2^{-\ell^{\beta n}}\right)$, they showed that the error exponent of F is given by

$$\frac{1}{\ell}\sum_{k=1}^{\ell}\log_\ell D_k, \tag{42}$$

where $\{D_1, \ldots, D_\ell\}$ are called partial distances of the matrix F, defined in [6]. On the basis of this result, they succeeded in obtaining upper and lower bounds of the best possible error exponents, and proved that error exponents can be made as close to 1 as possible by taking $\ell \to \infty$.

Theorem 1 can be extended to the case with an $\ell \times \ell$ matrix. Let $\{B_n\}$ be a sequence of i.i.d. random variables with $P(B_n = k) = 1/\ell$ for all $k \in \{1, \ldots, \ell\}$, and let $\mathbb{E}(F)$ and $\mathbb{V}(F)$ be the mean and variance of the random variables $\log_\ell D_{B_n}$. Applying the same argument as that in section 3 to the above random variables, one can prove that the error probability of successive cancellation decoding of polar codes, of codelength $N = \ell^n$, satisfies

$$P(\mathcal{E}_N) = o\left(2^{-\ell^{n\,\mathbb{E}(F)+t\sqrt{n\,\mathbb{V}(F)}}}\right), \tag{43}$$

for any t satisfying $Q(t) > R/I(W)$.

4.2. Channel with q-ary input

It is natural to consider channel polarization with a q-ary input memoryless channel with $q \geq 3$, rather than a binary-input channel. Şaşoğlu et al. [9] discussed the extension of channel polarization to channels with q-ary input, on the basis of the 2×2 matrix F given in (11), which is regarded as representing a linear transform over the ring $\mathbb{Z}/q\mathbb{Z}$ of residue classes modulo q. They proved the following theorem:

Theorem 2 *If q is a prime number, q-ary polar codes constructed with F as defined in* (11) *and successive cancellation decoding satisfies*

$$P(\mathcal{E}_N) = o\left(2^{-2^{n\beta}}\right) \tag{44}$$

for any $\beta < 1/2$.

If q is not a prime number, there are pathological cases in which channel polarization does not occur. In order to circumvent such difficulties, Şaşoğlu et al. [9] proposed introducing randomness in construction of polar codes, and proved the capacity-achieving property of the resulting codes.

Here, it seems more promising to consider fields rather than rings. We have given some sufficient conditions for an ℓ-input transformation (which can be nonlinear) on a set of q alphabets to induce channel polarization. An upper bound of error probability is also given, on the basis of which we have found that generator matrices of (generalized) Reed-Solomon codes can be good choices for providing larger error exponents. As an example, a 4×4 matrix over $\mathrm{GF}(q = 2^2)$ composed of a generator matrix of a generalized Reed-Solomon code over $\mathrm{GF}(2^2)$ has been shown to have the error exponent $\log 24/(4 \log 4) \approx 0.57312$, which is larger than the error exponent 0.51828 for the best possible $\ell \times \ell$ matrix over GF(2) for all $\ell \leq 16$, which was reported in [6]. Detailed arguments of the above results will be presented elsewhere [10].

5. Conclusion

We have reviewed channel polarization and polar codes, putting emphasis on our recent results. In this article, we have not mentioned several other important subjects related to channel polarization and polar codes. They include applications of polar codes to various problems of source coding, as well as exploration of more efficient decoding algorithms other than successive cancellation decoding, for which interested readers are encouraged to refer to, e.g., [11].

Acknowledgments

The author would like to thank Mr. Ryuhei Mori, Kyoto University, for his helpful comments. Support from the Grant-in-Aid for Scientific Research on Priority Areas (No. 18079010) is acknowledged.

References

[1] Arıkan E 2009 *IEEE Trans. Info. Theory,* **55** 3051–73
[2] Arıkan E and Telatar E 2009 *Proc. 2009 Int. Symp. Information Theory (Seoul, Korea)* pp 1493–5
[3] Mori R and Tanaka T 2009 *Proc. 32nd Symp. Information Theory and its Applications (SITA2009)* W43-2 (in Japanese).
[4] Tanaka T and Mori R 2010 Refined rate of channel polarization *Preprint* arXiv:1001.2067v1 [cs.IT]
[5] Hamed Hassani S and Urbanke R 2010 On the scaling of polar codes: I. The behavior of polarized channels *Preprint* arXiv:1001.2766v3 [cs.IT]
[6] Korada S B, Şaşoğlu E, and Urbanke R 2009 Polar codes: Characterization of exponent, bounds, and constructions *Preprint* arXiv:0901.0536v2 [cs.IT]
[7] Mori R and Tanaka T 2009 *Proc. 2009 Int. Symp. Information Theory (Seoul, Korea)* pp. 1496–1500
[8] Mori R and Tanaka T 2009 *IEEE Comm. Lett.*, **13** pp. 519–21
[9] Şaşoğlu E, Telatar E, and Arıkan E 2009 Polarization for arbitrary discrete memoryless channels *Preprint* arXiv:0908.0302v1 [cs.IT]
[10] Mori R and Tanaka T 2010 Channel polarization on q-ary discrete memoryless channels by arbitrary kernels *Preprint* arXiv:1001.2662v2 [cs.IT]
[11] Korada S B 2009 *Polar Codes for Channel and Source Coding* Doctoral Thesis, Ecole Polytechnique Fédèrale de Lausanne

International Workshop on Statistical-Mechanical Informatics 2010 IOP Publishing
Journal of Physics: Conference Series **233** (2010) 012019 doi:10.1088/1742-6596/233/1/012019

The solution space of metabolic networks: producibility, robustness and fluctuations

A. De Martino[1,2] **and E. Marinari**[2]
[1]CNR/IPCF, UoS Roma–Sapienza and [2]Dipartimento di Fisica, Sapienza Università di Roma, p.le A. Moro 2, 00185 Roma, Italy

E-mail: andrea.demartino@roma1.infn.it, enzo.marinari@roma1.infn.it

Abstract. By flux analysis one generically indicates a class of constraint-based approaches to the study of biochemical reaction networks concerned with the calculation of the flux configurations compatible with given stoichiometric and thermodynamic constraints. One of its main areas of application is the study of cellular metabolic networks. We briefly and selectively review the main approaches to this problem and then, building on recent work, we provide a characterization of the productive capabilities of the metabolic network of the bacterium *E.coli* in a specified growth medium in terms of the producible biochemical species. While a robust and physiologically meaningful production profile clearly emerges, the underlying constraints still allow for significant fluctuations in the net production even for key metabolites like ATP and, as a consequence, apparently lay the ground for different growth scenarios.

1. Introduction

At a rough conceptual level, living cells can be seen as devices that convey the free energy derived from the breakdown of nutrients (mostly sugars) into the chemical energy that fuels the production of the molecules required for survival and, when possible, growth and reproduction. The complex intracellular machinery that underlies the energy transduction is being increasingly unveiled at both the biochemical (reactions) and the regulatory level (enzymes, their corresponding genes, the genes' transcription factors, etc) through the massive genomic information available for different organisms [1]. In particular, the network of biochemical processes collectively known as 'metabolism' by which the nutrients are degraded and the housekeeping molecules are manufactured has been mapped in great detail for many bacteria and several eukaryotes [2]. While the study of the structure and dynamics of single metabolic pathways has a long standing history in biophysical chemistry [3, 4], the current wealth of data allows to analyze the behavior of cellular reaction networks at the scale of the whole genome. This is a crucial step both to shed light on the emergence of a metabolic phenotype from the underlying genotype and to formulate testable [5] predictions on a cell's adaptation and response to perturbations (the key to multiple biotechnological applications).

Ultimately, metabolic activity and capabilities are determined (or limited) by various constraints of chemical (stoichiometric), thermodynamic and regulatory origin. The uncertainty about their details however increases considerably as one passes from the stoichiometric to the regulatory level. Current theoretical approaches therefore try to infer the global organization of metabolism from simple schemes that implement explicitly only the best known restrictions,

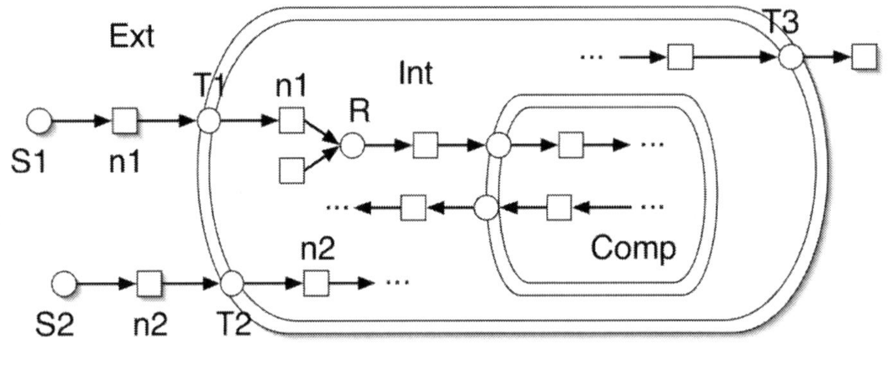

Figure 1. Basic scheme of a cellular metabolic network. Ext and Int denote respectively the exterior and the interior of the cell. Reactions (resp. metabolites) are denoted by circles (resp. squares). S1 and S2 are auxiliary fluxes supplying nutrients n1 and n2 to the environment. T1, T2 and T3 denote membrane transport reactions by which metabolites are taken in or expelled from the cell. R is an intracellular reaction. A cellular compartment Comp is also shown, together with the corresponding transport reactions connecting its interior with the cytosol.

making as few assumptions as possible on the rest. Even so, it turns out that some predictive and explanatory power on a cell's biochemical functioning can be achieved.

We begin by presenting a short selective review of stoichiometric models of metabolic networks as constraint-based systems. We then focus on one specific problem, that of metabolite producibility, characterizing the global metabolic output for the bacterium *E.coli* in a specified growth medium. While a robust and biologically significant production profile emerges, one still observes fluctuations in nutrient usage or in the output level of key metabolites that can be associated to variations in the overall flux of different metabolic pathways. This suggests that more metabolic phenotypes (with e.g. varying growth capabilities) might be compatible with the given environmental, thermodynamic and stoichiometric constraints. Finally we discuss some of the issues where statistical mechanics tools may have a significant impact in the near future. For the sake of simplicity we shall limit biochemical details to the necessary minimum and treat the reaction system as a standard input-output network. For metabolite abbreviations (as well as for stoichiometric data) regarding our case study, we shall follow the detailed genome-scale reconstruction of *E.coli*'s metabolic network described in [6]. The interested reader is referred to e.g. [3] for a thorough introduction to cellular metabolism.

2. Flux analysis: a bird's eye view

In essence (see Fig. 1), a cell's metabolic network can be seen as a set of interconnected chemical reactions coupled with a set of transport processes. For bacteria, this includes the membrane transport mechanisms by which nutrients are brought into the cell, and the intracellular reactions by which they are degraded and new biochemical species are produced. In organisms with compartmentalized cells (e.g. eukaryotes) one should also account for the transport of metabolites into and out of each compartment, i.e. for the cell's spatial structure. The basic and most reliable information on the, say, N reactions involving M chemical species is encoded in the $M \times N$ *stoichiometric matrix* Ξ, whose entry ξ_i^μ represents the stoichiometric coefficient with which species μ participates in reaction i. Ξ is a sparse, integer matrix and a sign convention is usually adopted to discern products ($\xi_i^\mu > 0$) from substrates ($\xi_i^\mu < 0$). The stoichiometric matrix often also contains a motivated assumption on reaction directionality to account for the fact that, while all reactions are in principle reversible, under physiological conditions some of them may occur in one direction only. We shall treat physiologically reversible reactions as two separate processes. In addition, we shall always assume that Ξ also includes external supply fluxes for the nutrients, i.e. auxiliary (free or fixed) reactions that provide the nutrient to the environment. To have an idea, in such a setting $N \simeq 1100$ and $M \simeq 700$ for the bacterium *E.coli*, whereas $N \simeq 1500$ and $M \simeq 900$ for the unicellular eukaryote *S.cerevisiæ*.

If we denote by $\boldsymbol{\nu} \geq \mathbf{0}$ the N-dimensional vector of reaction fluxes (the sign constraint arising from our treatment of reversibility), the time evolution of the vector $\mathbf{c}$ of metabolite concentrations is described simply by $\dot{\mathbf{c}} = \Xi\boldsymbol{\nu}$. Fluxes depend in principle on the membrane transport mechanisms or on the enzyme kinetics through various parameters like the rate constants $\mathbf{k}$ and $\mathbf{c}$ itself and in turn, through them, on temperature, activation energy, etc. If such details are known, then functional dependencies $\boldsymbol{\nu} = \boldsymbol{\nu}(\mathbf{c}, \mathbf{k}, \ldots)$ can be chosen according to the corresponding kinetics (e.g. Michaelis-Menten or Hill) and one can in principle solve the dynamical system for the concentrations, provided all underlying constants are known. Unfortunately, this is rarely the case in genome-scale networks (see [7] for a positive example). While methods have been developed to partially overcome this limitation and provide dynamical characterizations of metabolic activity (see e.g. [8, 9]), the standard modeling route assumes, on the basis of the timescale separation between chemical processes and genetic regulation, that metabolic networks operate in a non equilibrium steady state where $\dot{\mathbf{c}} = \mathbf{0}$. Note that flux vectors satisfying $\Xi\boldsymbol{\nu} = \mathbf{0}$ describe states in which intracellular metabolites obey Kirchhoff-type mass-balance conditions. If $N > M$, as it usually occurs in real metabolic networks, such vectors form a set of dimension $N - \text{rank}(\Xi)$ embedded in $\mathbb{R}_0^+$ that, in absence of additional constraints, contains equivalent feasible flux states of the network. In principle, this set should be explored uniformly to extract the relevant biological information. Unfortunately, Monte Carlo sampling becomes unaffordable as soon as the dimension of the solution space exceeds a few tens (see [10] for a feasible case; message-passing algorithms have been recently proposed as a working alternative for microbial cells [11]). In many cases it is however possible to select relevant configurations by imposing the maximization of an objective function normally represented as a linear combination of the fluxes with given coefficients $\boldsymbol{\alpha}$, i.e. $\boldsymbol{\alpha} \cdot \boldsymbol{\nu}$. This reduces the problem to a linear optimization one, namely

$$\max_{\mathbf{0} \leq \boldsymbol{\nu} \leq \boldsymbol{\nu}_{max}} (\boldsymbol{\alpha} \cdot \boldsymbol{\nu}) \quad \text{subject to} \quad \Xi\boldsymbol{\nu} = \mathbf{0} \tag{1}$$

The upper bounds on fluxes can be chosen to either model specific extracellular conditions (e.g. a certain nutrient's supply or intake cannot exceed a fixed level), to simulate known physiological limitations (e.g. the rate of a certain reaction cannot exceed a fixed level because of limited enzyme availability), or simply to impose well-defined bounds on the solution space. This scenario is the basis of various stoichiometry-based approaches such as Flux-Balance Analysis (FBA) [12]. Clearly, the vector $\boldsymbol{\alpha}$ contains the crucial biological assumptions. In some cases, it is grounded in experimental evidence. For instance, bacteria in rich environments appear to evolve, under selective pressure, to optimize their growth rate and reproduce as fast as possible [13]. A good proxy for a bacterium's growth capability lies in its ability to generate *biomass*, a combination of different metabolites in precise stoichiometric proportions (including the 20 proteinogenic amino acids, the molecular energy carrier ATP, water, key cofactors like nicotinamide adenine dinucleotide [nad], nicotinamide adenine dinucleotide phosphate [nadp] and coenzyme-A [coa] etc.) that is used to produce ADP, inorganic phosphate (PI), pyrophosphate (PPI) and water (h2o) [14, 6][1]. In brief, this reaction represents the cell's use of metabolic products in macromolecular processes (like building proteins, membranes, etc.) that are not accounted for in Ξ. The maximization of biomass yield is in these cases a widely used criterion implemented by adding the auxiliary biomass reaction to Ξ. In other cases, the choice may have physiological justifications. The need for energetically-efficient housekeeping e.g. in nutrient-limited conditions suggests that cells optimize the ATP yield, which requires to maximize the total flux of all ATP-producing reactions [15]. Other examples include the minimization of glucose consumption (a proxy for efficient nutrient usage [16]) or of the total flux of intracellular reactions (for maximal enzymatic efficiency [17]). We refer the reader to [18] for a comparison of the performance of several different objective functions for predicting the fluxes of the main carbon pathways in *E.coli* and to [19] for a study of objective function selection criteria.

This kind of approach can be extended to biologically more complicated situations like those induced by gene knock-outs that prevent the execution of certain reactions. Based on experimental evidences, such scenarios lead to consider different selection criteria, like the minimization of the overall flux rearrangements with respect to the wild type (by quadratic programming [20]) or the minimization of rearrangements in the large-flux backbone with respect to the wild type (by mixed linear-integer programming [21]). The former appears to be suited to capture the transient sub-optimal growth states that a bacterium takes on immediately after the perturbation, whereas the latter provides a better description of the states of fast growth that the knock-out organism reaches on longer time scales under selective pressure. However the dynamical structure of the response of metabolic networks to perturbations is far from being satisfactorily understood.

3. Producibility and Von Neumann's problem

An important problem related to flux analysis concerns the link between the network's structure and its productive capabilities. Following [22], a metabolite μ is said to be *producible* from a given set of nutrients (to be specified) if

$$\exists\, \boldsymbol{\nu} \geq \mathbf{0} \text{ such that } \Xi\boldsymbol{\nu} \geq 0 \ \& \ [\Xi\boldsymbol{\nu}]_{\mu} > 0 \tag{2}$$

(where $[\Xi\boldsymbol{\nu}]_{\mu} = \sum_{i=1}^{N} \nu_i \xi_i^{\mu}$) that is, if at least one flux vector exists allowing for a net production of μ irrespective of whether other metabolites are being also produced (nutrient usage may evidently never exceed its supply). Producible metabolites have the property that their concentrations can increase in a stationary flux state with the sole consumption of the nutrients, so that the cell is allowed to employ them for purposes other than metabolic (e.g. to form proteins, membranes, etc.). One would therefore expect that an organism's behavior in a given environment is related to its ability to produce the metabolites necessary for survival, growth etc. (e.g. biomass, ATP or else).

It is easy to understand that the possibility to actually produce a metabolite may be limited by the emergence of conservation laws from the stoichiometry. To clarify, note that M-dimensional vectors $\mathbf{z} \geq \mathbf{0}$ satisfying $\mathbf{z}^T\Xi = \mathbf{0}$ (or, positive semidefinite vectors from the left null-space of

[1] The biomass reaction reported in [6] reads as follows: (0.05)5mthf + (5.0E-5) accoa + (0.488) ala-L + (0.0010) amp + (0.281) arg-L + (0.229) asn-L + (0.229)asp-L + (45.7318)atp + (1.29E-4)clpn-EC + (6.0E-6)coa + (0.126)ctp + (0.087)cys-L + (0.0247)datp + (0.0254)dctp + (0.0254)dgtp + (0.0247)dttp + (1.0E-5)fad + (0.25)gln-L + (0.25)glu-L + (0.582)gly + (0.154)glycogen + (0.203)gtp + (45.5608)h2o + (0.09)his-L + (0.276)ile-L + (0.428)leu-L + (0.0084)lps-EC + (0.326)lys-L + (0.146)met-L + (0.00215)nad + (5.0E-5)nadh + (1.3E-4)nadp + (4.0E-4)nadph + (0.001935)pe-EC + (0.0276) peptido-EC + (4.64E-4)pg-EC + (0.176)phe-L + (0.21)pro-L + (5.2E-5)ps-EC + (0.035)ptrc + (0.205)ser-L + (0.0070)spmd + (3.0E-6)succoa + (0.241)thr-L + (0.054)trp-L + (0.131)tyr-L + (0.0030)udpg + (0.136)utp + (0.402)val-L → (45.5608)adp + (45.56035)h + (45.5628)pi + (0.7302)ppi. See [6] for a detailed list of metabolite abbreviations.

the stoichiometric matrix) define linear combinations of metabolites such that the corresponding weighted sums of their concentrations are constant over time [24]. Such conserved moieties[2] are abundant in real metabolic networks [25]. Clearly, an accumulation of metabolites belonging to one such pool, and hence their producibility, is ruled out by simple stoichiometric reasons. The duality of producibility and conservation can be exploited to identify (by linear programming) the growth media allowing for the production of specified sets of metabolites such as the biomass or slight modifications of it [23][3].

In order to evaluate the robustness of the cellular production profile (if any) emerging in given nutrient conditions one should study the set

$$\mathcal{V} = \{\boldsymbol{\nu} \text{ such that } \Xi\boldsymbol{\nu} \geq 0\} \tag{3}$$

(we shall henceforth assume that $\sum_i \nu_i = N$). Statistical sampling in this case turns out to be feasible, a possible route being suggested by work of J. Von Neumann. In order to present a rather simplified version of his model [27], let us re-define the stoichiometric matrix by separating the matrix $\mathbf{B}$ of input coefficients from the matrix $\mathbf{A}$ of output coefficients, so that $\mathbf{A} - \mathbf{B} = \Xi$. It is simple to see [28] that, given a constant $\rho > 0$, a flux vector $\boldsymbol{\nu}$ such that $\mathbf{A}\boldsymbol{\nu} \geq \rho\mathbf{B}\boldsymbol{\nu}$ describes a network state in which every species is being produced at a rate at least equal to ρ. In a generic input-output system specified by matrices $\mathbf{A}$ and $\mathbf{B}$, one can therefore ask what is the maximum value of ρ for which flux vectors satisfying

$$(\mathbf{A} - \rho\mathbf{B})\boldsymbol{\nu} \geq \mathbf{0} \tag{4}$$

exist. This value (denoted here as $\rho^\star$) represents the optimal productive performance allowed by the ways in which the available processes combine the metabolites. Depending on whether $\rho^\star$ is larger or smaller than 1, the system may have optimal states that are expanding or contracting. Correspondingly, the flux vector(s) satisfying the above set of constraints for $\rho = \rho^\star$ are the optimal flux states of the system. When $N, M \to \infty$ and $\mathbf{A}$ and $\mathbf{B}$ are taken to have non-negative entries but are otherwise unstructured random matrices, one can dissect the phase structure of the problem in detail by statistical mechanics tools, specifically using the replica trick (in a fully connected network where each reaction consumes and produces a finite fraction of the possible metabolites [28]) or the cavity method (in a finitely connected network where reactions use a finite number of substrates to generate a finite number of products [29]). The general lesson is that a critical value n_c of the structural parameter $n = N/M$ separates contracting from expanding regimes, with enhanced dilution increasing (resp. decreasing) $\rho^\star$ in the expanding (resp. contracting) phase. Moreover, in such cases a single flux vector satisfies conditions (4) when $\rho = \rho^\star$. In the biologically more sensible case in which $\mathbf{A}$ and $\mathbf{B}$ are real stoichiometric matrices, the situation is radically different. The mass balance conditions imposed by stoichiometry or, equivalently, the existence of conserved metabolic pools, imply $\rho^\star = 1$ (and force a finite volume of solutions in a random model with realistically structured input and output matrices [30]), so that the optimal solution space of (4) indeed coincides with (3). By exploiting the presence of ρ it is however possible to define a straightforward iterative algorithm that samples $\mathcal{V}$ uniformly. Its basic idea, see [29] for details, is to locate a solution of (4) for small ρ (easy) and track it as ρ is increased and the solution space shrinks, until $\rho^\star(= 1)$ is reached. Different "initial solutions" at small ρ lead to different final solutions at $\rho^\star$ and an unbiased sampling of (3) can be achieved. This ultimately allows for a complete and extensive characterization of producible metabolites and of the corresponding flux states in real cellular networks in a given environment at a modest computational cost even for genome-scale systems.

[2] Strictly, moieties are molecular subgroups (including functional groups) that are conserved over time in reaction networks, like e.g. the adenylate moiety in the pool formed by ATP, ADP and AMP.

[3] An interesting extension of producibility is *sustainability*, loosely defined (see [26]) as the property of being producible using producible metabolites as substrates besides nutrients.

4. Application to E.coli: production profiles and their fluctuations

The volume (3) generated by the stoichiometric matrix of the bacterium *E.coli* has been studied in [31], revealing that

(a) the predicted ranges of variability of the fluxes in $\mathcal{V}$ in specified extracellular conditions agree well with the (limited) experimental data available on the reaction rates inferred from ^{13}C-based experiments in a similar nutrient profile [5, 32];

(b) dynamically stiff variables, i.e. reactions with smaller allowed ranges, tend to correspond, via the associated enzyme, to *E.coli*'s phenomenologically essential genes, i.e. genes that are both necessary for the organism's survival and highly conserved across different bacterial species [33].

These results suggest that at least in some conditions metabolic networks may operate close to their optimal productive capacity and that the "shape" of $\mathcal{V}$ might contain useful information relating the metabolic phenotype to the underlying genotype. The natural question to ask now is whether a robust metabolite production profile emerges and how it correlates with the physiologically defined biomass (in suitable growth media). We therefore take a closer look at the set of produced metabolites for *E.coli* in a minimal growth medium, an environment with a tunable supply of a limited set of nutrients formed by inorganic phosphate (PI), molecular oxygen (O_2), sulfate (SO_4), carbon dioxide (CO_2), potassium ion (K^+), ammonia (NH_3) and a carbon source (glucose in this case) [34] that resembles closely the M9 medium widely used in the experimental literature [32]. This network consists of 1057 reactions involving 631 metabolites altogether (after pruning the trivial producibility constraints, see [31] for details).

For simplicity, we henceforth set $\boldsymbol{\lambda} = \boldsymbol{\Xi}\boldsymbol{\nu}$ and label the flux vectors from $\mathcal{V}$ as $\boldsymbol{\nu}_\alpha$, $\boldsymbol{\nu}_\beta$, etc. assuming to have sampled S such configurations. We shall furthermore write $\boldsymbol{\lambda}_\alpha = \boldsymbol{\Xi}\boldsymbol{\nu}_\alpha$.

The simplest way to check whether $\mathcal{V}$ defines a set of consistently produced metabolites is by studying the empirical covariance matrix $\mathbf{C}$ given by

$$C^{\mu\mu'} = \frac{1}{S}\sum_{\alpha=1}^{S} \lambda_\alpha^\mu \lambda_\alpha^{\mu'} \tag{5}$$

In the spirit of principal component analysis, $\mathbf{C}$'s eigenvalue spectrum contains much information on the collective properties of the system. In particular, it is possible to express the underlying "signals" $\{\boldsymbol{\lambda}_\alpha\}$ as weighted sums of the eigenvectors of $\mathbf{C}$. The presence of a large, isolated eigenvalue r_{max} indicates that $\lambda_\alpha^\mu \simeq \sqrt{r_{max}}\, V_{max}^\mu \eta_\alpha$, where $\mathbf{V}_{max}$ is the eigenvector corresponding to r_{max} and η_α is a unit-variance random variable. Hence r_{max} and its corresponding eigenvector may offer an effective, zero-order description of $\mathcal{V}$ in terms of the emergent production profile. The eigenvalue distribution for *E.coli* is shown in Fig. 2[4]. One clearly sees a giant eigenvalue separated by a roughly continuous spectrum, well described by the Marcenko-Pastur law with a point mass at $r = 0$ [35]. In metabolic terms, r_{max} represents a "network mode", a production profile that is common (to a first approximation) to all $\boldsymbol{\nu} \in \mathcal{V}$. The components of the eigenvector corresponding to the network mode (Fig. 2, bottom) point explicitly to some of the metabolites that are expected to form the core of the metabolic production. These turn out to include[5]

- biomass components (e.g. 5mthf, atp, fad, h2o, nad, nadh, nadp, nadph)
- products of the biomass reaction (adp, h, pi, ppi)
- end points of metabolic pathways (waste compounds like acetate [ac], co2 or lactate [lac])

[4] To remedy the uncertainty about relative fluctuations when $\lambda^\mu \ll 1$, one can set a threshold ϵ below which λ^μ is "effectively" zero. For the present case, values of $\epsilon \lesssim 0.01$ have been found to provide the same qualitative picture.
[5] See [6] for a list of metabolite abbreviations.

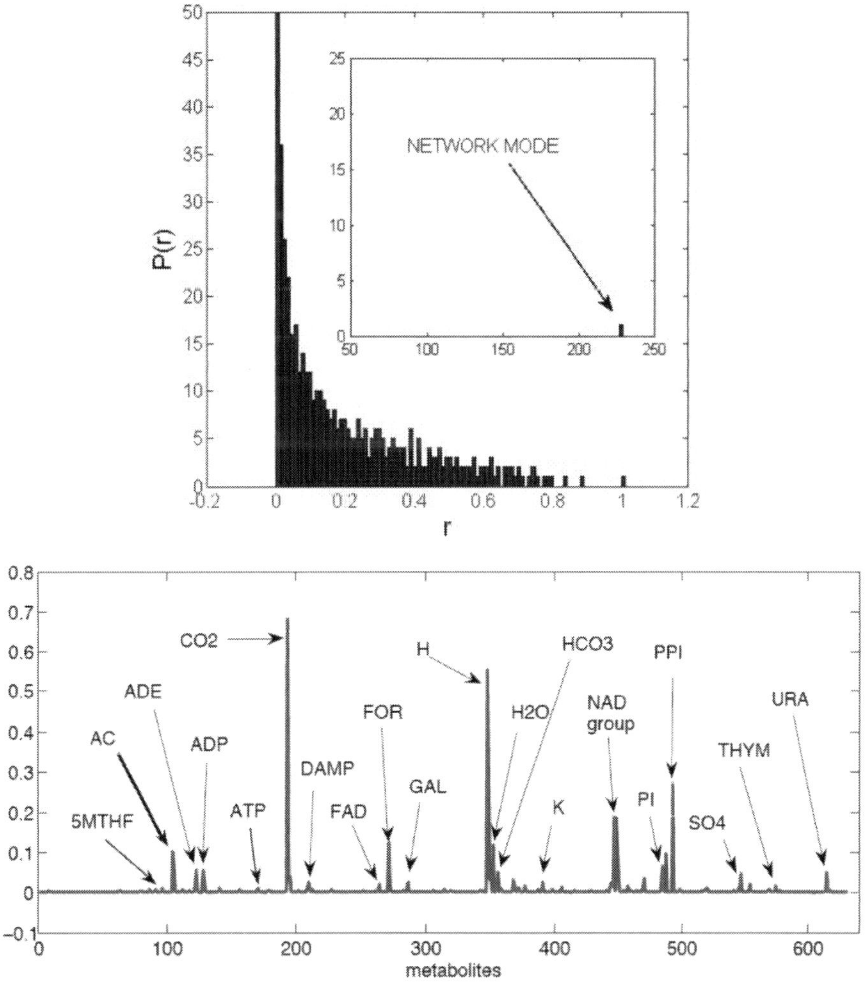

Figure 2. (Top) Histogram of the eigenvalues of the empirical correlation matrix **C**, Eq. (5), for *E.coli* in a minimal growth medium (see text). Inset: detail of the "network mode". (Bottom) Components of the eigenvector of **C** corresponding to the network mode, with some of the producible metabolites explicitly indicated (see [6] for a list of metabolite abbreviations).

Such a profile is largely recovered by studying the average of λ^{μ} over 500 flux configurations $\boldsymbol{\nu} \in \mathcal{V}$ for each μ, see Fig. 3. Focusing for clarity on the ones with the largest average producibility we find multiple biomass components (5mthf, amp, atp, coa, fad, h2o, leu-L, nad, nadh, nadp, nadph, ppi) and biomass products (adp, h, pi and ppi), supporting a growth scenario; typical waste products of metabolic activity (such as ac and co2, end point of oxidative phosphorylation); and metabolites with clear metabolic roles like formate [for] (central for h production), bicarbonate [hco3] and potassium ion [k] (helping to maintain a stable intracellular condition), succinate [succinate] (a key node of the Krebs cycle). In addition to these, a host of producible species appear, though with smaller λ^{μ}, including all amino acids (with the notable exception of leucine, which appears to has a large average net output).

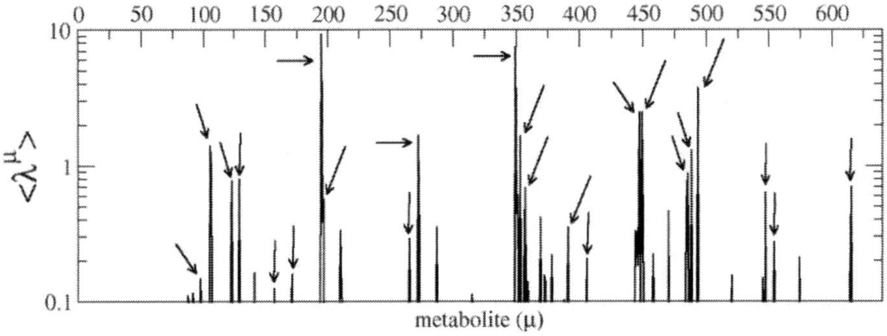

Figure 3. Value of λ^μ averaged over 500 flux states $\boldsymbol{\nu} \in \mathcal{V}$ versus μ (metabolite, horizontal axis) for *E.coli* (631 metabolites) in a minimal growth medium (see text). Arrows mark, from left to right, the following metabolites (see [6] for a list of metabolite abbreviations): 5mthf, ac, ade, adp, amp, atp, co2, coa, fad, for, h, h2o, hco3, k, leu-L, the nad group (nad, nadh, nadp, nadph), nh4, pi, ppa, ppi, so4, succ and ura.

For most compounds, however, one finds $[\Xi\boldsymbol{\nu}]_\mu = 0$ for each $\boldsymbol{\nu} \in \mathcal{V}$. As said above, $[\Xi\boldsymbol{\nu}]_\mu = 0$ if μ belongs to a conserved metabolic pool. Interestingly, such pools do not seem to exhaust the list of unproducible metabolites, as metabolite producibility turns out to be a significantly fluctuating property. To evaluate how the production profile of each metabolite varies across solutions in $\mathcal{V}$ one can calculate, for each μ and for each pair of (distinct) flux vectors $\boldsymbol{\nu}_\alpha, \boldsymbol{\nu}_\beta \in \mathcal{V}$, the overlap

$$q^\mu_{\alpha\beta} = \frac{2\lambda^\mu_\alpha \lambda^\mu_\beta}{(\lambda^\mu_\alpha)^2 + (\lambda^\mu_\beta)^2} \tag{6}$$

When the above quantity is averaged over pairs $\alpha \neq \beta$ (properly accounting for the case where at least one $\lambda^\mu_\alpha \ll 1$), one obtains an index (we shall denote it by q^μ) that is closer to 1 the smaller are the fluctuations in λ^μ for metabolite μ. The distribution of q^μ is shown in Fig. 4. One sees that the production rate of many key metabolites including co2, atp, amp, the nad group, etc. may oscillate considerably from solution to solution in $\mathcal{V}$. This holds for intracellular species (e.g. ATP or biomass components) as well as for extracellular ones, so that even the amount of nutrients necessary for the cell to maintain its productive capability in an environment where a specific group of metabolites is present may have a large allowed range.

It is instructive to inspect two particular solutions out of those sampled from $\mathcal{V}$, namely those with the largest and smallest production of ATP. Denoting these respectively by $\boldsymbol{\nu}_{max}$ and $\boldsymbol{\nu}_{min}$, in Fig. 5 we show, component by component, the quantities $\Delta\boldsymbol{\lambda} = \boldsymbol{\lambda}_{max} - \boldsymbol{\lambda}_{min}$ and $\Delta\boldsymbol{\nu} = \boldsymbol{\nu}_{max} - \boldsymbol{\nu}_{min}$. Analyzing the production profiles it turns out that the state with larger ATP production is closely associated to the network mode, i.e. it comes with a significant production of biomass components and of metabolic waste like ac and co2. On the contrary the state with reduced ATP production is associated to decreased co2 output and high lactate output. This is consistent with an increase of the flux through glycolisis, a central, anaerobic carbon pathway with limited ATP productive efficiency that leads to the generation of lactate. Large ATP output is instead achievable by strengthening aerobic pathways that employ oxidative phosphorylation (with higher ATP-production efficiency) with the concomitant production of co2. Note however that oxygen is available in the cell's environment, so it is reasonable to expect that aerobic solutions will dominate (as the the network mode also shows). The oxygen intake indeed does not fluctuate significantly across the solution space and co2 production, while not tightly constrained, appears to be consistently large. These different states are thus unlikely

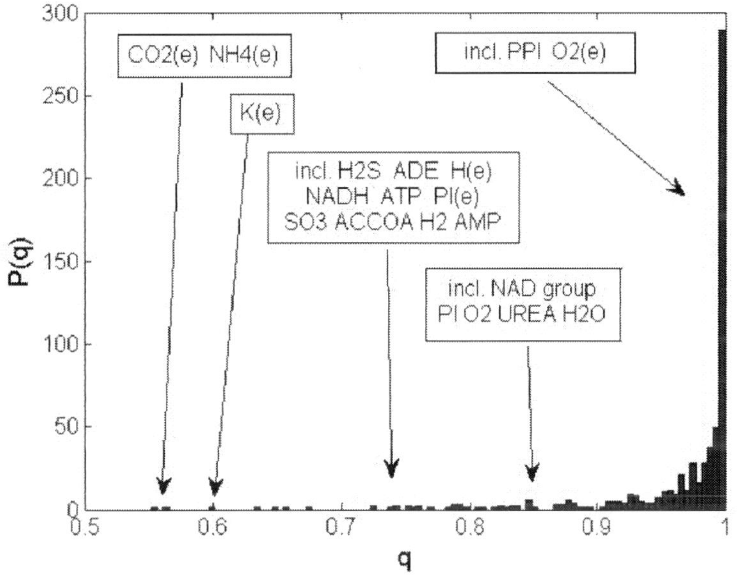

Figure 4. Distribution of $q^\mu = \langle q^\mu_{\alpha\beta} \rangle$ (average over pairs of distinct solutions from $\mathcal{V}$). The index e (extracellular) denotes metabolites that also serve as nutrients. Arrows mark groups of metabolites (as shown) falling in the corresponding range of q.

to represent genuinely different metabolic phenotypes operating in aerobic and, respectively, anaerobic modes, but more simply signal the existence of viable flux rearrangements (compatible with all specified constraints) with increased weight for anaerobic pathways relative to the more efficient aerobic ones. While a more careful analysis is needed to map the exact locations of the flux rearrangements, the overall difference in the organization of flux between the two states is indeed significant. From a biological perspective, these states may correspond to different growth properties, that could perhaps be identified by studying distribution of growth rates of bacterial colonies in a fixed environment. On the other hand, this shows that a sharper selection of production profiles, and thus a better identification of metabolic objectives, requires constraints that are still not included in the available theories.

5. Discussion

Metabolic networks control, directly or indirectly, many of the most basic tasks cells must accomplish, from the synthesis of amino acids, to the maintenance of osmotic balance with the exterior, to the response to environmental shifts. A possible key to improve our understanding of their organization lies in our opinion in the identification of the physical, stiochiometric, thermodynamic or regulatory factors that intrinsically limit their productive potential and determine the overall outcome of metabolic activity. Constraint-based approaches provide a simple mathematical frameworks where many of the emerging properties of metabolism can be analyzed quantitatively. We have shown here that while the constraints that are normally employed are sufficient to describe the main physiological aspects of a cell's growth performance in a certain medium, they may be still insufficient to infer precise metabolic objective functions by which one could capture, e.g., objective shifts under varying nutrient conditions. Otherwise it would be important to understand whether the observed fluctuations have biological counterparts. In addition to the problems presented here, the methods of

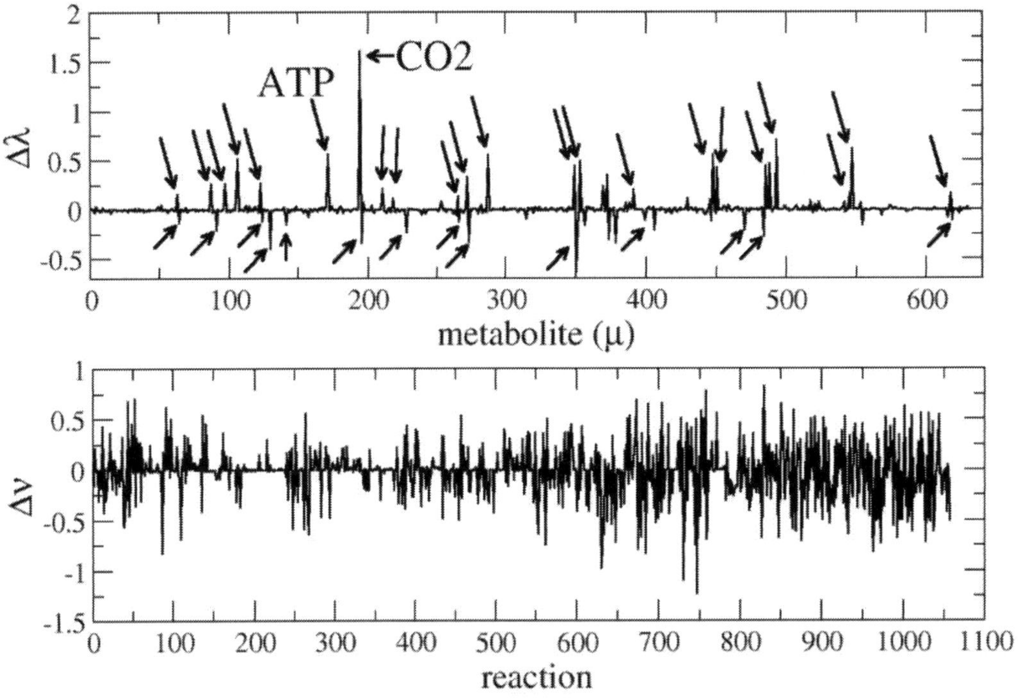

Figure 5. (Top) Component by component difference in $\boldsymbol{\lambda}$ and (Bottom) in $\boldsymbol{\nu}$ between the solution with the largest and smaller production rate of ATP. The number of metabolites is 631, that of reactions 1057. Arrows in the top figure mark, from left to right, the following metabolites: in the positive half, 3mob, 5aizc, 5mthf, ac, ade, atp, co2, damp, dcyt, fad, for, gal, h, h2o, k(e), nad group, nh4, pi(e), ppi, so3, so4(e) and urea; in the negative half 3mop, 5caiz, ade(e), adp, akg, co2(e), dhf, fadh2, for(e), h(e), lac, pant-R, pi and urea(e).

statistical mechanics developed for the analysis of constraint-satisfaction problems may prove crucial to address many of the important questions arising in this field, both for the structural and the dynamical level [36].

Acknowledgments
This work was supported by the IIT Seed Project DREAM.

References
[1] Liolios K *et al.* 2008 *Nucl. Acids Res.* **36** (Database issue) D475
[2] Palsson B O 2006 *Systems Biology: Properties of Reconstructed Networks* (New York: Cambridge University Press)
[3] Heinrich R and Schuster S 1996 *The regulation of cellular systems* (New York: Chapman & Hall)
[4] Beard B A and Qian H 2008 *Chemical biophysics: Quantitative analysis of cellular systems* (Cambridge: Cambridge University Press)
[5] Sauer U 2006 *Mol. Syst. Biol.* **2** 62
[6] Reed J L *et al.* 2003 *Genome Biology* **4** R54
[7] Jamshidi N *et al.* 2001*Bioinformatics* **17** 286
[8] Mahadevan R, Edwards J S and Doyle F J 2002 *Biophys J.* **83** 1331
[9] Steuer R *et al.* 2006 *Proc. Natl. Acad. Sci. USA* **103** 11868
[10] Wiback S J *et al.* 2004 *J. Theor. Biology* **228** 437
[11] Braunstein A, Mulet R and Pagnani A 2008 *BMC Bioinformatics* **9** 240

[12] Kauffman K, Prakash P and Edwards J S 2003 *Curr. Opin. Biotech.* **14** 491
[13] Ibarra R U, Edwards J S and Palsson B O 2002 *Nature* **420** 186
[14] Edwards J S and Palsson B O 2000 *Proc. Natl. Acad. Sci. USA* **97** 5528
[15] Ramakrishna R *et al.* 2001 *Am J. Physiol. Regul. Integr. Comp. Physiol.* **280** R695
[16] Oliveira A P, Nielsen J and Forster J 2005 *BMC Microbiol.* **5** 39
[17] Holzhütter HG 2004 *Eur. J. Biochem.* **271** 2905
[18] Schuetz R, Kuepfer L and Sauer U 2007 *Mol. Syst. Biol.* **3** 119
[19] Knorr A L, Jain R and Srivastava R 2007 *Bioinformatics* **23** 351
[20] Segre D, Vitkup D and Church G M 2002 *Proc. Natl. Acad. Sci. USA* **99** 15112
[21] Shlomi T, Berkman O and Ruppin E 2005 *Proc. Natl. Acad. Sci. USA* **102** 7695
[22] Imielinski M *et al.* 2005 *Bioinformatics* **21** 2008
[23] Imielinski M *et al.* 2006 *Biophys. J.* **90** 2659
[24] Famili I and Palsson BO 2003 *Biophys. J.* **85** 16
[25] Nikolaev E V, Burgard A P and Maranas C D 2005 *Biophys. J.* **88** 37
[26] Kruse K and Ebenhoeh O 2008 *Genome Inform.* **20** 91
[27] Von Neumann J 1945 *Rev. Econ. Stud.* **13** 1
[28] De Martino A and Marsili M 2005 *J. Stat. Mech.* L09003
[29] De Martino A *et al.* 2007 *J. Stat. Mech.* P05012
[30] De Martino A, Martelli C and Massucci F 2009 *Europhys. Lett.* **85** 38007
[31] Martelli C *et al.* 2009 *Proc. Natl. Acad. Sci. USA* **106** 2607
[32] Emmerling M *et al.* 2002 *J. Bacteriol* **184** 152
[33] Gerdes S Y *et al.* 2003 *J. Bacteriol* **185** 5673
[34] Almaas E *et al.* 2004 *Nature* **427** 839
[35] Bouchaud J P and Potters M 2009 Financial Applications of Random Matrix Theory: a short review *Preprint* arXiv:0910.1205
[36] De Martino A *et al.* 2009 Optimal flux states, reaction replaceability and response to knockouts in human red blood cell *Preprint* arXiv:0907.2824

On the operating point of cortical computation

Robert Martin, Marcel Stimberg, Klaus Wimmer and Klaus Obermayer

Bernstein Center for Computational Neuroscience Berlin and School of Electrical Engineering and Computer Science, Technische Universität Berlin, FR 2-1, Franklinstr. 28/29, D-10587 Berlin, Germany

E-mail: oby@cs.tu-berlin.de

Abstract. In this paper, we consider a class of network models of Hodgkin-Huxley type neurons arranged according to a biologically plausible two-dimensional topographic orientation preference map, as found in primary visual cortex (V1). We systematically vary the strength of the recurrent excitation and inhibition relative to the strength of the afferent input in order to characterize different operating regimes of the network. We then compare the map-location dependence of the tuning in the networks with different parametrizations with the neuronal tuning measured in cat V1 *in vivo*. By considering the tuning of neuronal dynamic and state variables, conductances and membrane potential respectively, our quantitative analysis is able to constrain the operating regime of V1: The data provide strong evidence for a network, in which the afferent input is dominated by strong, balanced contributions of recurrent excitation and inhibition, operating *in vivo*. Interestingly, this recurrent regime is close to a regime of "instability", characterized by strong, self-sustained activity. The firing rate of neurons in the best-fitting model network is therefore particularly sensitive to small modulations of model parameters, possibly one of the functional benefits of this particular operating regime.

1. Introduction

The major tasks of primary visual cortex (V1) is computing a representation of simple features present in the visual field. Most prominently, its neurons show sensitivity to orientations presented in their receptive fields, and are organized in a topographic map by, amongst others, their orientation preference. Consequently, preferred orientation varies smoothly, with the exception of occasional singularities, the so-called pinwheels. Early models of orientation processing in V1 [1], which combined the center-surround receptive fields of lateral geniculate nucleus to give rise to orientation selectivity, have been shown to be over-simplistic [2, 3]. Yet, even today a debate remains regarding the contribution of afferent and of recurrent excitatory and inhibitory influences, respectively, to the orientation selectivity of V1 [4, 5]. Information processing in cortex changes dramatically with this "cortical operating regime", i. e. depending on the relative strengths of the afferent and recurrent inputs [6, 7].

For instance, a theoretical study [7] demonstrated that in a Mexican-hat like operating regime, networks can express sharp orientation tuning that is largely independent of map position. Mexican-hat like connectivity patterns arising from broader inhibition have been routinely used in theoretical studies of the visual system [8]. However, recent experimental studies [9] seem to indicate that excitatory and inhibitory short-range connections spread over an approximately similar range, thereby not supporting a Mexican-hat like connectivity pattern. Yet, provided inhibition is fast (or conversely, excitation is slow, as in NMDA synapses with a large time constant), Mexican-hat like interaction patterns do occur for similar or even shorter-range inhibition [7]. The study by Kang et al. [7] relied on a mean-field network and allowed analytical treatment under certain simplifying assumptions. Here, we study a similar network but with spiking units, incorporating conductance-based synapses and spike-frequency adaption, thus allowing for a richer, non-linear behavior of the system.

2. The models

In our study we consider two different kinds of network models that both model a patch of approximately $1.56 \times 1.56\,\mathrm{mm}^2$ of V1: (1) a firing rate model using a meanfield approximation, similar to the network in [7], and (2) a spiking network consisting of Hodgkin-Huxley type units. Both models share a common network structure where all cells have a spatial position on a two-dimensional grid. The neurons' preferred orientations are determined by their position in an artificial orientation map and realized by the tuned afferent input (Fig. 1). Pinwheel distance and the spread of recurrent connections of the map is calibrated to match experimental data [9]. Recurrent connections within the map follow a spatially isotropic two-dimensional Gaussian profile with similar spatial extent for excitatory and inhibitory connections, consistent with *in vivo* data [9]. This leads to characteristic differences in the recurrent input a cell receives, depending on its map position: Cells close to a singularity of the orientation map (*pinwheel centers*) receive input from cells with very heterogeneous preferred orientations, whereas cells in the linear zones of the *iso-orientation domains* receive input from cells with very similar preferred orientations. In order to avoid boundary effects, we used periodic boundary conditions.

The activity of the firing-rate units in the mean field models is given by:

$$\tau_{\mathrm{E}_i}\frac{dm_{\mathrm{E}_i}(\vec{r},t)}{dt} = -m_{\mathrm{E}_i}(\vec{r},t) + \big[I_{\mathrm{Aff}}(\vec{r}) + S_{\mathrm{EE}}\alpha_1 I^{\mathrm{E}_1}_{\mathrm{rec}}(\vec{r},t) + S_{\mathrm{EE}}(1-\alpha_1)I^{\mathrm{E}_2}_{\mathrm{rec}}(\vec{r},t) - S_{\mathrm{EI}}I^{\mathrm{I}}_{\mathrm{rec}}(\vec{r},t)\big]^+, \quad (1)$$

$$\tau_{\mathrm{I}}\frac{dm_{\mathrm{I}}(\vec{r},t)}{dt} = -m_{\mathrm{I}}(\vec{r},t) + \big[I_{\mathrm{Aff}}(\vec{r}) + S_{\mathrm{IE}}\alpha_1 I^{\mathrm{E}_1}_{\mathrm{rec}}(\vec{r},t) + S_{\mathrm{IE}}(1-\alpha_1)I^{\mathrm{E}_2}_{\mathrm{rec}}(\vec{r},t) - S_{\mathrm{II}}I^{\mathrm{I}}_{\mathrm{rec}}(\vec{r},t)\big]^+, \quad (2)$$

where $i \in \{1,2\}$ is an index for the two excitatory populations, $\vec{r}$ is a vector describing a unit's position in its layer, $\tau_{\mathrm{E}_1} = 5\,\mathrm{ms}$ and $\tau_{\mathrm{E}_2} = 50\,\mathrm{ms}$ denote the excitatory synaptic time constants, $\tau_{\mathrm{I}} = 5\,\mathrm{ms}$ denotes the inhibitory time constant and $[\ldots]^+$ is the rectification function. The parameters $S_{..}$ denote the weight of the connections between the pre-synaptic excitatory and inhibitory populations and the postsynaptic excitatory (S_{EE} and S_{EI}) and inhibitory populations (S_{IE} and S_{II}). The recurrent input from the excitatory populations is additionally weighted by $\alpha_1 = 0.4$ and $1-\alpha_1 = 0.6$, respectively, for the two populations of excitatory neurons, i.e. the total excitatory recurrent input is a mixture from the fast (40%) and the slow population (60%). The recurrent input from a population $j \in \{\mathrm{E}_1, \mathrm{E}_2, \mathrm{I}\}$ is given by

$$I^j_{\mathrm{rec}}(\vec{r},t) = \int d\vec{r}\,'\varphi(\vec{r},\vec{r}\,')m_j, \quad (3)$$

where the function φ,

$$\varphi(\vec{r},\vec{r}\,') = \frac{1}{\sqrt{2\pi}\sigma}\exp\left(-\frac{(\vec{r}-\vec{r}\,')^2}{2\sigma^2}\right), \quad (4)$$

defines the connectivity profile.

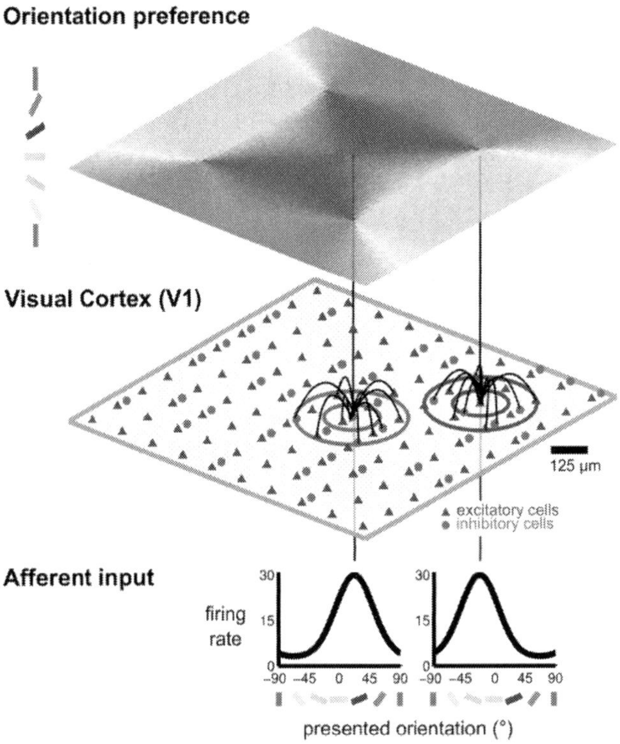

Figure 1. Network architecture. The cartoon sketches the architecture of both network model classes: A layer of excitatory (blue triangles) and inhibitory neurons (green circles) receives afferent as well as lateral input. In the Hodgkin-Huxley networks, the number of inhibitory cells is one third of the number of excitatory cells. Cells are placed on a grid (inhibitory neurons in random grid positions) of 50×50 (Hodgkin-Huxley networks), respectively 64×64 (firing rate networks) positions, modeling a patch of cortex $1.56 \times 1.56\,\mathrm{mm}^2$ in size (see scale bar). Examples for lateral connections are indicated for an inhibitory neuron in an iso-orientation domain (lines connecting to the neuron in the center) and an excitatory cell close to a pinwheel center (lines connecting to the neuron at the right). The preferred orientation of each neuron depends on its position in the artificial orientation map with four pinwheels (top; see connecting lines for the two example cells). A circular Gaussian tuning curve with standard deviation 27.5° (bottom) determines the afferent input rate for each neuron, depending on the presented orientation and the orientation preference of the cell.

The activity of the Hodgkin-Huxley units in the spiking network is given by:

$$C_{\mathrm{m}} \frac{dV_{\mathrm{m}}}{dt} = -g_{\mathrm{L}} \left(V_{\mathrm{m}} - E_{\mathrm{L}}\right) - \sum_{\mathrm{int}} I_{\mathrm{int}} - I_{\mathrm{syn}} - I_{\mathrm{bg}}, \tag{5}$$

where I_{syn}, I_{int} and I_{bg} denote the voltage-dependent synaptic, intrinsic and background currents, g_{L} and E_{L} denote the leak conductance and its reversal potential, C_{m} denotes the membrane capacitance, and t denotes time. We included three voltage dependent intrinsic currents: a fast Na^+ current and a delayed-rectifier K^+ current for the generation of action potentials, and a slow non-inactivating K^+ current responsible for spike frequency adaptation. For parameters and the functional forms of the activating and inactivating variables, see [10]. The peak conductance of the non-inactivating K^+-current is multiplied by the factor 0.1 for inhibitory neurons in order to reduce their spike-frequency adaptation compared to the excitatory neurons. The synaptic background current I_{bg} is determined, independently for each cell, from an excitatory and an inhibitory background conductance, which follow an Ornstein-Uhlenbeck-like stochastic process [11]. The synaptic current for neurons of the population $j \in \{\mathrm{E,I}\}$ is composed of the input received from $N_{\mathrm{Aff}} = 20$ afferent excitatory, $N_{\mathrm{E}} = 100$ recurrent excitatory and $N_{\mathrm{I}} = 50$ recurrent inhibitory neurons. The currents are given by

$$I^j_{\mathrm{syn}}(t) = \Bigg[\underbrace{\frac{\bar{g}^{\mathrm{Aff}}_j}{N_{\mathrm{Aff}}} g^{\mathrm{Aff}}(t)}_{\text{afferent excitatory}} + \underbrace{\frac{\bar{g}_{j\mathrm{E}}}{N_{\mathrm{E}}} \big(\alpha g_{\mathrm{E1}}(t) + (1-\alpha) g_{\mathrm{E2}}(t)\big)}_{\text{recurrent excitatory}} \Bigg] (V_{\mathrm{m}} - E_{\mathrm{e}}) \\ - \underbrace{\frac{\bar{g}_{j\mathrm{I}}}{N_{\mathrm{I}}} g_{\mathrm{I}}(t)(V_{\mathrm{m}} - E_{\mathrm{i}})}_{\text{recurrent inhibitory}}, \quad (6)$$

where $g(t)$ are the time-dependent conductances, E_{e} and E_{i} are the synaptic reversal potentials, $\alpha = 0.7$ determines the number of fast (70%) versus slow (30%) receptors of recurrent excitatory synapses, and $\bar{g}$ are scale factors (peak conductances; for values see [12]).

3. Operating regimes

Following [7] we define the operating regime of a meanfield network by properties of its "cortical feedback kernel" $D(k)$. For fixed connectivity φ, $D(k)$ depends on the strength of the recurrent connections only:

$$D(k) = S_{\mathrm{EE}}\tilde{\varphi}(k) - S_{\mathrm{EI}}S_{\mathrm{IE}}\tilde{\varphi}^2(k) - S_{\mathrm{II}}\tilde{\varphi}(k) + S_{\mathrm{II}}S_{\mathrm{EE}}\tilde{\varphi}^2(k). \quad (7)$$

The different operating regimes ("phases") shown in Fig. 2A are characterized by certain properties of $D(k)$: Unstable if $D(k) > 1$ for any k; Feedforward (FF) if $|D(k)| < 0.5$ for all k; Excitatory dominated (EXC) if $D(k) > 0$ for all k and $D_{\max} = D(0)$; Recurrent (REC) if $D_{\max} > 0.5$ and $1 - D(0) > 2(1 - D_{\max})$; Inhibitory dominated (INH) if $D(k) < 0$ for some k and $D_{\max} < 0.5$.

Although this analytical classification is not possible in a Hodgkin-Huxley model, we can define similar operating regimes numerically (Fig. 2B). We base this definition on the excitatory and inhibitory currents a cell receives when stimuli with varying orientations are presented, normalizing all currents with respect to the afferent input current at the preferred orientation. This leads to the same operating regimes (Fig. 2C): FF, if the sum of the absolute recurrent excitatory and inhibitory current is less than 0.3; EXC, if the excitatory current is larger than the inhibitory current for all presented orientations; REC, if the excitatory current is larger than the inhibitory currents for some orientations and lower for others; INH, if the excitatory current is lower than the inhibitory current for all presented orientations. We refer to the network as "unstable" if the model neurons show strong responses (average firing rate exceeds 100 Hz) and remain at high firing rates if the afferent input is turned off, i. e. the network shows self-sustained activity. In this regime, the model neurons lose their orientation tuning.

4. Characterizing the operating regimes

Naturally, networks in different operating regimes differ in their behavior. Firstly and quite predictably, as recurrent excitation increases (i. e. moving upward in the phase plot), peak firing rates increase. Moreover, the rate of change in the firing rate increases, too, as parameters approach the unstable region (movement towards top left in the phase plot).

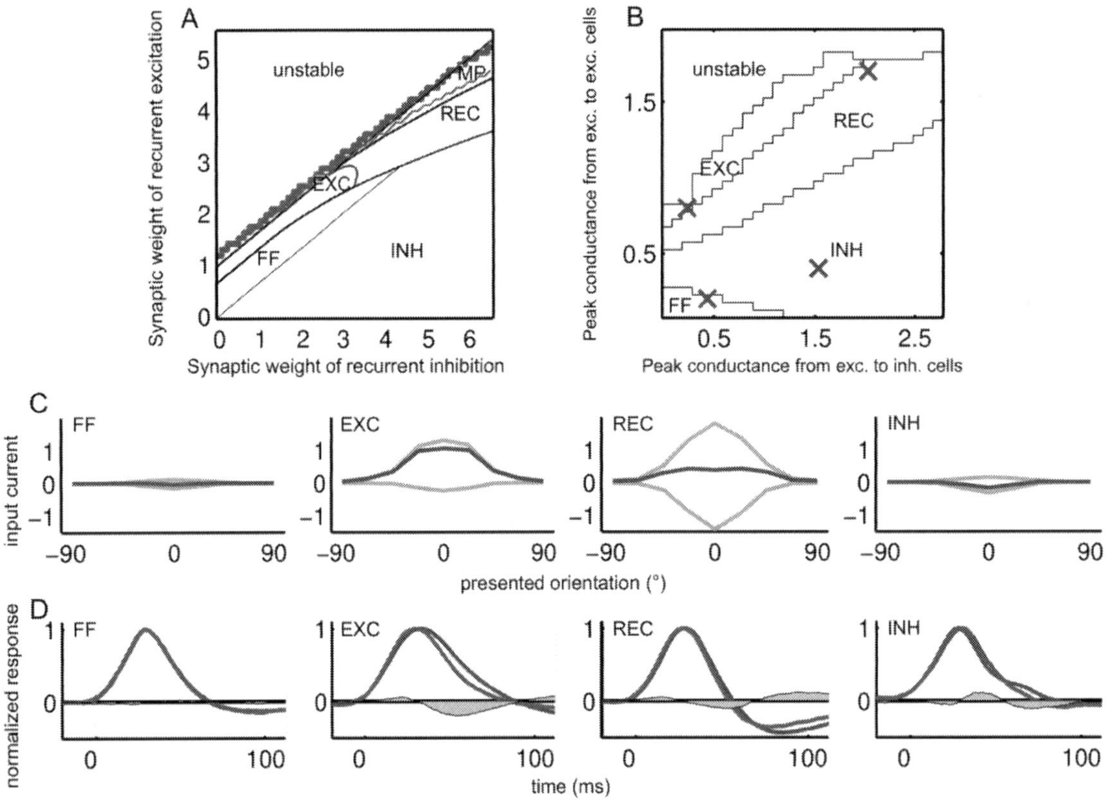

Figure 2. Operating regimes and their characteristics. (A) Operating regimes of a meanfield network model as a function of the synaptic weights of recurrent excitation (S_{EE}) and inhibition ($S_{\mathrm{EI}} \times S_{\mathrm{IE}}$). Black lines denote analytically obtained borders of the operating regimes [7]: FF – feed-forward, EXC – recurrent excitatory dominated, INH – recurrent inhibitory dominated, REC – strong recurrent excitation and inhibition (cf. phases II and III in [7]) and MP – marginal phase. The borders to the unstable region (thick red line) and to the MP (thin red line) were also determined numerically, defining the network to operate in the MP when for an almost untuned input (mean $A = 1$, modulation $B = 0.001$) the maximum and the minimum activity in the network differ by more than 5%. (B) Operating regimes of a Hodgkin-Huxley network model as a function of the peak conductance of synaptic excitatory connections to excitatory ($\overline{g}_{\mathrm{EE}}$) and inhibitory ($\overline{g}_{\mathrm{IE}}$) neurons. Conductances are given as multiples of the afferent peak conductance of excitatory neurons ($\overline{g}_{\mathrm{E}}^{\mathrm{Aff}}$). The figure summarizes simulation results for 38×28 different values of $\overline{g}_{\mathrm{EE}}$ and $\overline{g}_{\mathrm{IE}}$. (C) Feedback kernels for selected operating regimes in the Hodgkin-Huxley model. Gray lines depict the recurrent excitatory and inhibitory currents, respectively, relative to the cells' preferred orientation, averaged across cells for the example models marked in B, normalized to the afferent current received at their preferred orientation. Blue lines depict the corresponding total cortical feedback kernels. (D) Time-course of responses in the Hodgkin-Huxley model, determined using a reverse correlation paradigm, at the cells' preferred orientation averaged over all pinwheel (map OSI < 0.3; red lines) and iso-orientation domain ($0.6 <$ map OSI < 0.9; blue lines) cells, normalized to a peak of one for the example models marked in B. Shaded areas show the difference between the averaged responses of pinwheel and orientation domain cells.

Secondly, the operating regime influences not only activity, but also temporal dynamics of a cell's response, as illustrated in Fig. 2D. The employed reverse correlation paradigm (for details see [13]) moreover illustrates characteristic differences between cells at different map positions (Fig. 2D): In the FF and the REC regime, both cells close to pinwheel centers and cells in orientation domains show very similar temporal responses. In the EXC regime on the other hand, the response of both populations is markedly prolonged, but the effect is even more pronounced in iso-orientation domain cells. Although less clear, the opposite relationship is present in the INH regime, where cells close to pinwheel centers show a more prolonged response. Further differences in the variability found in pinwheel- and iso-orientation domain cells are discussed elsewhere [12].

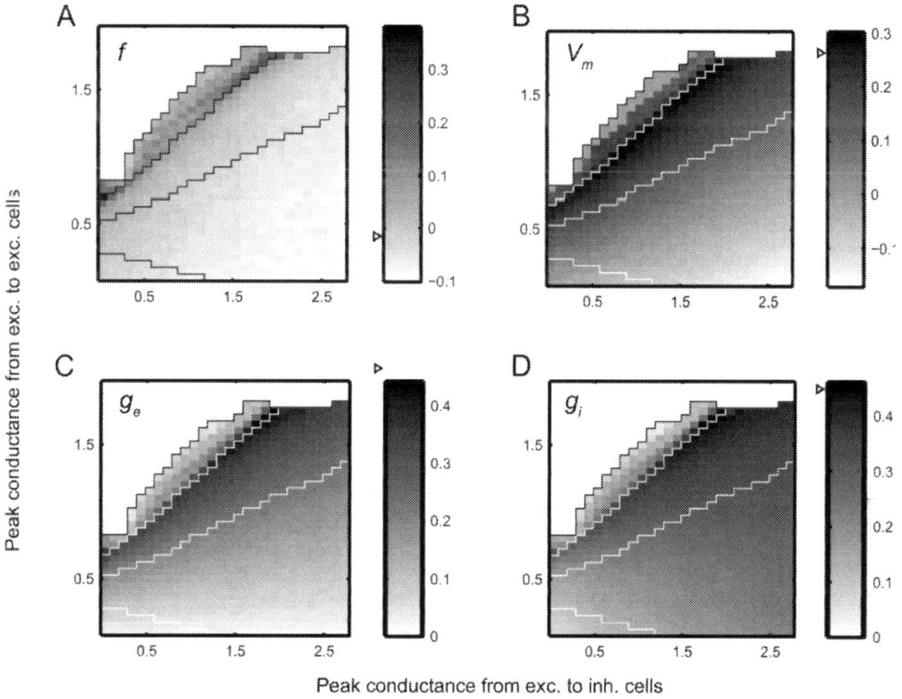

Figure 3. Location dependence of orientation tuning in the network models. The figure shows the slope values of the OSI-OSI regression lines (in gray values) as a function of the peak conductance of synaptic excitatory connections to excitatory ($\overline{g}_{\mathrm{EE}}$) and inhibitory ($\overline{g}_{\mathrm{IE}}$) neurons, separately (A) for the spike rate f, (B) for the membrane potential V_m, (C) for the total synaptic excitatory conductance g_e and (D) for the total synaptic inhibitory conductance g_i. Conductances are given as multiples of the afferent peak conductance of excitatory neurons ($\overline{g}_{\mathrm{E}}^{\mathrm{Aff}}$). Thin lines denote the borders of the different operating regimes (cf. Figure 2). The arrows at the colorbars indicate the values of the respective experimental measurements in [9].

Thirdly and most importantly for the analysis presented below, the operating regime also has a profound effect on the tuning of the neurons' firing rate f, membrane potential V_m and their excitatory and inhibitory input conductances g_e and g_i. Again, these properties change in a way that is strongly related to the localization of a neuron, either in the region close to a pinwheel center or in an iso-orientation domain. For the further analysis, we describe the strength of tuning of a any property for a particular cell using the orientation selectivity index (OSI). The dependence of the OSI on a cell's position in the orientation map between pinwheel and iso-orientation domain for any of these properties is described using a linear fit with a certain slope (see Appendix for further details). Plotting the slopes, Fig. 3 illustrates how strongly the tuning of f, V_m, g_e and g_i varies between these two regions, with positive values indicating a sharper tuning of the variable in iso-orientation domains and negative values vice versa. The rising values show that for all four quantities, the strength of their linear dependence on map position tends to increase as the regime gets closer to the unstable region (top left). In other words, as the regime becomes increasingly recurrently/excitatory dominated, the recurrent contribution leads to more pronounced tuning in the iso-orientation domain than in pinwheel regions. Note the exception in the excitatory regimes close to instability: consistent with the observation of the previous section, here the map-dependence of the tuning almost vanishes (slope approaching zero).

The map-dependence of orientation tuning in different regimes points to one handle on the underconstrained problem of finding the actual biological operating regime, as is illustrated below.

5. Searching biologically plausible operating regimes

Recent experimental and theoretical studies have investigated how the orientation tuning of a cell's spike output depends on its position in the orientation preference map [7, 14, 15], but none of these has thus far been able to conclude the debate about the cortical operating regime. The wide range of models operating in different regimes that are discussed in the literature are an indication that models of V1 orientation selectivity are underconstrained. We show now that studying the specific location dependence of the tuning of internal neuronal properties can provide further constraints to find the appropriate cortical operating regime.

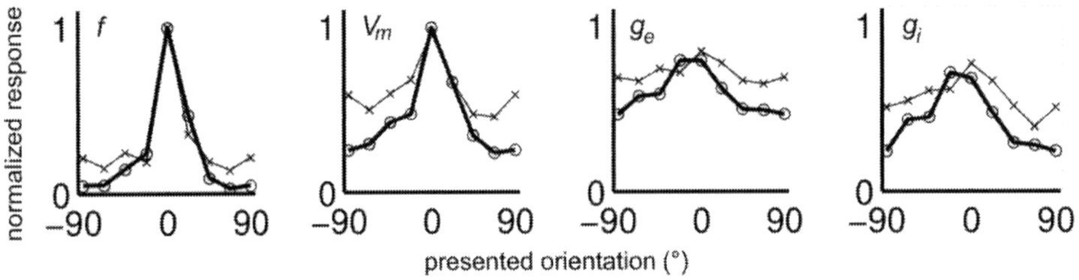

Figure 4. Tuning curves for experimental data from [9]. Mean responses across cells are shown for the firing rate (f), the membrane potential (V_m), the total excitatory (g_e) and total inhibitory conductance (g_i), separately for cells in iso-orientation domains ($0.6 <$ map OSI < 0.9, thick lines) and cells close to pinwheel centers (map OSI < 0.3, thin lines). For each cell, responses were individually aligned to its preferred orientation and normalized to its maximum response. To allow comparison of the magnitude of g_i and g_e responses, both types of conductances were normalized to the maximum g_i response.

The data originates from intracellular recordings of cat V1 [9], combined with optical imaging. This allowed to measure *in vivo* the output (firing rate f) of neurons, their input (excitatory and inhibitory conductances g_e and g_i) and further a state variable (membrane potential V_m) of a neuron as a function of its position in the orientation map. Note that the latter three properties have hitherto not been available and could thus far not constrain the operating regime. Fig. 4 shows population averages of the experimentally observed tuning curves of each of these properties for neurons around pinwheel centers and for neurons within iso-orientation domains.

Note that the tuning of V_m as well as the tuning of g_e and g_i vary strongly with map location. Specifically, these properties are more sharply tuned for neurons within an iso-orientation domain, where the neighboring neurons have very similar orientation preferences, as compared to neurons close to a pinwheel center, where the neighboring neurons show a broad range of orientation preferences. The tuning of the firing rate f is on the other hand largely invariant with map location.

6. Finding the biologically plausible operating regime

We analyzed the dependence of the orientation tuning properties on the operating regimes and compared them to the experimental data. For every combination of $\overline{g}_{\mathrm{EE}}$ and $\overline{g}_{\mathrm{IE}}$, we simulated the responses of neurons in the network model to oriented stimuli in order to measure the orientation tuning of f, V_m, g_e and g_i. The OSI of each of the four quantities can then be plotted against the map OSI (varying from 0 close to pinwheel centers to 1 in an iso-orientation domain; see Appendix for details) to reveal the dependence of the tuning on map location. The slope of the linear regression of this OSI-OSI dependence was used to characterize the different operating points of the network. This slope captures the specific map-location dependence of the tuning width for each of the four measured quantities f, V_m, g_e and g_i.

Fig. 3 shows the slopes for the tuning of f, V_m, g_e and g_i as a function of $\overline{g}_{\mathrm{EE}}$ and $\overline{g}_{\mathrm{IE}}$ of the respective Hodgkin-Huxley network models (gray scale). Model networks with strong recurrent excitation (large values of $\overline{g}_{\mathrm{EE}}$), as in the REC regime, predict steeper slopes than networks with less recurrent excitation. As the experimentally found values (arrows at the color bar) indicate rather steep values for the slopes, V_m, g_e and g_i constrain the network parameters to a highly recurrent regime. However, yet closer to the line of instability the map-dependence of the tuning almost vanishes, with flat slopes in the strongly excitatory regimes, where tuning is lost regardless of map location. Also, the strongly inhibitory dominated regimes (large values of $\overline{g}_{\mathrm{IE}}$) at the bottom right corner of Fig. 3 are of interest. Here, the slope of the location dependence becomes negative for the tuning of firing rate f and membrane potential V_m. Such a sharpening of the tuning close to pinwheel centers in an inhibition dominated regime has been observed in other theoretical studies operating in an inhibition dominated regime [14]. Note that the slope of the tuning of the firing rate (Fig. 3A) is flat – and thus consistent with the experimental data – for most parts of the parameter space.

We can now quantitatively compare the slopes of the OSI-OSI relationship found *in vivo*, as indicated by the arrows in the colorbars of Fig. 3, to the model data. For each of the four measured quantities and for each point in the phase plane we determine the likelihood that a network operating at this point has generated the OSI-OSI relationship found *in vivo*. Fig. 5A shows the product of the four normalized likelihood functions. This "joint likelihood" peaks in a strongly recurrent regime, and we can thus conclude that the experimental data constrains the network to operate in an operating regime with both, strong recurrent excitation and inhibition, approximately balanced with one another, but dominating the afferent input. The normalized likelihood for the firing rate f (Fig. 5B) alone, on the other hand, has high values for most points in the phase space, illustrating the difficulty in determining the correct operating regime by just considering this one property.

7. Discussion

Although much is known about the anatomy of lateral connections in cat V1, the *strengths* of synapses formed by short-range connections are largely unknown. In this study, we numerically explore the operating regimes of a spiking neural network with different strengths of their recurrent connections. We then use intracellular physiological measurements to constrain the strengths of these connections, and find that neither feed-forward dominated nor recurrent excitatory- or inhibitory-dominated networks are consistent with the tuning properties observed *in vivo*. We therefore conclude that the cortical network in cat V1 operates in a regime with a dominant recurrent influence that is balanced between inhibition and excitation.

The analysis presented here focuses on the steady state the network reaches when presented with one non-changing orientation. Interestingly, a comparable operating regime has been

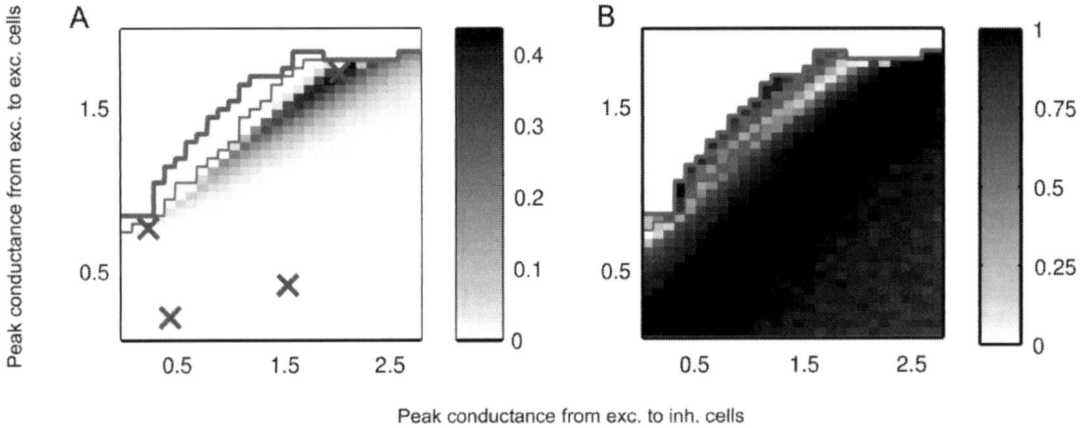

Figure 5. Likelihood of regimes. (A) Product of the individual likelihoods for V_m, f, g_e and g_i for the slopes of the regression lines in the OSI-OSI plots. (B) Likelihood for the slopes of the OSI-OSI relationship of the firing rate response V_m.
Both plots are shown as a function of the peak conductance of synaptic excitatory connections to excitatory ($\overline{g}_{\mathrm{EE}}$) and inhibitory ($\overline{g}_{\mathrm{IE}}$) neurons. Gray values denote the likelihood (scale bars at the right); conductances are given as multiples of the afferent peak conductance of excitatory neurons ($\overline{g}_{\mathrm{E}}^{\mathrm{Aff}}$). The area above the thin red line corresponds to parameter values for which the model neurons exhibit untuned responses (all firing rate OSIs are below 0.3). The area above the thick red line corresponds to parameter values for which the model network becomes "unstable", i.e. model cells fire with an average firing rate above 100 Hz, independent of the stimulus orientation. The likelihood was not evaluated for the unstable regime. The figure summarizes simulation results for 38×28 different values of $\overline{g}_{\mathrm{EE}}$ and $\overline{g}_{\mathrm{IE}}$.

indicated in an analysis of the dynamic properties of orientation tuning in cat V1 [13].

The finding that tuning properties of cat V1 are best explained by a network operating in a recurrent regime is robust against variation of other parameters not considered here, e. g. the relative range of the recurrent inhibition and the peak conductances $\overline{g}_{\mathrm{II}}$ and $\overline{g}_{\mathrm{EI}}$ (data not shown). Nevertheless, the network architecture is based on a range of basic assumptions, notably that all neurons in the network receive tuned input of equal sharpness. Explicit inclusion of location dependence of the input tuning might well lead to tuning properties compatible with the experimental data in different operating regimes. However, there is no evidence supporting such a location dependence of the afferent input, and therefore assuming location-independent input seemed the most prudent basis for this analysis.

In the recurrent regime, neurons responded to their preferred stimulus with a 1.8-fold increased firing rate compared to the feed-forward regime. Previous modeling studies have already attributed such an amplification to the cortical network [8,16], and so has experimental work [17].

Our analysis demonstrates that the cortical operating point seems to be close to instability. Such a network is very sensitive to changes in its governing parameters, e. g. small changes in connection strengths lead to large changes in the overall firing rate. In the "best fitting" operating regime, a 10% change in firing rate, which is of similar magnitude as observed firing rate changes under attention in macaque V1 [18], is easily achieved by increasing $\overline{g}_{\mathrm{EE}}$ by just 2%. It therefore seems plausible that one benefit of being in such a regime is the possibility of significantly changing the "operating point" of the network through only small adjustments of the underlying parameters. Candidates for such an adjustment could be contextual modulations, adaptation or attentional effects, which may all be explainable as temporal shifts of the operating point, for example to make the cortical response particularly sensitive to small changes in afferent or feedback signals. Experimental data, for example detailing spatial and temporal contextual interactions [19], may allow to further investigate the functional implications of the cortical operating regime. For example, Dragoi et al. [20] found pronounced adaptation-induced short-term plasticity close to pinwheel centers, which was explained by a network model operating in a strongly recurrent, balanced regime [21]. Center-surround interactions [22], too, are compatible with such a regime [23].

The peculiar location of the cortical operating point close to the line of instability is reminiscent of the theory of "computing at the edge of chaos" [24], which has attracted some attention in the past decade. Theoretical studies demonstrated abstract networks to accomplish more challenging computations near the border between ordered to chaotic dynamics [25]. Moreover, such networks are responsive to inputs over a wider dynamic range [26]. These theoretical considerations are complemented by *in vivo* data showing recurrent excitation and inhibition in balance [27], which there, too, has been interpreted as a facilitator for easy transition between different cortical states. Thus, our findings here add to the evidence that the recurrent network contributes to maintaining an operating point which allows easy switching between different cortical states.

Appendix A. Simulation

Appendix A.1. The Hodgkin-Huxley network model

The network contained $50 \times 50 = 2500$ excitatory cells arranged on a grid and 833 inhibitory neurons (25%) placed at random grid locations. Recurrent excitatory conductances were modeled as arising to 70% from fast (AMPA-like) and to 30% from slow (NMDA-like) receptors. Presynaptic spike were transferred to the postsynaptic neuron with a delay, drawn from a Gaussian distribution. We used exponential models for the synaptic conductances originating from $GABA_A$-like inhibitory and AMPA-like excitatory synapses [28]. Slow NMDA-like excitatory synapses were modeled by the difference of two exponentials. The Poisson input spike trains exclusively trigger fast, AMPA-like excitatory synapses. Further details and model parameters can be found in [12].

The network usually settled into a steady state within a few hundred milliseconds. Thus, to measure orientation tuning curves of f, V_m, g_e and g_i, the response of the network to inputs with different orientations was computed for 1.5 s with 0.25 ms resolution. We then calculated the average f, V_m, g_e and g_i for every cell in an interval between 0.5 s and 1.5 s.

Appendix A.2. The firing-rate network model

The firing-rate network is similar to the network in [7] with two exceptions: (1) We include self-inhibition ($S_{II} \neq 0$) in our model, and (2) we use the same spatial connectivity function φ for excitatory and inhibitory connections, consistent with experimental observations [9]. Further details and model parameters can be found in [12].

Appendix B. Orientation selectivity index (OSI) and OSI-OSI slopes

We analyze orientation tuning for an arbitrary property $R(\phi_i)$ in response to a stimulus of orientation ϕ_i using the orientation selectivity index [29], which is given by

$$\mathrm{OSI} = \sqrt{\left(\sum_{i=1}^{N} R(\phi_i)\cos(2\phi_i)\right)^2 + \left(\sum_{i=1}^{N} R(\phi_i)\sin(2\phi_i)\right)^2} \Big/ \sum_{i=1}^{N} R(\phi_i). \tag{B.1}$$

For all measurements, eight stimulus orientations $\phi_i \in \{-67.5, -45, -22.5, 0, 22.5, 45, 67.5, 90\}$ were presented. The OSI is then a measure of tuning sharpness ranging from 0 (unselective) to 1 (perfectly selective). In addition, the OSI was used to characterize the sharpness of the recurrent input a cell receives based on the orientation preference map. To calculate this *map OSI*, we estimated the local orientation preference distribution by binning the orientation preference of all pixels within a radius of 250 µm around a cell into bins of 10° size; the number of cells in each bin replaces $R(\phi_i)$. The map OSI ranges from almost 0 for cells close to pinwheel centers to almost 1 in the linear zones of the iso-orientation domains. The dependence of each tuning property on the local map OSI was then described by a linear regression line using the least squares method.

Acknowledgments

We thank James Schummers, David Lyon, Jorge Mariño and Mriganka Sur for sharing experimental data and discussing the modeling work presented in this study. Support was contributed by grant 01GQ0414 from the BMBF.

References

[1] Hubel D H and Wiesel T N 1962 *J. Physiol.* **160** 106
[2] Sompolinsky H and Shapley R 1997 *Curr. Opin. Neurobiol.* **7** 514
[3] Ferster D and Miller K D 2000 *Annu. Rev. Neurosci.* **23** 441
[4] Martin K A C 2002 *Curr. Opin. Neurobiol.* **12** 418
[5] Teich A F and Qian N 2006 *J. Neurophysiol.* **96** 404
[6] Ben-Yishai R, Bar-Or R L and Sompolinsky H 1995 *Proc. Natl. Acad. Sci. USA* **92** 3844
[7] Kang K, Shelley M and Sompolinsky H 2003 *Proc. Natl. Acad. Sci. USA* **100** 2848
[8] Somers D, Nelson S and Sur M 1995 *J. Neurosci.* **15** 5448
[9] Mariño J, Schummers J, Lyon D C, Schwabe L, Beck O, Wiesing P, Obermayer K and Sur M 2005 *Nat. Neurosci.* **8** 194
[10] Destexhe A and Paré D 1999 *J. Neurophysiol.* **81** 1531
[11] Destexhe A, Rudolph M, Fellous J and Sejnowski T 2001 *Neuroscience* **107** 13
[12] Stimberg M, Wimmer K, Martin R, Schwabe L, Mariño J, Schummers J, Lyon D C, Sur M and Obermayer K 2009 *Cereb. Cortex* **19** 2166
[13] Schummers J, Cronin B, Wimmer K, Stimberg M, Martin R, Obermayer K, Koerding K and Sur M 2007 *Frontiers in Neuroscience* **1** 145
[14] McLaughlin D, Shapley R, Shelley M and Wielaard D J 2000 *Proc. Natl. Acad. Sci. USA* **97** 8087
[15] Nauhaus I, Benucci A, Carandini M and Ringach D L 2008 *Neuron* **57** 673
[16] Douglas R J, Koch C, Mahowald M, Martin K A and Suarez H H 1995 *Science* **269** 981
[17] Sharon D and Grinvald A 2002 *Science* **295** 512
[18] McAdams C J and Maunsell J H 1999 *J. Neurosci.* **19** 431
[19] Schwartz O, Hsu A and Dayan P 2007 *Nat. Rev. Neurosci.* **8** 522
[20] Dragoi V, Rivadulla C and Sur M 2001 *Nature* **411** 80
[21] Wimmer K and Obermayer K 2009 Synaptic plasticity changes orientation tuning and provides a mechanistic explanation for the tilt aftereffect in network models of V1 *2009 Neuroscience Meeting Planner* (Society for Neuroscience) online
[22] Schwabe L, Obermayer K, Angelucci A and Bressloff P C 2006 *J. Neurosci.* **26** 9117
[23] Stimberg M and Obermayer K 2009 The local recurrent network influences surround modulation in a model of V1 *2009 Neuroscience Meeting Planner* (Society for Neuroscience) online
[24] Langton C G 1990 *Physica D: Nonlinear Phenomena* **42** 12
[25] Bertschinger N and Natschläger T 2004 *Neural Comput.* **16** 1413
[26] Kinouchi O and Copelli M 2006 *Nature Physics* **2** 348
[27] Haider B, Duque A, Hasenstaub A R and McCormick D A 2006 *J. Neurosci.* **26** 4535
[28] Destexhe A, Mainen Z F and Sejnowski T J 1998 Kinetic models of synaptic transmission *Methods in neuronal modeling* ed Koch C and Segev I (Cambridge, MA: MIT Press) pp 1–25 2nd ed
[29] Swindale N V 1998 *Biol. Cybern.* **78** 45

International Workshop on Statistical-Mechanical Informatics 2010 IOP Publishing
Journal of Physics: Conference Series **233** (2010) 012021 doi:10.1088/1742-6596/233/1/012021

Doubly sparse factor models for unifying feature transformation and feature selection

Kentaro Katahira[1,2,3], Narihisa Matsumoto[4], Yasuko Sugase-Miyamoto[4], Kazuo Okanoya[1,3] and Masato Okada[1,2,3]

[1] ERATO, Okanoya Emotional Information Project, Japan Science Technology Agency, Saitama, Japan
[2] Graduate School of Frontier Sciences, The University of Tokyo, Kashiwa, Chiba, Japan
[3] RIKEN Brain Science Institute, Saitama, Japan
[4] Human Technology Research Institute, National Institute of Advanced Industrial Science and Technology, Ibaraki, Japan.

E-mail: okada@k.u-tokyo.ac.jp

Abstract. A number of unsupervised learning methods for high-dimensional data are largely divided into two groups based on their procedures, i.e., (1) feature selection, which discards irrelevant dimensions of the data, and (2) feature transformation, which constructs new variables by transforming and mixing over all dimensions. We propose a method that both selects and transforms features in a common Bayesian inference procedure. Our method imposes a doubly automatic relevance determination (ARD) prior on the factor loading matrix. We propose a variational Bayesian inference for our model and demonstrate the performance of our method on both synthetic and real data.

1. Introduction

Procedures to reduce dimensionality are crucial for extracting meaningful structures from multidimensional data to classify and visualize data. Statistical methods to reduce dimensionality are largely divided into two groups, depending on their procedures, i.e., feature selection (e.g., the wrapper approach [1]) and feature transformation. Principal component analysis (PCA) and factor analysis (FA) are examples of the methods with feature transformation. Feature selection discards irrelevant dimensions of given data [2], while feature transformation synthesizes new variables by transforming all dimensions. Feature selection is more effective than feature transformation if most dimensions are irrelevant to interesting features. If most dimensions are relevant to interesting features and have correlations with other dimensions, feature transformation gives us a more efficient representation of the data. However, there are no systematic ways to determine which procedure is better. Furthermore, both procedures can be used if one wants to discard irrelevant noisy features and synthesize new variables by transforming the remaining features. Let us consider the case in which the activities of many neurons are simultaneously recorded from animals in electrophysiological experiments, and the activities of some neurons are not relevant to stimuli or behavioral tasks. In such a situation, it may be beneficial to eliminate irrelevant neurons from analysis and synthesize new variables by transforming the activities of remaining neurons to visualize the activities of many neurons.

We propose a method that unifies both procedures. The method is based on a probabilistic formulation of generative models and automatic relevance determination (ARD) technique obtained from hierarchical Bayesian approaches. We assumes that interesting (relevant) features are generated from the linear transformation of low-dimensional generative models, such as a single Gaussian distribution, a mixture of Gaussians (MoGs), or hidden Markov models (HMMs), and irrelevant features are generated from Gaussian noise models which are independent of other features. ARD in our framework serves two purposes. The first purpose is to absorb irrelevant features into a noise model, and the second is to suppress irrelevant low-dimensional features [3, 4]. Our method selects and transforms features within a common probabilistic model framework. Hence, we can directly compare among the method with feature selection, the one with feature transformation, and the one with both procedures by a method of model selection within a Bayesian framework. We derive a variational Bayesian algorithm for our model. We demonstrate the effectiveness of our method on both synthesized data and real data sets including neuron data recorded from monkeys.

2. Doubly Sparse Factor (DSF) model

We assume that a D-dimensional observation vector $\mathbf{y}$ is generated from the following generative model

$$\mathbf{y} = \mathbf{V}\mathbf{W}\mathbf{x} + \mu_b + \epsilon, \tag{1}$$

where $\mathbf{V}$ is a $D \times D$ diagonal matrix $\mathrm{diag}(v_1, ..., v_D)$ and $\mathbf{W}$ is a $(D \times M)$ matrix (here, we assume $D \geq M$). The product $\mathbf{VW}$ corresponds to the factor loading matrix in FA terminology. A M-dimensional latent variable $\mathbf{x}$ is generated from an "interesting" low-dimensional factor model that captures the overall structure of the given data. μ_b is a D-dimensional bias parameter and ϵ is a D-dimensional noise vector. We assume that ϵ obeys a zero-mean non-isotropic Gaussian distribution, as in FA:

$$\epsilon \sim \mathcal{N}(\mathbf{0}, \Psi^{-1}), \quad \Psi = \mathrm{diag}(\psi_1, ..., \psi_D)$$

We assume each ψ_i can have a different value as FA. Each element of the diagonal matrix, $\mathbf{V}$, serves as a "gate" that determines whether relevant factor model $\mathbf{x}$ projects onto the feature, or the data variance in the feature is only attributed to the noise model, ϵ. If v_i has a non-zero value, the model selects the former, otherwise the latter. The matrix, $\mathbf{W}$, can be regarded as an ordinary factor loading matrix. If all elements in a column are zero, the corresponding dimension in latent variable $\mathbf{x}$ is suppressed. Therefore, $\mathbf{W}$ also serves as a filter that determines whether each dimension of $\mathbf{x}$ is projected onto the data (through being multiplied by $\mathbf{V}$).

2.1. Factor model

The factor model of latent variable $\mathbf{x}$ represents the low-dimensional data structure in which we are interested. If we use a Gaussian distribution with zero mean and unit covariance, the corresponding model can be regarded as an extension of conventional FA. We can also use more complex models. We use the finite mixture of Gaussians (MoGs) in this paper for clustering purposes, defined as

$$\mathbf{x}_n|\mathbf{z}_n \sim \prod_{k=1}^{K}[\mathcal{N}(\mathbf{x}_n|\mu_k, (\Lambda_k)^{-1})]^{z_{nk}}, \qquad \mathbf{z}_n|\pi \sim \prod_{k=1}^{K}\pi_k^{z_{nk}}$$

where $\mathcal{N}(\cdot)$ denotes the Gaussian distribution, K is the number of components and $\pi = \{\pi_k\}_{k=1}^{K}$ represents the mixing proportions. The hidden variable $\mathbf{z}_n = \{z_{nk}\}_{k=1}^{K}$ indicates that if $z_{nk} = 1$,

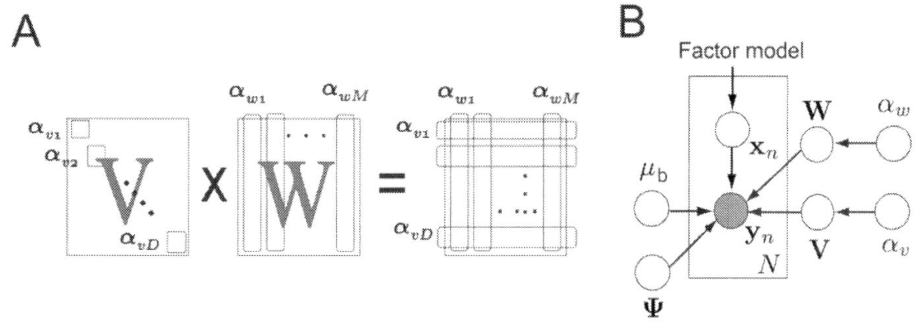

Figure 1: (A) Schematic of doubly ARD. The hyperparameters α_{vi} induce the corresponding row of the effective factor loading matrix $\mathbf{VW}$ to have a zero value, while the hyperparameters α_{wi} induce the corresponding column to have a zero value. (B) Graphical model of DSF model. Arrows represent conditional dependencies. Factor model for $\mathbf{x}_n$ can be a single Gaussian or a mixture of Gaussians. Only $\mathbf{x}_n$ in this graphical model connects to the variables in factor models.

nth data are generated by the kth component, otherwise $z_{nk} = 0$. By using this model, we can achieve clustering in low-dimensional space. Furthermore, we can jointly optimize feature selection, feature transformation, and MoGs, since all are within a common probabilistic model framework (i.e., we do not use a two-stage procedure, such that we first use various static techniques to reduce dimensionality, such as FA or PCA, or static techniques of feature selection before applying clustering techniques to the reduced space). We can also use other interesting models depending on the purpose, i.e., for time series data, we may use hidden Markov models or Gaussian processes [5].

2.2. Doubly automatic relevance determination

We try to impose double sparsity on the effective factor loading matrix, $\mathbf{VW}$. As the first is row-wise sparsity, irrelevant features are not projected from the factor model, and as the second is component-wise sparsity, the effective dimensions of the latent factor model become as small as possible. To achieve this, we impose an ARD prior on $\mathbf{V}$ and $\mathbf{W}$ as

$$p(\mathbf{V}|\alpha_v) = \prod_{i=1}^{D} \left(\frac{\alpha_{vi}}{2\pi}\right)^{\frac{1}{2}} \exp\left\{-\frac{1}{2}\alpha_{vi}v_i^2\right\},$$
$$p(\mathbf{W}|\alpha_w) = \prod_{i=1}^{M} \left(\frac{\alpha_{wi}}{2\pi}\right)^{\frac{D}{2}} \exp\left\{-\frac{1}{2}\alpha_{wi}||\mathbf{w}_i||^2\right\},$$

where $\mathbf{w}_i$ is the ith column of $\mathbf{W}$. Although one could adopt the type II maximum-likelihood approach that optimizes α_v and α_w so that the marginal likelihood is maximized, here, we use a hierarchical prior as in [4]:

$$p(\alpha_v) = \prod_{i=1}^{D} \mathcal{G}(\alpha_{vi}|a_{\alpha v0}, b_{\alpha v0}), \quad p(\alpha_w) = \prod_{i=1}^{M} \mathcal{G}(\alpha_{wi}|a_{\alpha w0}, b_{\alpha w0}).$$

where $\mathcal{G}(\cdot)$ denotes the Gamma distribution. If $\langle\alpha_{vi}\rangle \to \infty$ (here, $\langle\cdot\rangle$ denotes the posterior mode), the corresponding feature, i, is neglected from the factor model and only explained by the noise model. If $\langle \alpha_{wi} \rangle \to \infty$, the corresponding dimension of latent variable i is effectively eliminated.

There are a number of approaches that impose sparsity prior on the factor loading matrix [4, 6]. Our original idea is to use $\mathbf{V}$ and to also impose ARD prior on it. From this, we can doubly impose the sparsity requirements on the row and column of the effective factor loading matrix. We may use component-wise ARD as in [6], but this does not directly induce the elimination of each feature and each latent dimension. One may wonder if we can directly impose row-wise and column-wise precision hyperparameters on the factor loading matrix like

$$p(\mathbf{W}|\alpha_v, \alpha_w) = \prod_{i=1}^{D} \prod_{j=1}^{M} \left(\frac{\alpha_{vi}\alpha_{wj}}{2\pi} \right)^{\frac{1}{2}} \exp \left\{ -\frac{1}{2} \alpha_{vi}\alpha_{wj} w_{ij}^2 \right\}.$$

We can also derive tractable inference for this model, but we found that for this model, α_{vi}'s and α_{wi}'s compensated each other, and the effect weakened.

In summary, our goals of the use of sparsity prior are twofold: (1) determine the number of dimension of the low dimensional latent factors and (2) select the relevant features (in original data set) to which the latent factors (mixture of Gaussians) project. The first one is achieved by imposing row-wise sparsity on the factor loading matrix, while the second one is done by imposing column-wise sparsity on the factor loading matrix. However, it is difficult to impose both sparsity on the same matrix. To achieve this, in addition to the factor loading matrix $\mathbf{W}$, we used the diagonal matrix $\mathbf{V}$ with ARD prior which induces row-wise sparsity of the effective factor loading matrix.

To achieve an analytically tractable inference based on the variational Bayesian inference, we also use conjugate priors for the other parameters, which are specified in a forthcoming paper.

2.3. Variational Inference

For tractable inference, we adopt variational approximation for the posteriors of all parameters and all latent variables. We denote all model parameters and latent variables as $\mathcal{Z} = \{\mathbf{V}, \mathbf{W}, \mathbf{X}, \alpha_W, \alpha_V, \theta_X\}$, where $\mathbf{X} = \{\mathbf{x}_n\}_{n=1}^N$ and θ_X denotes variables in a factor model (in the MoG case, $\theta_X = \{\mathbf{Z}, \mu, \mathbf{\Lambda}, \pi\}$ where $\mathbf{Z} = \{\mathbf{z}_n\}_{n=1}^N$). We approximate the posteriors over these variables by the factorized distribution, $q(\mathcal{Z}) = q(\mathbf{V})q(\mathbf{W})q(\mathbf{X})q(\theta_X)$. For MoG, $q(\theta_X)$ is further factorized as $q(\theta_X) = q(\pi)q(\mathbf{Z})q(\mu, \mathbf{\Lambda})$. To obtain the best approximation within this restricted distribution, we seek $q(\mathcal{Z})$ that minimizes the Kullback-Leibler (KL) divergence from $q(\mathcal{Z})$ to the true posterior, $p(\mathcal{Z}|\mathbf{Y})$:

$$\mathrm{KL}(q(\mathcal{Z})||p(\mathcal{Z}|\mathbf{Y})) = \log p(\mathbf{Y}) + \mathcal{F}, \tag{2}$$

where $\mathcal{F}$ is called the variational free energy, which gives the lower bound for the log marginal likelihood, $\log p(X)$. Minimizing the KL divergence is equivalent to maximizing $\mathcal{F}$. Maximization is obtained by iterating the general update rule (cf. [7])

$$q(\mathcal{Z}_i) \leftarrow \frac{1}{C_{\mathcal{Z}_i}} \exp \left(\langle \ln p(\mathbf{Y}, \mathcal{Z}) \rangle_{q(\mathcal{Z}_{\setminus i})} \right) \tag{3}$$

where index $\mathcal{Z}_i$ indicates the individual variable of $\mathcal{Z}$, $\setminus i$ denotes those omitting $\mathcal{Z}_i$, and $C_{\mathcal{Z}_i}$ is a normalization constant. Here, we show the specific update rules related to doubly ARD;

$$q(\mathbf{V}) \leftarrow \prod_i^D \mathcal{N}(v_i|m_v^{(i)}, \sigma_v^{(i)2}), \quad q(\mathbf{W}) \leftarrow \prod_i^D \mathcal{N}(\tilde{\mathbf{w}}_i|m_w^{(i)}, \Sigma_w^{(i)})$$

$$q(\alpha_W) \leftarrow \prod_{i=1}^M \mathcal{G}(\alpha_{wi}|\tilde{a}_{\alpha w}, \tilde{b}_{\alpha wi}), \quad q(\alpha_V) \leftarrow \prod_{i=1}^D \mathcal{G}(\alpha_{vi}|\tilde{a}_{\alpha v}, \tilde{b}_{\alpha vi}),$$

where $\tilde{\mathbf{w}}_i$ denotes a column vector corresponding to the i-th row of $\mathbf{W}$. Here, we have defined

$$m_v^{(i)} = \sigma_v^{(i)2} \langle \psi_i \rangle \langle w_i^T \rangle \sum_{n=1}^{N} \langle \mathbf{x}_n \rangle (y_{ni} - \langle \mu_{bi} \rangle), \quad \sigma_v^{(i)2} = \left(\langle \psi_i \rangle \mathrm{Tr}(\langle w_i w_i^T \rangle \sum_{n=1}^{N} \langle \mathbf{x}_n \mathbf{x}_n^T \rangle) + \langle \alpha_{vi} \rangle \right)^{-1},$$

$$m_w^{(i)} = \langle \psi_i \rangle \langle v_i \rangle \Sigma_w^{(i)} \sum_{n=1}^{N} (\mathbf{y}_n - \langle \mu_{bi} \rangle) \langle \mathbf{x}_n \rangle, \quad \Sigma_w^{(i)} = \left(\langle v_i^2 \rangle \langle \psi_i \rangle \sum_{n=1}^{N} \langle \mathbf{x}_n \mathbf{x}_n^T \rangle + \mathrm{diag} \langle \alpha_W \rangle \right)^{-1},$$

and

$$\tilde{a}_{\alpha v} = a_{\alpha v0} + \frac{1}{2}, \quad \tilde{b}_{\alpha vi} = b_{\alpha v0} + \frac{\langle v_i^2 \rangle}{2}, \quad \tilde{a}_{\alpha w} = a_{\alpha w0} + \frac{D}{2}, \quad \tilde{b}_{\alpha wi} = b_{\alpha w0} + \frac{\langle ||\mathbf{w}_i||^2 \rangle}{2},$$

where T denotes the transposition.

3. Related work

Our model is closely related to the independent factor analysis (IFA) [8]. The IFA models each latent factor by single dimensional MoG, but our method models all dimensions of the factor model by a single, but correlated, MoG model. The purpose of IFA was to find factors that were mutually independent, but our aim with this method is to find a low-dimensional manifold to represent the entire data structure, and we do not want any dimensions of the factor to be independent. Choudrey and Roberts [9] proposed a mixture of independent component analyzers and formulated variational Bayesian inference for the models. They also imposed an ARD prior on the factor loading matrix, but only in a column-wise manner (hence, not for feature selection).

Our model includes a standard generative model for high dimensionality data. If we restrict the matrix, $\mathbf{V}$, to a unit matrix and the factor model to a unit Gaussian distribution, we can recover FA (and its variational Baysian variant). If we further restrict the noise model to an isotropic Gaussian, we can recover the probabilistic PCA (PPCA) [10].

Mixtures of principal component analyzers [11] or mixtures of factor analyzers [12] are powerful methods of modeling density in high-dimensional data. Each component in these methods has individual factor loadings so that different components lie in different subspaces. Although this is a reasonable assumption for density modeling, it can be difficult to look at the overall data structure solely based on one single figure. Our method restricts the latent data to be on a single subspace, hence we can readily capture the entire structure of the data.

4. Demonstration on synthetic data

Here we demonstrate the performance of DSF on two synthetic data sets to illustrate how our method works.

4.1. Single Gaussian factor model

First, we used a single Gaussian factor model to clarify how it differs from related methods. We generated 200 samples from a zero mean and unit covariance Gaussian with a dimension of $M = 3$. We projected these data into four-dimension features by multiplying the factor loading matrix, $W = [e_1, e_2, e_3]$, where $e_1 = (4, 4, 4, 4)^T$, $e_2 = (3, 3, -3, -3)^T$, $e_3 = (2, -2, 2, -2)^T$. Four additive features were generated independently from the zero-mean Gaussian distribution having standard deviations of (4, 3, 2, 1). As these features have no correlation with other elements, we want them to be absorbed into the noise model.

Figure 2 compares a Hinton diagram of the factor loading matrix for different methods. For DSF, we used the effective factor loading, $\langle \mathbf{V} \rangle \langle \mathbf{W} \rangle$. The principal components of PCA

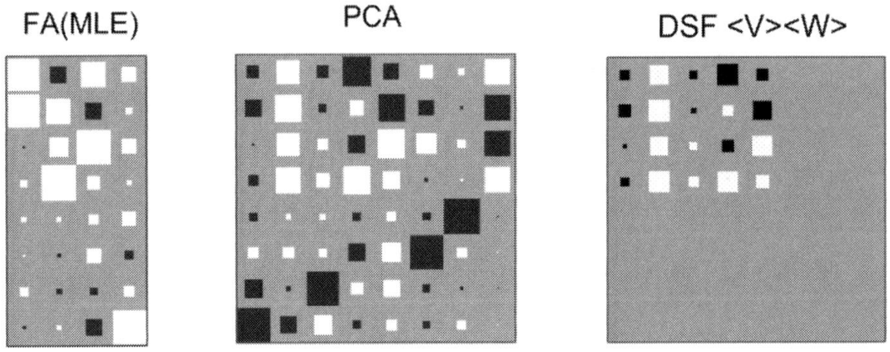

Figure 2: Hinton diagrams of factor loading matrices for different methods fitted to a synthesized data set generated from a single Gaussian factor model (M=3). Four dimensions, which correspond to the bottom four rows, were generated from an independent Gaussian distribution.

correspond to the direction with largest variance; hence, it is largely affected by irrelevant features. As the bottom four features had comparable variance with interesting features, the four columns of $\mathbf{W}$ (first, third, sixth, seventh) found the direction with the bottom four directions. As FA with maximum likelihood estimation (MLE) absorbs the bottom four directions into the noise model, the corresponding lows of factor loading took a small value. However, due to the small sample size and overfitting of MLE, they deviated from zero. This overfitting would reduce predictive ability. For DSF, we can see that the bottom four features are discarded by ARD. In addition, three dimensions of the latent factor were discarded. The number of remaining dimensions did not match exactly that of the generative model in this example. There were nearly parallel columns (first and second dimensions and third and fifth dimensions).

4.2. MoG factor model

Next, we demonstrate the performance of our model in a clustering problem. To show how our method seeks a low-dimensional factor model that is suitable for clustering, let us consider a simple case where there are originally three dimensions, but one factor dimension is sufficient for clustering. The data were generated from a mixture of Gaussians. Figure 3 shows the scatter plots for the data. As the feature y_3 obeyed a single Gaussian, it was irrelevant to clustering. We wish to discard this feature. The features, y_1 and y_2, exhibit a bimodal distribution and hence provide useful information for clustering. However, we can further reduce the number of effective dimensions for clustering by constructing a new variable, $y' = y_1 + y_2$ (broken line). The Hinton diagram in Figure 3B shows that our DSF achieved this. Figure 3C represents the posterior means of the factor model, $\langle \mathbf{x} \rangle$. We can see that $\langle \mathbf{x} \rangle$ is distributed in only the $\langle x_1 \rangle$ direction. The ellipses represent the contours of Gaussian components.

5. Demonstration on Real Data

5.1. Wine Data

Next, we demonstrate the performance of DSF on a wine data set extracted from the UCI machine-learning repository. The wine data had three labels for wine, and thirteen features and 178 samples. We did not use the class labels during the inference procedure. We used an MoG with $K = 3$ (equal to the number of wine labels) components for the factor model. As all the features in original data set had information for clustering (as we confirmed by plotting the distributions of all features), no features were eliminated by ARD for the original data set. These results clearly differed from those of Dy and Brodley [2], in which the number of selected features was seven (with the trace criterion) or three (with the likelihood criterion). This difference arises

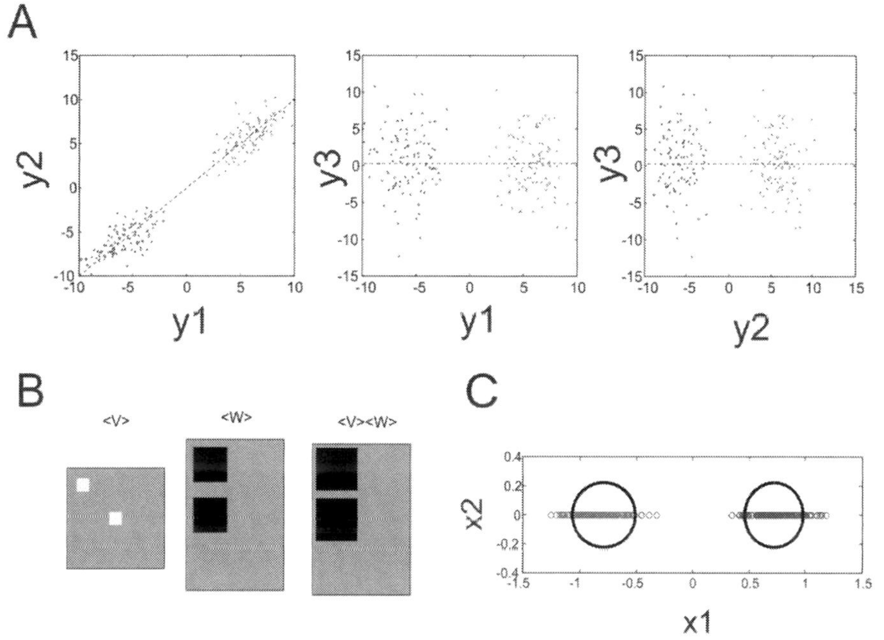

Figure 3: Results from fitting model to synthetic data described in Section 4.2. (A) Scatter plots of synthesized data. (B) Hinton diagrams of $\langle \mathbf{V} \rangle$, $\langle \mathbf{W} \rangle$, and $\langle \mathbf{V} \rangle \langle \mathbf{W} \rangle$. (C) Scatter plot of $\langle \mathbf{X} \rangle$.

from the goal of the model fitting. In contrast to Dy and Bordley's method, which treats only a subset of the features, our model tries to model the entire data distribution, while "irrelevant" features are absorbed into the noise model. Thus, if the class label significantly affected the distribution of each feature, the feature was not discarded. For major data sets, the features that are thought to relate to data classification are collected. However, there are many cases where some features provide no information for clustering. To simulate such cases, we shuffled the order of D_s randomly chosen features in the wine data set (here, D_s denotes the number of shuffled features). This procedure reserves the distribution of each feature, but destroys the correlation with other features.

The Hinton diagram for this shuffled data ($D_s = 4$) is shown in Figure 4A. While the model without ARD had finite values at all elements, our model with row-wise ARD (on $\mathbf{V}$) and column-wise ARD (on $\mathbf{W}$) eliminated six rows (including the shuffled four features) and five columns. The model with only column-wise ARD reduced the effective dimension of the latent variable, but did not remove the irrelevant features. Figure 4B shows how this differences affects the quality of clustering. Figure 4B compares the clustering label error of various versions of our model including (1) without ARD, (2) only column-wise ARD, (3) only row-wise ARD, and (4) no reduced dimensionality, i.e., the VB-MoG is directly applied to all data dimensions. We can see that DSF with both ARDs outperformed the others, suggesting that including feature selection (through row-wise ARD) and eliminating latent dimensions (through column-wise ARD) improve performance.

Based on the factor distribution $\langle \mathbf{x}_n \rangle$ we can visualize the entire structure of the data as shown in Figure 4C. We see that the model without ARD or with ARD only on $\mathbf{W}$ used only two Gaussian components. This affects the correct clustering rate in Figure 4B. We should note that these scatter plots represent a posterior mode of each $\mathbf{x}_n$, and are not a subspace of the

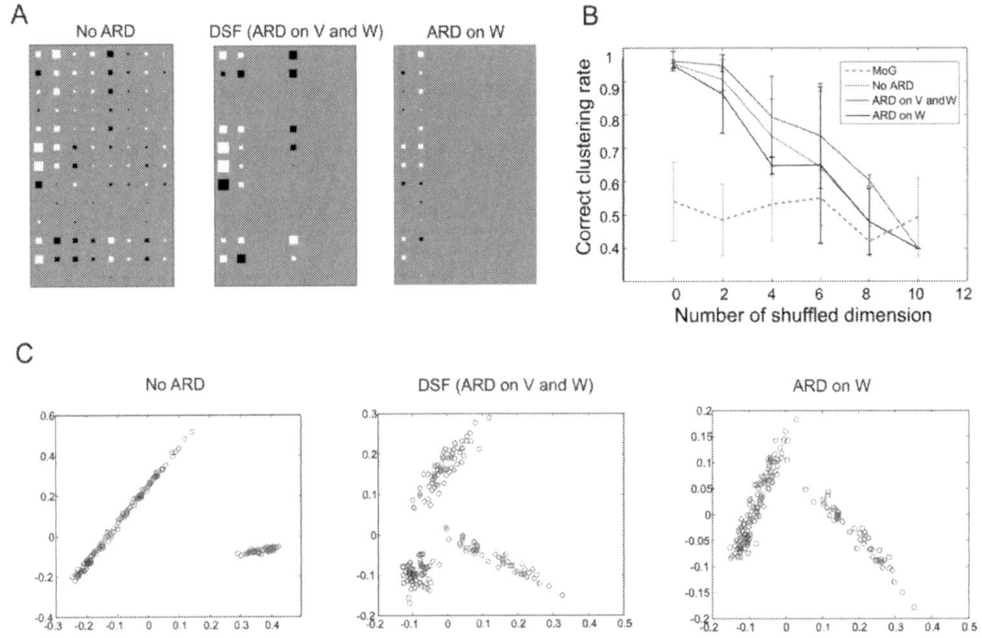

Figure 4: Results of model fitting to wine data. (A) Hinton diagram of $\langle \mathbf{V} \rangle \langle \mathbf{W} \rangle$ of different models. (B) Correct classification rates for varying the number of shuffled dimension. The error bars indicate standard deviation over five different choices of shuffled features. (C) The scatter plot of $\langle \mathbf{X} \rangle$ for the data with the number of shuffled dimension $D_s = 6$. Colors represent the most likely component label.

original space, i.e., it was largely affected by the factor model. Furthermore, the effective factor loading matrix $\mathbf{VW}$ is not guaranteed to be orthonormalized; hence, the scale of the factor dimension in the original data space can differ, and different dimensions can mutually correlate with each other even if they are independent in the original data space. If we wish to eliminate the disadvantages, we can use the orthonormalization of $\langle \mathbf{V} \rangle \langle \mathbf{W} \rangle$ by using the singular value decomposition [5], although it is not examined here.

5.2. Neural data from inferior temporal cortex

Next we applied the proposed method to neural activity data collected from the inferior temporal cortex of two macaque monkeys (Macaca fuscata) [13]. Single-unit activities were recorded while the monkeys performed a fixation task. Visual stimuli were 12 human faces of three models with four expressions (neutral, happy, surprised, angry), 16 monkey faces of four models with four expressions (neutral, pout-lips, full-open mouthed, mid-open mouthed) and 10 simple geometric shapes. We used the single-unit activities of 45 neurons that were tested using all 38 stimulus sets. We calculated a population activity vector for each stimulus by averaging the spike counts over a number of trials and by smoothing using a Gaussian filter with a variance of 10 ms [14].

We used MoG as a factor model for clustering the population activity vectors. The number of Gaussian components were set from $K = 2$ to $K = 7$. We selected the number of Gaussian components that gave the highest variational free energy (lower bound on log-marginal likelihood). The results obtained in the time window [140 ms, 190 ms] are shown in Figure 5. $K = 4$ was selected for this time window. Figure 5A shows the Hinton diagram of the posterior mean for the effective factor loading matrix $\mathbf{VW}$. Of the 45 rows, values of all the elements

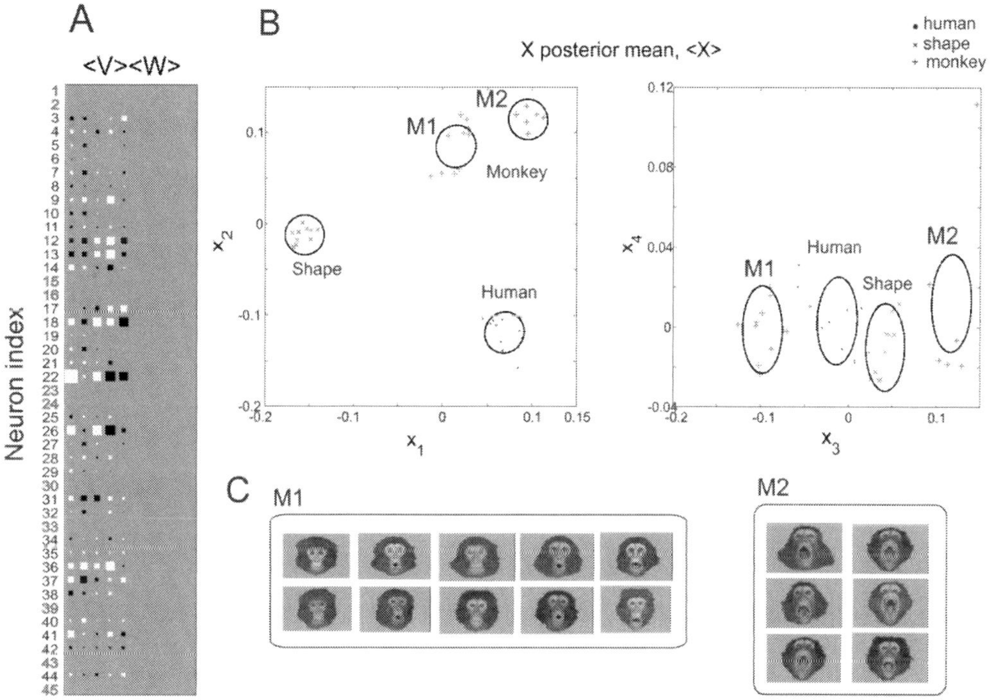

Figure 5: Results of model fitting to spike data from the inferior temporal cortex. (A) Hinton diagram of $\langle \mathbf{V} \rangle \langle \mathbf{W} \rangle$. The indices of eliminated neurons are shown in red. (B) The scatter plot of posterior mean, $\langle \mathbf{X} \rangle$. Colors indicate the most likely component label. The ellipses represent the one standard deviation contour of Gaussian components.

were close to zero for 11 rows, showing that the data of 11 neurons were effectively eliminated. Five columns remained, showing that the number of used dimensions of latent space was five. The posterior mean for $\mathbf{x}$ and alignment of Gaussian components are shown in Figure 5B. The population activity vectors were separated into clusters corresponding to the global category, i.e., human faces, monkey faces, and shapes. Two sub-clusters (denoted as "M1" and "M2") were assigned for the monkey faces. The members of M1 and M2 are shown in Figure 5C. Sub-cluster M2 includes all four full open-mouthed faces and two mid-open-mouthed faces. M1 includes the remaining monkey faces. We compared the clustering results with the results reported in Matsumoto et al. [14]. The same cluster as M2 was observed. However, M1 was further separated into two clusters, human faces were separated into two clusters, and simple shapes were separated into two clusters also [Fig. 6 in 14]. Our algorithm tends to select the model with fewer components partly because we used an independent noise term, ϵ, that compensates the decrease in likelihood with fewer number of clusters, and because there is a penalty for increases in the number of model parameters.

Our results suggest that 34 of the 45 analyzed neurons and five latent dimensions are sufficient in categorizing the neural responses according to the global category and one fine category, i.e. monkey facial expressions. We should emphasize that our algorithm automatically set these values in a single model fitting process.

6. Discussion

In this study, we proposed a novel use for ARD to combine the selection and transformation of features within the context of unsupervised learning. Our method can extract "interesting" structures from multi-dimensional data while discarding irrelevant features. "Irrelevant" features within our framework are defined as those that can be regarded as having been generated from unimodal Gaussian noise independent of other features. Their "interestingness" is determined by the choice of the factor model. Our framework is general in the sense that we can replace the factor model with various generative models. Although we have demonstrated the use of finite mixture models for the factor model, we can instead use non-parametric Bayes models such as the Dirichlet mixture model. Variational inference for the combined model would be possible [15, 16].

As is typical for VB, our model often suffers from slow convergence due to the coupling of parameters. Combining Parameter eXpanded-VB (PX-VB) [17] with our model would help accelerate convergence. Our framework is restricted to cases where the data can be seen as a linear transformation of the low-dimensional manifold. Furthermore, we considered data with real values. Our future work is to extend our approach to non-negative or discrete data with exponential family distributions (c.f., [18]) and to incorporate non-linear transformation with kernel methods.

Acknowledgments

We thank Toshiyuki Tanaka and Yoshiyuki Kabashima for helpful discussions. This work was partially supported by Grants-in-Aid No. 20240020, 20650019, 18079003, 18079008, 20300096, and 21119528 from the Ministry of Education, Culture, Sports, Science, and Technology of the Japanese Government.

References

[1] Kohavi R and John G 1997 *Artif. Intell.* **97** 273–324
[2] Dy J G and Brodley C E 2004 *J. Mach. Learn. Res.* **5** 845–889
[3] Bishop C 1999 *Adv. Neural Inf. Process. Syst.* **11** 382–388
[4] Bishop C 1999 *Proc. Int. Conf. on Artificial Neural Networks* vol 1 (Conf. Publ. No. 470)
[5] Yu B M, Cunningham J P, Santhanam G, Ryu S I, Shenoy K V and Sahani M 2009 *Adv. Neural Inf. Process. Syst.* **21** 1881–1888
[6] Archambeau C and Bach F 2009 *Adv. Neural Inf. Process. Syst.* **21** 73–80
[7] Bishop C 2006 *Pattern Recognition and Machine Learning* (Springer New York.)
[8] Attias H 1999 *Neural Comput.* **11** 803–851
[9] Choudrey R and Roberts S 2003 *Neural Comput.* **15** 213–252
[10] Tipping M and Bishop C 1999 *J. Roy. Stat. Soc.* B **61** 611–622
[11] Tipping M and Bishop C 1999 *Neural Comput.* **11** 443–482
[12] Ghahramani Z and Hinton G 1997 *University of Toronto Technical Report CRG-TR-96-1*
[13] Sugase Y, Yamane S, Ueno S and Kawano K 1999 *Nature* **400** 869–873
[14] Matsumoto N, Okada M, Sugase-Miyamoto Y, Yamane S and Kawano K 2005 *Cerebral Cortex* **15** 1103
[15] Blei D and Jordan M 2006 *Bayesian Analysis* **1** 121–144
[16] Kurihara K, Welling M and Vlassis N 2007 *Adv. Neural Inf. Process. Syst.* **19** 761–768
[17] Qi Y A and Jaakkola T S 2007 *Adv. Neural Inf. Process. Syst.* **19** 1097–1104
[18] Mohamed S, Heller K and Ghahramani Z 2009 *Adv. Neural Inf. Process. Syst.* **21** 1089–1096

International Workshop on Statistical-Mechanical Informatics 2010 IOP Publishing
Journal of Physics: Conference Series **233** (2010) 012022 doi:10.1088/1742-6596/233/1/012022

A haplotype inference method based on sparsely connected multi-body Ising model

Masashi Kato, Qian Ji Gao, Hiroshi Chigira, Hiroyuki Shindo and Masato Inoue

Department of Electrical Engineering and Bioscience, (Graduate) School of Advanced Science and Engineering, Waseda University, 3–4–1, Okubo, Shinjuku-ku, Tokyo 169–8555, Japan

E-mail: masato.inoue@eb.waseda.ac.jp

Abstract. Statistical haplotype inference is an indispensable technique in the field of medical science. The method usually has two steps: inference of haplotype frequencies and inference of diplotype for each subject. The first step can be done by using the expectation-maximization (EM) algorithm, but it incurs an unreasonably large calculation cost when the number of single-nucleotide polymorphism (SNP) loci of concern is large. In this article, we describe an approximate probabilistic model of haplotype frequencies. The model is constructed by using several distributions of nearby local SNPs. This approximation seems good because SNPs are generally more strongly correlated when they are close to one another on a chromosome. To implement this approach, we use a log linear model, the Walsh-Hadamard transform, and a combinatorial optimization method. Artificial data suggested that the overall haplotype inference of our method is good if there are nine or more local consecutive SNPs. Some minor problems should be dealt with before this method can be applied to real data.

1. Introduction

The human genome project [1, 2] revealed typical DNA sequences of the human genome. Its successor, the human haplotype map project [3], is revealing the genetic variation of the human genome for each haplotype block. Such an understanding is important for medical science because genetic variations directly or indirectly affect many diseases. The type of each haplotype block can be determined from a specific combination of alleles. The most common type of human genetic variation is the single-nucleotide polymorphism (SNP), which exists in about 0.1% bases of the human genome. Genome-wide gene-disease association studies [4] are also revealing the relationship between genetic variation and common diseases. The outcomes of such studies will greatly contribute to the development of individualized medicine.

Statistical haplotype inference is an indispensable technique for individualized medicine. However, the haplotype inference problem arises because we cannot easily distinguish which of the two alleles at a specific polymorphic locus originated from the father or the mother. Thus, if we want to determine the type of a certain haplotype block including several polymorphic loci, we have to infer two combinations (phases) of alleles. Haplotype inference is generally possible if we have enough data from genetically closely related individuals.

Various haplotype inference methods have been developed [5, 6, 7, 8, 9, 10, 11, 12], and these are mainly based on SNP data. Haplotype inference usually involves two steps. The first step infers the haplotype frequencies in the population. The second step infers the diplotype (combination of two haplotypes) for each subject by using the haplotype frequencies inferred in the first step. The expectation-maximization (EM) algorithm is often used in the first step to infer the maximum likelihood (ML) estimator or the maximum a posteriori (MAP) solution of parameters based on a certain genetic model. However, the EM algorithm suffers from an unreasonably large calculation cost, because the number of haplotypes of concern grows as $\mathcal{O}(2^N)$, where N denotes the number of SNP loci. Numerous approximations have been devised to overcome this difficulty.

In this manuscript, we introduce an approximate model of haplotype frequencies. Specifically, we consider local clusters of SNP loci that are nearby each other on a certain chromosome. If we restrict the size of each cluster to less than N, say K, we can guarantee an upper limit to the calculation cost. After the EM algorithm determines the haplotype frequencies for each cluster, we combine those distributions into one distribution. The idea behind this method is straightforward, and the method combines common techniques of statistical mechanics and informatics, i.e., Ising model, log linear model, multi-body interactions, fast Walsh-Hadamard transform, and combinatorial optimization.

2. Model

2.1. Hardy-Weinberg equilibrium model

We assume a Hardy-Weinberg equilibrium model for a given multilocus genotype data set. This model is equivalent to the following formulation. We define an N-dimensional binary random vector $\boldsymbol{x} \equiv [x_1, x_2, ..., x_N] \in \{+1, -1\}^N$ (Fig. 1), where N denotes the number of SNP loci. Each dimension of $\boldsymbol{x}$ specifies the allele (major/minor) of the SNP, and the configuration of $\boldsymbol{x}$ specifies one haplotype. Hereafter, we call each x_l a spin. This random vector $\boldsymbol{x}$ is assumed to follow a distribution,

$$P(\boldsymbol{x}|\boldsymbol{\theta}) \equiv \theta_{\boldsymbol{x}}, \quad \sum_{\boldsymbol{x}} \theta_{\boldsymbol{x}} = 1, \quad \theta_{\boldsymbol{x}} \geq 0, \tag{1}$$

where $\boldsymbol{\theta} \equiv [\theta_{+1,+1,...,+1}, \theta_{-1,+1,..,+1}, ..., \theta_{-1,-1,...,-1}] \in \mathcal{R}^{2^N}$ denotes the haplotype frequencies in the population. The degree of freedom of $\boldsymbol{\theta}$ is $2^N - 1$ since the dimensionality of $\boldsymbol{\theta}$ is 2^N, and we have one normalization restriction. This distribution $P(\boldsymbol{x}|\boldsymbol{\theta})$ can represent any distribution of $\boldsymbol{x}$, although it is parameterized. We independently sample $\boldsymbol{x}$ from this distribution $2I$ times and obtain $\boldsymbol{X} \equiv [\boldsymbol{x}^{(1,1)}, \boldsymbol{x}^{(1,2)}, ..., \boldsymbol{x}^{(I,1)}, \boldsymbol{x}^{(I,2)}]$, where I denotes the number of subjects, and the combination of $\boldsymbol{x}^{(i,1)}$ and $\boldsymbol{x}^{(i,2)}$ denotes the diplotype of the i-th subject for $i = 1, 2, ..., I$. An N-dimensional ternary random vector $\boldsymbol{y} \equiv [y_1, y_2, ..., y_N] \in \{+1, 0, -1\}^N$ is used to represent the multilocus genotype data. The multilocus genotype data of the i-th subject are defined using the diplotype:

$$\boldsymbol{y}^{(i)} \equiv \frac{1}{2}\left(\boldsymbol{x}^{(i,1)} + \boldsymbol{x}^{(i,2)}\right). \tag{2}$$

We also use $\boldsymbol{Y} \equiv [\boldsymbol{y}^{(1)}, .., \boldsymbol{y}^{(I)}]$. Note that we have assumed that the observation of the multilocus genotype data is performed noiselessly.

The joint distribution of $\boldsymbol{X}$ and $\boldsymbol{Y}$ can be expressed as

$$P(\boldsymbol{X}, \boldsymbol{Y}|\boldsymbol{\theta}) = \prod_{i=1}^{I} \delta_{\boldsymbol{y}^{(i)} = \frac{1}{2}(\boldsymbol{x}^{(i,1)} + \boldsymbol{x}^{(i,2)})} \theta_{\boldsymbol{x}^{(i,1)}} \theta_{\boldsymbol{x}^{(i,2)}} \tag{3}$$

where δ_c denotes the indicator function that returns 1 if the given condition c is true or 0 otherwise.

2.2. Haplotype inference

A common way of haplotype inference is two-step inference. First, we determine the most probable haplotype frequencies by maximum likelihood estimation,

$$\boldsymbol{\theta}^* \equiv \arg\max_{\boldsymbol{\theta}} P(\boldsymbol{Y}|\boldsymbol{\theta}), \tag{4}$$

where $P(\boldsymbol{Y}|\boldsymbol{\theta})$ is the marginal distribution of $P(\boldsymbol{X}, \boldsymbol{Y}|\boldsymbol{\theta})$. The $\boldsymbol{\theta}^*$ can be usually determined by using the EM algorithm. Nevertheless, the EM algorithm is not applicable if N is large because the dimensionality of $\boldsymbol{\theta}$ grows as $\mathcal{O}(2^N)$. Empirically, the maximum likelihood estimator $\boldsymbol{\theta}^*$ is quite sparse, i.e., the number of non-zero elements of $\boldsymbol{\theta}^*$ is at most $2I$. Thus, we can reduce the calculation cost if we use an efficient algorithm such as GrEM [12]. In this study, we reduce the calculation cost to within an acceptable value by using an approximation.

Second, we determine the most probable diplotypes that have the maximum probability:

$$\boldsymbol{X}^* \equiv \arg\max_{\boldsymbol{X}} P(\boldsymbol{X}|\boldsymbol{Y}, \boldsymbol{\theta}^*). \tag{5}$$

This step is not hard if the effective dimensionality of $\boldsymbol{\theta}^*$ is small.

2.3. Kth-order nearest-neighbor Ising model

Here, we consider a useful distribution that approximates $P(\boldsymbol{x}|\boldsymbol{\theta}^*)$ with very few parameters compared with 2^N. Because we are dealing with SNP data on a chromosome, the distribution of alleles at a certain SNP locus is considerably dependent on those at nearby SNP loci, but almost independent of those at far SNP loci. Accordingly, we can construct a model by merging several 'local' distributions.

Specifically, we consider all sets of consecutive K $(< N)$ spins, e.g., $\boldsymbol{\nu} \equiv \{1,2,3\}, \{2,3,4\}, \{3,4,5\}, ..., \{N-2, N-1, N\}, \{N-1, N, 1\}, \{N, 1, 2\}$ for the case of $K = 3$, where $\boldsymbol{\nu}$ denotes a set of ordinal spin numbers. A loop formation, e.g., $\{N-1, N, 1\}$, is employed in order to make the cardinalities of the spins the same although the loop formation is not a good model for sequential SNP data on a chromosome. We focus on the spins specified by $\boldsymbol{\nu}$ and the simultaneous distribution $P_{\boldsymbol{\nu}}(\boldsymbol{X}_{\boldsymbol{\nu}}, \boldsymbol{Y}_{\boldsymbol{\nu}}|\boldsymbol{\theta}_{\boldsymbol{\nu}})$.

The EM algorithm can determine the optimal haplotype frequencies for each local distribution. Recall that when N is large, the algorithm suffers from the large dimensionality of $\boldsymbol{\theta}$, 2^N, but if we can reduce the dimensionality of $\boldsymbol{x}$ from N to K, and K is not large, we can use the EM algorithm to infer the maximum likelihood estimator,

$$\hat{\boldsymbol{\theta}}_{\boldsymbol{\nu}} \equiv \arg\max_{\boldsymbol{\theta}_{\boldsymbol{\nu}}} P_{\boldsymbol{\nu}}(\boldsymbol{Y}_{\boldsymbol{\nu}}|\boldsymbol{\theta}_{\boldsymbol{\nu}}), \tag{6}$$

with high probability. Note that $P_{\boldsymbol{\nu}}(\boldsymbol{x}_{\boldsymbol{\nu}}|\hat{\boldsymbol{\theta}}_{\boldsymbol{\nu}})$ is not the marginal distribution of $P(\boldsymbol{x}|\boldsymbol{\theta}^*)$.

Next, we define the approximate distribution of $P(\boldsymbol{x}|\boldsymbol{\theta}^*)$, as

$$P(\boldsymbol{x}|\hat{\boldsymbol{\theta}}) \propto \sqrt[K]{\prod_{\boldsymbol{\nu}} P_{\boldsymbol{\nu}}(\boldsymbol{x}_{\boldsymbol{\nu}}|\hat{\boldsymbol{\theta}}_{\boldsymbol{\nu}})}, \tag{7}$$

(Fig. 2) where the proportion symbol means that normalization is required. The root of the K-th power is taken because each dimension of $\boldsymbol{x}$ is multiplied K times. This root is necessary for making a good approximation of $P(\boldsymbol{x}|\boldsymbol{\theta}^*)$, but has no effect on the final result of $\hat{\boldsymbol{X}}$ in Eq. (8).

Finally, we determine the most probable diplotypes by using $\hat{\boldsymbol{\theta}}$,

$$\hat{\boldsymbol{X}} \equiv \arg\max_{\boldsymbol{X}} P(\boldsymbol{X}|\boldsymbol{Y}, \hat{\boldsymbol{\theta}}). \tag{8}$$

Equations (7) and (8) look difficult, but the log linear model that we introduce later enables both equations to be managed. To be more exact, we use $\hat{\boldsymbol{J}}$ instead of $\hat{\boldsymbol{\theta}}$, where $\hat{\boldsymbol{J}}$ is the parameter of the log linear model. The dimensionality of $\hat{\boldsymbol{J}}$ is also 2^N, but the effective dimensionality should be quite small.

3. Method

3.1. EM algorithm

Equations (4) and (6) can be approximately and numerically solved by using the EM algorithm if the dimensionality of $\boldsymbol{\theta}$ is small. Here, we shall explain the EM algorithm only for the case of Eq. (4); its extension to the case of Eq. (6) is easy. The EM algorithm starts with an arbitrarily chosen parameter value $\boldsymbol{\theta}^{(0)}$ and updates it until it converges.

$$\boldsymbol{\theta}^{(t+1)} \equiv \arg\max_{\boldsymbol{\theta}} \langle \ln P(\boldsymbol{X}, \boldsymbol{Y}|\boldsymbol{\theta})\rangle_{P(\boldsymbol{X}|\boldsymbol{Y},\boldsymbol{\theta}^{(t)})}\,. \tag{9}$$

This update equation can be solved analytically, resulting in

$$\theta_{\boldsymbol{x}}^{(t+1)} = \frac{1}{2I}\sum_{i=1}^{I} \frac{\sum_{\boldsymbol{x}',\boldsymbol{x}''} \delta_{\boldsymbol{y}^{(i)}=\frac{1}{2}(\boldsymbol{x}'+\boldsymbol{x}'')}\theta_{\boldsymbol{x}'}^{(t)}\theta_{\boldsymbol{x}''}^{(t)}(\delta_{\boldsymbol{x}'=\boldsymbol{x}}+\delta_{\boldsymbol{x}''=\boldsymbol{x}})}{\sum_{\boldsymbol{x}',\boldsymbol{x}''} \delta_{\boldsymbol{y}^{(i)}=\frac{1}{2}(\boldsymbol{x}'+\boldsymbol{x}'')}\theta_{\boldsymbol{x}'}^{(t)}\theta_{\boldsymbol{x}''}^{(t)}}. \tag{10}$$

The EM algorithm guarantees that the parameter converges to one of the extremals of the likelihood function, in this case, $P(\boldsymbol{Y}|\boldsymbol{\theta})$, and usually this value coincides with one of the local maxima. Several trials with different initial values should be performed to obtain one of the global maxima.

3.2. Log linear model

Almost all of the binary vector distributions can be exactly expressed with a log linear model. Here, we shall only explain the expression for $P(\boldsymbol{x}|\boldsymbol{\theta})$; it is easy to extend it to the case of $P_\nu(\boldsymbol{x}_\nu|\boldsymbol{\theta}_\nu)$. By using the parameter $\boldsymbol{J} \in \mathcal{R}^{2^N}$, we can express $P(\boldsymbol{x}|\boldsymbol{\theta})$ in the form

$$Q(\boldsymbol{x}|\boldsymbol{J}) \equiv \exp\left(\sum_{\mu\subseteq\{1,2,..,N\}} J_\mu \prod_{l\in\mu} x_l\right). \tag{11}$$

For example, in the case of $N = 3$, $\boldsymbol{J} = [J_0, J_1, J_2, J_{1,2}, J_3, J_{1,3}, J_{2,3}, J_{1,2,3}]$ and

$$Q(\boldsymbol{x}|\boldsymbol{J}) = \exp\left(J_0 + J_1x_1 + J_2x_2 + J_3x_3 + J_{1,2}x_1x_2 + J_{1,3}x_1x_3 + J_{2,3}x_2x_3 + J_{1,2,3}x_1x_2x_3\right), \tag{12}$$

where we have used a simplified expression, e.g., $J_{1,2}$ instead of $J_{\{1,2\}}$. We also used J_0 instead of $J_{\{\}}$, for convenience. J_0 acts as the normalization term. The degree of freedom of $\boldsymbol{J}$ is $2^N - 1$, which is the same as that of $\boldsymbol{\theta}$. The log linear expression $Q(\boldsymbol{x}|\boldsymbol{J})$ is generally equivalent to the ordinal expression $P(\boldsymbol{x}|\boldsymbol{\theta})$ except that the former cannot express any distributions including zero probability ($\theta_{\boldsymbol{x}} = 0 \;\; \exists\boldsymbol{x}$); such cases are measure zero, and we can avoid them by setting their probability to be a small positive.

3.3. Walsh-Hadamard transform

The parameter $\boldsymbol{J}$ of the equivalent distribution $Q(\boldsymbol{x}|\boldsymbol{J})$ is easy to determine if we know the parameter $\boldsymbol{\theta}$ of $P(\boldsymbol{x}|\boldsymbol{\theta})$. Basically, we have 2^N simultaneous linear equations, $Q(\boldsymbol{x}|\boldsymbol{J}) = P(\boldsymbol{x}|\boldsymbol{\theta})$.

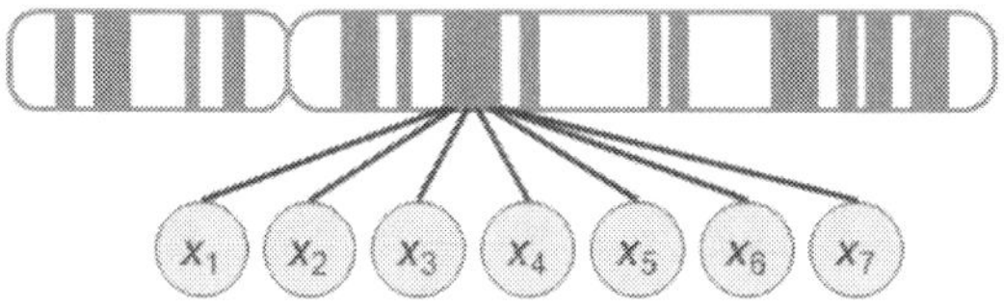

Figure 1. Schema of locations of $x_1, .., x_N$ on a chromosome. We consider N SNP loci on the same chromosome; they are numbered sequentially $x_1, .., x_N$.

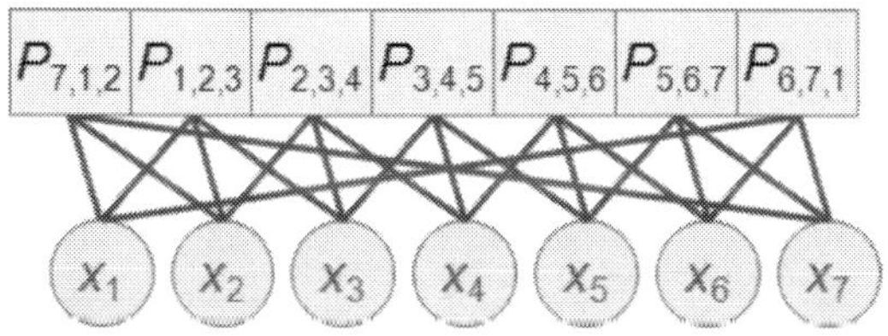

Figure 2. Factor graph expression of model $P(\boldsymbol{x}|\hat{\boldsymbol{\theta}})$ for the case of $N = 7, K = 3$. $P_{7,1,2}, ...$ denote factors $(P_{\boldsymbol{\nu}}(\boldsymbol{x}_{\boldsymbol{\nu}}|\hat{\boldsymbol{\theta}}_{\boldsymbol{\nu}}))$ of K consecutive spins. In this model, the first and last spins are assumed to form a loop in order to make the cardinality of each spin the same.

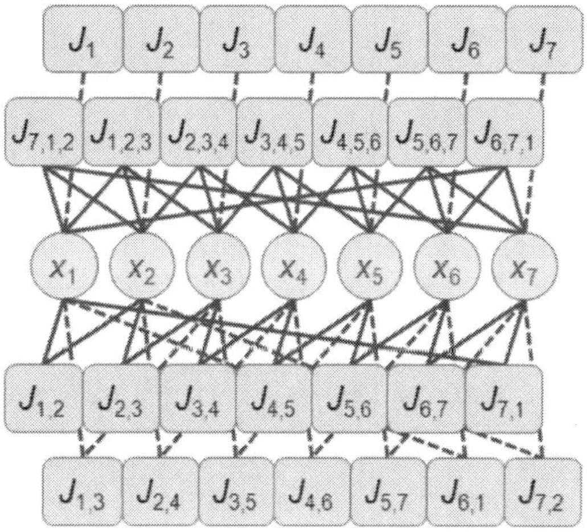

Figure 3. Factor graph expression of log linear model $Q(\boldsymbol{x}|\hat{\boldsymbol{J}})$ in the case of $N = 7, K = 3$. The number of parameters except J_0 is $N2^{K-1}$.

For example, in the case of $N = 3$, we have

$$\begin{cases} Q(+1,+1,+1|\boldsymbol{J}) = \theta_{+1,+1,+1} \\ Q(-1,+1,+1|\boldsymbol{J}) = \theta_{-1,+1,+1} \\ Q(+1,-1,+1|\boldsymbol{J}) = \theta_{+1,-1,+1} \\ Q(-1,-1,+1|\boldsymbol{J}) = \theta_{-1,-1,+1} \\ Q(+1,+1,-1|\boldsymbol{J}) = \theta_{+1,+1,-1} \\ Q(-1,+1,-1|\boldsymbol{J}) = \theta_{-1,+1,-1} \\ Q(+1,-1,-1|\boldsymbol{J}) = \theta_{+1,-1,-1} \\ Q(-1,-1,-1|\boldsymbol{J}) = \theta_{-1,-1,-1} \end{cases}, \tag{13}$$

and the solution is

$$\begin{pmatrix} J_0 \\ J_1 \\ J_2 \\ J_{1,2} \\ J_3 \\ J_{1,3} \\ J_{2,3} \\ J_{1,2,3} \end{pmatrix} = 2^{-3} \begin{pmatrix} +1 & +1 & +1 & +1 & +1 & +1 & +1 & +1 \\ +1 & -1 & +1 & -1 & +1 & -1 & +1 & -1 \\ +1 & +1 & -1 & -1 & +1 & +1 & -1 & -1 \\ +1 & -1 & -1 & +1 & +1 & -1 & -1 & +1 \\ +1 & +1 & +1 & +1 & -1 & -1 & -1 & -1 \\ +1 & -1 & +1 & -1 & -1 & +1 & -1 & +1 \\ +1 & +1 & -1 & -1 & -1 & -1 & +1 & +1 \\ +1 & -1 & -1 & +1 & -1 & +1 & +1 & -1 \end{pmatrix} \begin{pmatrix} \ln\theta_{+1,+1,+1} \\ \ln\theta_{-1,+1,+1} \\ \ln\theta_{+1,-1,+1} \\ \ln\theta_{-1,-1,+1} \\ \ln\theta_{+1,+1,-1} \\ \ln\theta_{-1,+1,-1} \\ \ln\theta_{+1,-1,-1} \\ \ln\theta_{-1,-1,-1} \end{pmatrix}. \tag{14}$$

The general solution of $\boldsymbol{J}$ is

$$\boldsymbol{J} = 2^{-N}\boldsymbol{M}\ln\boldsymbol{\theta}, \tag{15}$$

where the vectors $\boldsymbol{J}$ and $\boldsymbol{\theta}$ are treated as column vectors such as in Eq. (14), and $\ln\boldsymbol{\theta}$ denotes the element-wise natural logarithm. $\boldsymbol{M}$ denotes the Walsh-Hadamard matrix, and the element of the i-th row and j-th column of the matrix is defined as

$$(-1)^{w_{i,j}}, \qquad w_{i,j} \equiv \sum_{k=0}^{K-1} \delta_{(i-1)\&(j-1)\&2^k \neq 0}, \tag{16}$$

where & denotes a bitwise logical product. Although it is not used in this paper, the inverse transformation has a similar formulation:

$$\boldsymbol{\theta} = \exp \boldsymbol{MJ}, \tag{17}$$

where $\exp \boldsymbol{MJ}$ denotes the element-wise exponent. The calculation cost of Eqs. (15) or (17) is apparently $\mathcal{O}((2^N)^2)$, but it falls to $\mathcal{O}(2^N \log 2^N)$ if we use a fast Walsh-Hadamard transform [13].

3.4. Determination of $\hat{\boldsymbol{J}}$

So far we have explained the EM algorithm and Walsh-Hadamard transform from which we can determine $\hat{\boldsymbol{J}}_{\boldsymbol{\nu}}$ for each local distribution specified by $\boldsymbol{\nu}$ such that

$$Q(\boldsymbol{x}_{\boldsymbol{\nu}}|\hat{\boldsymbol{J}}_{\boldsymbol{\nu}}) \equiv P(\boldsymbol{x}_{\boldsymbol{\nu}}|\hat{\boldsymbol{\theta}}_{\boldsymbol{\nu}}). \tag{18}$$

The dimensionality of $\hat{\boldsymbol{\theta}}_{\boldsymbol{\nu}}$ and $\hat{\boldsymbol{J}}_{\boldsymbol{\nu}}$ is 2^K. Accordingly, we can determine $\hat{\boldsymbol{J}}$ such that

$$Q(\boldsymbol{x}|\hat{\boldsymbol{J}}) \equiv P(\boldsymbol{x}|\hat{\boldsymbol{\theta}}), \tag{19}$$

(Fig. 3) by simply calculating

$$\hat{\boldsymbol{J}} = \frac{1}{K}\sum_{\boldsymbol{\nu}} \hat{\boldsymbol{J}}'_{\boldsymbol{\nu}}, \tag{20}$$

where $\hat{\boldsymbol{J}}'_{\boldsymbol{\nu}}$ is a sparse expression of $\hat{\boldsymbol{J}}_{\boldsymbol{\nu}}$; i.e., the dimensionality of $\hat{\boldsymbol{J}}'_{\boldsymbol{\nu}}$ is 2^N, and 2^K elements of $\hat{\boldsymbol{J}}'_{\boldsymbol{\nu}}$ are specified by the corresponding values in $\hat{\boldsymbol{J}}_{\boldsymbol{\nu}}$; the other elements are zero. Only the J_0 element of $\hat{\boldsymbol{J}}$ is the exception to Eq. (20); it must be calculated as a normalizing term, but we do not use J_0 values hereafter. The dimensionality of $\hat{\boldsymbol{J}}$ is 2^N but the effective dimensionality without J_0 is $N2^{K-1}$. For example, in the case of $N = 5$ and $K = 3$, $\boldsymbol{\nu}$ is $\{1,2,3\},\{2,3,4\},\{3,4,5\},\{4,5,1\},\{5,1,2\}$, and each $Q(\boldsymbol{x}_{\boldsymbol{\nu}}|\hat{\boldsymbol{J}}_{\boldsymbol{\nu}})$ has the form

$$\begin{aligned}
Q_{1,2,3}(\boldsymbol{x}_{1,2,3}|\hat{\boldsymbol{J}}_{1,2,3}) &= \exp(J_0+J_1x_1+J_2x_2+J_{1,2}x_1x_2+J_3x_3+J_{1,3}x_1x_3+J_{2,3}x_2x_3+J_{1,2,3}x_1x_2x_3)\\
Q_{2,3,4}(\boldsymbol{x}_{2,3,4}|\hat{\boldsymbol{J}}_{2,3,4}) &= \exp(J_0+J_2x_2+J_3x_3+J_{2,3}x_2x_3+J_4x_4+J_{2,4}x_2x_4+J_{3,4}x_3x_4+J_{2,3,4}x_2x_3x_4)\\
Q_{3,4,5}(\boldsymbol{x}_{3,4,5}|\hat{\boldsymbol{J}}_{3,4,5}) &= \exp(J_0+J_3x_3+J_4x_4+J_{3,4}x_3x_4+J_5x_5+J_{3,5}x_3x_5+J_{4,5}x_4x_5+J_{3,4,5}x_3x_4x_5)\\
Q_{4,5,1}(\boldsymbol{x}_{4,5,1}|\hat{\boldsymbol{J}}_{4,5,1}) &= \exp(J_0+J_4x_4+J_5x_5+J_{4,5}x_4x_5+J_1x_1+J_{4,1}x_4x_1+J_{5,1}x_5x_1+J_{4,5,1}x_4x_5x_1)\\
Q_{5,1,2}(\boldsymbol{x}_{5,1,2}|\hat{\boldsymbol{J}}_{5,1,2}) &= \exp(J_0+J_5x_5+J_1x_1+J_{5,1}x_5x_1+J_2x_2+J_{5,2}x_5x_2+J_{1,2}x_1x_2+J_{5,1,2}x_5x_1x_2).
\end{aligned} \tag{21}$$

Consequently, $Q(\boldsymbol{x}|\hat{\boldsymbol{J}})$ has the form

$$\begin{aligned}
Q(\boldsymbol{x}|\hat{\boldsymbol{J}}) = \exp(&J_0 + J_1x_1 + J_2x_2 + J_3x_3 + J_4x_4 + J_5x_5\\
&+ J_{1,2}x_1x_2 + J_{2,3}x_2x_3 + J_{3,4}x_3x_4 + J_{4,5}x_4x_5 + J_{5,1}x_5x_1\\
&+ J_{1,3}x_1x_3 + J_{2,4}x_2x_4 + J_{3,5}x_3x_5 + J_{4,1}x_4x_1 + J_{5,2}x_5x_2\\
&+ J_{1,2,3}x_1x_2x_3 + J_{2,3,4}x_2x_3x_4 + J_{3,4,5}x_3x_4x_5 + J_{4,5,1}x_4x_5x_1 + J_{5,1,2}x_5x_1x_2).
\end{aligned} \tag{22}$$

Since the effective dimensionality of $\hat{\boldsymbol{J}}$ is small, $Q(\boldsymbol{x}|\hat{\boldsymbol{J}})$ does not generally coincide with $P(\boldsymbol{x}|\boldsymbol{\theta}^*)$ but approximates it well with a reasonably small number of non-zero parameters.

3.5. Personalized objective functions

Here, we determine the objective function in Eq. (8). If each set of data $\{\boldsymbol{x}^{(i,1)}, \boldsymbol{x}^{(i,2)}, \boldsymbol{Y}^{(i)}\}$ is independent among subjects, as defined in Eq. (3), Eq. (8) can be reduced to the form of personalized objective functions,

$$[\hat{\boldsymbol{x}}^{(i,1)}, \hat{\boldsymbol{x}}^{(i,2)}] = \arg\max_{[\boldsymbol{x}', \boldsymbol{x}'']} \delta_{\boldsymbol{y}^{(i)} = \frac{1}{2}(\boldsymbol{x}' + \boldsymbol{x}'')} \hat{\theta}_{\boldsymbol{x}'} \hat{\theta}_{\boldsymbol{x}''}. \tag{23}$$

Furthermore, if we replace $P(\boldsymbol{x}|\hat{\boldsymbol{\theta}})$ with $Q(\boldsymbol{x}|\hat{\boldsymbol{J}})$ and consider that $\boldsymbol{x}^{(i,1)}$ and $\boldsymbol{x}^{(i,2)}$ have permutation symmetry, a simplified personalized objective function can be written in the form

$$\hat{\boldsymbol{x}}^{(i,1)} = \arg\max_{\boldsymbol{x}} Q(\boldsymbol{x}|\hat{\boldsymbol{J}}^{(i)}). \tag{24}$$

$Q(\boldsymbol{x}|\hat{\boldsymbol{J}}^{(i)})$ is also a log linear model whose parameter, $\hat{\boldsymbol{J}}^{(i)}$, satisfies

$$Q(\boldsymbol{x}|\hat{\boldsymbol{J}}^{(i)}) \equiv Q(\boldsymbol{x}|\hat{\boldsymbol{J}})\; Q(2\boldsymbol{y}^{(i)} - \boldsymbol{x}|\hat{\boldsymbol{J}}). \tag{25}$$

For example, in the case of Eq. (22) and $\boldsymbol{y}^{(i)} = [+1, 0, -1, 0, +1]$, we obtain

$$Q(\boldsymbol{x}|\hat{\boldsymbol{J}}^{(i)}) \propto \exp\left(2(J_{2,4} + J_{2,3,4}) x_2 x_4\right) \tag{26}$$

where $J_{2,4}$ and $J_{2,3,4}$ are elements of $\hat{\boldsymbol{J}}$. Note that we have omitted the normalization term J_0. In this example, the $J_{2,4}$ element of $\hat{\boldsymbol{J}}^{(i)}$ can be determined as $2(J_{2,4} + J_{2,3,4})$ of $\hat{\boldsymbol{J}}$, and the other elements except J_0 are zero.

In the next subsection, we maximize Eq. (24) over $\boldsymbol{x}$, and the effective space of $\boldsymbol{x}$ is 2^h, where h denotes the number of zeros in $\boldsymbol{y}^{(i)}$. As shown by Eq. (26), odd-order terms in $\hat{\boldsymbol{J}}^{(i)}$ generally vanish. Thus, at least two optimal configurations of $\boldsymbol{x}$ must exist.

3.6. Maximization of objective function

We propose a simple maximization algorithm to determine all the configurations of $\boldsymbol{x}$ that give the maximum value of the objective function. Suppose that $N = 7, K = 3$ and the objective function $f(\boldsymbol{x})$ has the form of

$$\begin{aligned} f(\boldsymbol{x}) \equiv & \sum_{i=1}^{N} f_{i,i+1,i+2}(x_i, x_{i+1}, x_{i+2}) \\ = & f_{1,2,3}(x_1, x_2, x_3) + f_{2,3,4}(x_2, x_3, x_4) + f_{3,4,5}(x_3, x_4, x_5) + f_{4,5,6}(x_4, x_5, x_6) \\ & + f_{5,6,7}(x_5, x_6, x_7) + f_{6,7,1}(x_6, x_7, x_1) + f_{7,1,2}(x_7, x_1, x_2). \end{aligned} \tag{27}$$

The maximization problem of Eq. (24) is naturally equivalent to the maximization of this function. We maximize the argument vector $\boldsymbol{x}$ in an element-wise manner, e.g., $x_1, x_2, ..., x_N$. The order of spins to be maximized is arbitrary but it affects the calculation cost.

First, we focus on x_1 and separate the terms of $f(\boldsymbol{x})$ into two parts:

$$f(\boldsymbol{x}) = g_1(x_1; x_2, x_3, x_6, x_7) + h_1(x_4, x_5; x_2, x_3, x_6, x_7), \tag{28}$$

where

$$g_1(x_1; x_2, x_3, x_6, x_7) = f_{1,2,3} + f_{6,7,1} + f_{7,1,2} \tag{29}$$

$$h_1(x_4, x_5; x_2, x_3, x_6, x_7) = f_{2,3,4} + f_{3,4,5} + f_{4,5,6} + f_{5,6,7} \tag{30}$$

g_1 consists of the terms including the focused spin x_1. Also, g_1 is a function of x_1 having parameters of linked spins x_2, x_3, x_6, x_7. On the other hand, h_1 consists of the residual terms.

Next, we maximize g_1 with respect to x_1. Specifically, we list up all the configuration of spins x_1, x_2, x_3, x_6, x_7 and numerically calculate the value of g_1 for each configuration. Accordingly, we have 2^5 configurations with g_1 values. From these, we choose the optimal x_1 for each condition of x_2, x_3, x_6, x_7,

$$\hat{x}_1 \equiv \arg\max_{x_1} g_1(x_1; x_2, x_3, x_6, x_7), \tag{31}$$

$$\hat{g}_1 \equiv \max_{x_1} g_1(x_1; x_2, x_3, x_6, x_7). \tag{32}$$

Now, we have the following 2^4 if-then items:

$$\text{if } [x_2, x_3, x_6, x_7] = [+1, +1, +1, +1], \text{ then } x_1 = \hat{x}_1 \text{ and } g_1 = \hat{g}_1, \tag{33}$$

$$\vdots$$

Next, we maximize $f(\boldsymbol{x})$ with respect to x_2. Again, we separate the terms of $f(\boldsymbol{x})$ into two parts,

$$f(\boldsymbol{x}) = g_2(x_2; \hat{x}_1, x_3, x_4, x_6, x_7) + h_2(x_5; \hat{x}_1, x_3, x_4, x_6, x_7), \tag{34}$$

where

$$g_2(x_2; \hat{x}_1, x_3, x_4, x_6, x_7) = \hat{g}_1 + f_{2,3,4} \tag{35}$$

$$h_2(x_5; \hat{x}_1, x_3, x_4, x_6, x_7) = f_{3,4,5} + f_{4,5,6} + f_{5,6,7}. \tag{36}$$

g_2 consists of $\hat{g}_1$ and the terms including x_2. h_2 consists of the residual terms. Then, we list up all the configurations of the spins that are used by the g_2 but not the spins that have already been maximized: i.e., the list is x_2, x_3, x_4, x_6, x_7 (since x_1 is already maximized). Note that we already have 2^4 if-then items above. So it is enough for us to just add two cases, $x_4 = +1$ and $x_4 = -1$, to each item. After numerically calculating g_2 and determining the optimal x_2, we have 2^4 if-then items, e.g.,

$$\text{if } [x_3, x_4, x_6, x_7] = [+1, +1, +1, +1], \text{ then } [x_1, x_2] = [\hat{x}_1, \hat{x}_2] \text{ and } g_2 = \hat{g}_2 \tag{37}$$

Next, we maximize $f(\boldsymbol{x})$ with respect to x_3. If we proceed in the previous manner, g_3 is supposed to be $\hat{g}_2 + f_{3,4,5}$. However, we also add terms that consist of only the spins appeared in $\hat{g}_2 + f_{3,4,5}$. Since the spins of x_3, x_4, x_5, x_6, x_7 are appeared, the terms of $f_{4,5,6}$ and $f_{5,6,7}$ are also added to g_3. This extension of the definition of g_3 is also applicable to g_1 and g_2; it does not contradict the definitions of g_1 and g_2. Accordingly, we can adopt this extended definition for every g_i. As a result, we obtain

$$f(\boldsymbol{x}) = g_3(x_3; \hat{x}_1, \hat{x}_2, x_4, x_5, x_6, x_7), \tag{38}$$

where

$$g_3(x_3; \hat{x}_1, \hat{x}_2, x_4, x_5, x_6, x_7) = \hat{g}_2 + f_{3,4,5} + f_{4,5,6} + f_{5,6,7}. \tag{39}$$

Then, we add the conditions of $x_5 = +1$ and $x_5 = -1$ to each if-then item and maximize g_3 resulting 2^4 if-then items, e.g.,

$$\text{if } [x_4, x_5, x_6, x_7] = [+1, +1, +1, +1], \text{ then } [x_1, x_2, x_3] = [\hat{x}_1, \hat{x}_2, \hat{x}_3] \text{ and } g_3 = \hat{g}_3. \tag{40}$$

Next, we maximize $f(\boldsymbol{x})$ with respect to x_4. Because g_3 consists of all the terms of $f(\boldsymbol{x})$, no new terms are added to g_4. Thus, we obtain

$$f(\boldsymbol{x}) = g_4(x_4; \hat{x}_1, \hat{x}_2, \hat{x}_3, x_5, x_6, x_7), \tag{41}$$

where

$$g_4(x_4; \hat{x}_1, \hat{x}_2, \hat{x}_3, x_5, x_6, x_7) = \hat{g}_3. \tag{42}$$

Then, we maximize g_4 and obtain 2^3 if-then items, e.g.,

$$\text{if } [x_5, x_6, x_7] = [+1, +1, +1], \text{ then } [x_1, x_2, x_3, x_4] = [\hat{x}_1, \hat{x}_2, \hat{x}_3, \hat{x}_4] \text{ and } g_4 = \hat{g}_4. \tag{43}$$

In the same manner, we maximize g_5 and obtain 2^2 if-then items. Again, we maximize g_6 and obtain 2^1 if-then items. Finally, we maximize g_7 and obtain 2^0 if-then item. To be more exact, the last if-then item no longer has the if-clause,

$$\text{if (nothing), then } [x_1, x_2, x_3, x_4, x_5, x_6, x_7] = [\hat{x}_1, \hat{x}_2, \hat{x}_3, \hat{x}_4, \hat{x}_5, \hat{x}_6, \hat{x}_7] \text{ and } g_7 = \hat{g}_7. \tag{44}$$

This configuration $[\hat{x}_1, \hat{x}_2, \hat{x}_3, \hat{x}_4, \hat{x}_5, \hat{x}_6, \hat{x}_7]$ is the optimal configuration to give the maximum value of $f(\boldsymbol{x})$, and the maximum value of $f(\boldsymbol{x})$ is $\hat{g}_7$.

So far, we explained an example in which the number of the optimal configurations is only one. If the number is more than one, and we like to obtain all such configurations, we can easily extend the above algorithm to retain all the (locally) optimal if-then items at each maximization step. In the case that the number of the optimal configurations is quite large, e.g., $\mathcal{O}(2^N)$, we can avoid memory overflow by adopting the idea of a linked list to remember the spin configurations in the then-clause.

This algorithm is suboptimal compared with the junction tree algorithm [14]. One of the reasons is that we do not optimize the order of the spins to be maximized. The order affects the number of spins appeared in g_i, and the number of spins affects the calculation cost of enumerating all the spin configurations. Nevertheless, the proposed algorithm has two advantages: it is simpler than the junction tree algorithm, and it can determine all the optimal spin configurations.

3.7. *Calculation costs*

Let us summarize the calculation costs of the proposed method. We have to perform for the EM algorithm N times. In each run, $\hat{\boldsymbol{\theta}}_\nu$ costs $O(I2^K)$. Hence, the total cost of EM algorithm is $O(IN2^K)$. The fast Walsh-Hadamard transform is also performed N times. For each run, $\hat{\boldsymbol{J}}_\nu$ is determined from $\hat{\boldsymbol{\theta}}_\nu$, whose cost is $O(K2^K)$. Hence, the total cost of the fast Walsh-Hadamard transform is $O(NK2^K)$. The determination of $\hat{\boldsymbol{J}}$ from $\hat{\boldsymbol{J}}_\nu$ costs $O(N2^K)$. The determination of each $\hat{\boldsymbol{J}}^{(i)}$ costs $O(N2^{K-1})$, so the total cost of determining $\{\hat{\boldsymbol{J}}^{(i)}\}$ is $O(IN2^{K-1})$. The maximization of $Q(\boldsymbol{x}|\hat{\boldsymbol{J}}^{(i)})$ costs $O(N2^{2K-1})$ for each subject. Hence, the total cost of the maximization is $O(IN2^{2K-1})$. Consequently, the total cost of the overall method is $O(IN2^{2K-1})$, or more simply, $O(IN2^{2K})$.

4. Results

We validated the proposed method by using artificial data sets. The artificial data sets were created according to Niu et al. [8]; i.e., we generated data sets by using the Hardy-Weinberg equilibrium, single point mutation, and recombination criteria. More specifically, we randomly prepared each allele of 30 ancestors and then mated them randomly to create successive

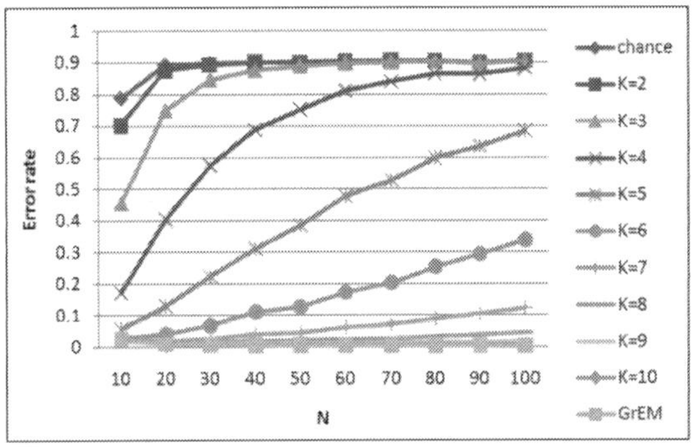

Figure 4. Average error rates of 100 subjects over 100 artificial data sets under the conditions of $N = 10, 20, .., 100$ and $K = 2, 3, .., 10$. Because some $\boldsymbol{y}^{(i)}$ contain a small number of or none of zeros, the chance level is not 1 but rather about 0.9. Error rates were almost zero when the proposed method with $K \geq 9$ or the GrEM method was used.

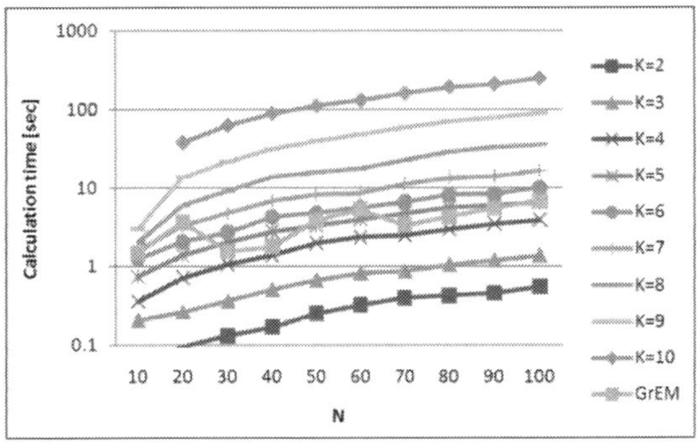

Figure 5. Average times for the same sets as in Fig. 4. The effect of N on the time cost was linear, while that of K was exponential. The case of $N = 10, K = 10$ is omitted because the model assumes $K < N$. The time cost of GrEM largely depends on the complexity of given data sets; thus, it tends to be more unstable than the cost of the proposed method.

generations with a point mutation rate of 10^{-5} per single meiosis and a cross-over rate of 10^{-3} per single meiosis (it may happen between every neighboring SNP loci). The growth rate for the first two generations was 2.0 and 1.05 for the remaining generations. Each artificial data set was constructed by random sampling from the 101st generation. $\boldsymbol{Y}$ was generated from the sampled data $\boldsymbol{X}$.

Next, we inferred $\boldsymbol{X}$ from $\boldsymbol{Y}$ by using the proposed method and validated the average error rate under several conditions. We judged that only the cases of $\hat{\boldsymbol{x}}^{(i,1)} = \boldsymbol{x}^{(i,1)}$ and $\hat{\boldsymbol{x}}^{(i,2)} = \boldsymbol{x}^{(i,2)}$ or $\hat{\boldsymbol{x}}^{(i,1)} = \boldsymbol{x}^{(i,2)}$ and $\hat{\boldsymbol{x}}^{(i,2)} = \boldsymbol{x}^{(i,1)}$ are correct or otherwise there was an error for each subject. The number of subjects was fixed at $I = 100$. We chose the number of SNP loci to be $N = 10, 20, ..., 100$ and the number of the order of multi-body interactions to be $K = 2, 3, ..., 10$.

Figure 4 shows the average error rates over 100 artificial data sets for each condition. For small K, the error rates were almost at the chance level. For large K, the error rates were almost zero. The approximate model was satisfactory for these data sets if $K \geq 9$. To confirm that the maximum likelihood estimator $\boldsymbol{\theta}^*$ gives a correct inference of $\hat{\boldsymbol{X}}$ in these data sets, we also performed the GrEM method [12]; this method is an efficient extension of the EM algorithm with no approximation. It can determine $\boldsymbol{\theta}^*$ with high probability. Indeed, the GrEM method could infer the correct $\hat{\boldsymbol{X}}$ in almost all the data sets.

Figure 5 shows the average calculation times for each condition. As expected, the time cost was roughly proportional to N and the exponential order of K. To be more exact, the time cost was about $O(2^{1.12K})$ regarding K.

5. Discussion

We shall discuss the reason why the error rates are large when K is small. The main reason is the limited expression ability of the low-dimensional model. For example, suppose that $N = 3$ and the true haplotype frequencies are given by

$$\begin{aligned} &\theta_{+1,+1,+1} = \theta_{+1,-1,-1} = \theta_{-1,+1,-1} = \theta_{-1,-1,+1} = 0.15, \\ &\theta_{-1,+1,+1} = \theta_{+1,-1,+1} = \theta_{+1,+1,-1} = \theta_{-1,-1,-1} = 0.1. \end{aligned} \tag{45}$$

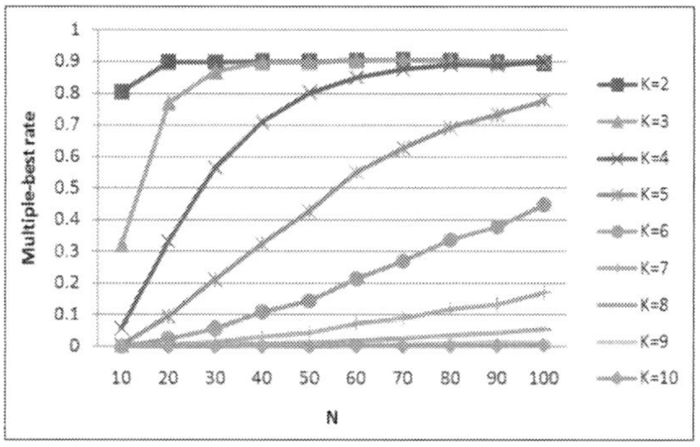

Figure 6. Average rates of 'multiple-best' solutions (depicted in the same manner as that in Fig. 4). 'Multiple-best' denotes that the number of optimal solutions for each personalized object function is more than two.

Surprisingly, no two-body model ($K = 2$)

$$Q(\boldsymbol{x}|\boldsymbol{J}) = \exp(J_0+J_1x_1+J_2x_2+J_3x_3+J_{1,2}x_1x_2+J_{2,3}x_2x_3+J_{3,1}x_3x_1) \quad (46)$$

can express this simple distribution.

Figure 6 shows the rates of the cases in which each personalized objective function in Eq. (24) has more than two optimal solutions. Note that each personalized objective function has at least two optimal solutions because $\hat{\boldsymbol{x}}^{(i,1)}$ and $\hat{\boldsymbol{x}}^{(i,2)}$ are symmetrical. If the number of optimal solutions is more than two, inferring the correct haplotype is difficult; i.e., the probability of correct inference is at most $2/|\mathcal{L}|$, where $|\mathcal{L}|$ denotes the number of optimal solutions. The graph of Fig. 6 is similar to that of Fig. 4. This suggests that one of the major reasons for the high error rates with low K owes to the 'multiple-best' solutions. It further suggests that the approximation of the model with low K is not suitable for this haplotype inference. The determination of the optimal K is one of our future tasks.

Further management may be needed to deal with real data sets. One foreseeable problem is that of missing data. For example, the situation that $y_l^{(i)}$ is missing means the genotype datum of the l-th SNP locus from the i-th subject may be -1, 0, or $+1$. Usually, data is missing from about 1% of real data. Another problem could be how to choose good spins for each local distribution. This is because the rates of the major allele of an SNP differ from one another. An SNP is informative if the rate is about a half, whereas it is less informative if the rate is almost one. We did not encounter such a problem in our study because we created the artificial data by assuming the rate was a half. Extending the method from SNP data to copy-number variations (CNVs) data and single-nucleotide variation in a copy unit (SNVC) data [15, 16, 17] will be another important task.

6. Conclusion

In this study, we approximate the model of an N-dimensional binary distribution by merging N K-dimensional binary distributions. Each distribution is constructed using K consecutive spins. We infer the most probable haplotypes for each subject by solving a combinatorial optimization problem in an N-dimensional binary space. The total calculation cost of the proposed method is $O(IN2^{2K})$, where I, N, and K denote the number of subjects, number of SNP loci, and number of SNP loci of each local distribution. If we restrict the value of K, we can guarantee a polynomial time cost. Although our model needs to be validated on real data sets, we found that $K \geq 9$ gave satisfactory results on artificial data sets.

Acknowledgments
This work was partially supported by a Grant-in-Aid for Scientific Research on Priority Areas (No. 18079012) and by the "High-Tech Research Center" Project of the Japanese Ministry of Education, Culture, Sports, Science and Technology.

References
[1] International Human Genome Sequencing Consortium 2000 *Nature* **409** 860-921
[2] International Human Genome Sequencing Consortium 2006 *Nature* **440** 1194-8
[3] International HapMap Consortium 2005 *Nature* **437** 1299-320
[4] Pearson T A and Manolio T A *J. Am. Med. Assoc.* **299** 1335-44
[5] Clark A G 1990 *Mol. Biol. Evol.* **7** 111–22
[6] Excoffier L and Slatkin M 1995 *Mol. Biol. Evol.* **12** 921–7
[7] Stephens M, Smith N J and Donnelly P 2001 *Am. J. Hum. Genet.* **68** 978–89
[8] Niu T H, Qin Z H S, Xu X P and Liu J S 2002 *Am. J. Hum. Genet.* **70** 157–69
[9] Qin Z H S, Niu T H and Liu J S 2002 *Am. J. Hum. Genet.* **71** 1242–7
[10] Stephens M and Donnelly P 2003 *Am. J. Hum. Genet.* **73** 1162–9
[11] Xing E P, Jordan M I and Sharan R 2007 *J. Comput. Biol.* **14** 267–84
[12] Shindo H, Chigira H, Tanaka J, Kamatani N and Inoue M 2008 *J. Hum. Genet.* **53** 747–56
[13] Fino B J and Algazi V R 1976 *IEEE Trans. Comput.* **25** 1142–6
[14] Lauritzen S L and Spiegelhalter D J 1988 *J. Roy. Stat. Soc.* B **50** 157-224
[15] Kato M, Nakamura Y and Tsunoda T 2008 *Bioinformatics* **24** 1645–6
[16] Kato M, Nakamura Y and Tsunoda T 2008 *Am. J. Hum. Genet.* **83** 157–69
[17] Shindo H, Chigira H, Nagaoka T, Kamatani N and Inoue M 2009 *J. Phys. Conference Series* **197** 012009

Maximum *a posteriori* X-ray computed tomography using graph cuts

Shin-ichi Maeda[1], **Wataru Fukuda**[1], **Atsunori Kanemura**[1,2] **and Shin Ishii**[1]

[1] Graduate School of Informatics, Kyoto University, Kyoto 611-0011, Japan
[2] Department of Electrical Engineering, University of California, Santa Cruz, CA 95064, USA

E-mail: {fukuda,ichi,atsu-kan,ishii}@sys.i.kyoto-u.ac.jp

Abstract. We develop maximum *a posteriori* (MAP) method for X-ray computed tomography (CT). We present a mixture prior to represent the knowledge that the human body is composed of a finite number of material kinds whose CT values are roughly known in advance. The tomographic image and material classes are simultaneously estimated in an alternating manner, where a graph cut algorithm is used to minimize the MAP objective function. Experiments show that the proposed algorithm performs better than the existing methods in severe situations where samples are limited or metals are inserted into the body.

1. Introduction

Computed tomography (CT) reconstructs tomographic images from their projections, and X-ray CT is a widely used technique for e.g. medical diagnosis and noninvasive tests. A number of CT algorithms have been proposed in the literature [1–8], but there still exist difficulties that must be resolved. This article addresses two difficulties, namely, CT under fewer projections [9; 10] and metal artifact reduction (MAR) [4–7]. Our approach is based on the knowledge that the objects to be reconstructed are composed of a finite number of material kinds, e.g. air, fat, normal tissue (muscle), bone, metal, and so on, and that their CT values (X-ray attenuation coefficients) are roughly known in advance.

The background behind the difficulty of fewer projections is the need to avoid overdoses of radiation. Exposure to X-ray can cause cancer, and recent studies [9; 10] have reported the cancer risk may be higher than that expected; therefore, X-ray exposure should be minimized. However, the reduced exposure to radiation results in low signal-to-noise ratio (SNR), making the observed projection data noisy due to the fluctuation of the detected number of photons. One way to deal with low SNR is to use a good regularization on or *a priori* knowledge of the objects to be reconstructed. The classical filtered back projection (FBP) or maximum likelihood (ML) methods for CT do not utilize such prior knowledge, and they produce unstable solutions to this ill-posed problem of CT. In contrast, maximum *a posteriori* estimation incorporates *a priori* knowledge of the objects as a prior probability, and this is the approach we adopt in this study. Our knowledge is expressed as a mixture distribution that reflects the finite nature of the objects' material kinds.

The need for the second difficulty, MAR, is based on the fact that high density objects such as metal prostheses and dental fillings attenuate X-ray so strongly that the projections passing through them have almost no information. If we reconstruct the tomographic image from strongly attenuated projections by using a naive algorithm such as FBP, streak or star artifacts will be present in the reconstructed image [4–7]. The most popular approach to coping with MAR is the projection completion (PC) method [4], which tries to recover the lost information by detecting and interpolating the metal regions in the sinogram (projected domain) before applying FBP. A weak point of the PC method is that it is not easy to determine the metal regions correctly, especially when two or more metals are inserted. Moreover, the PC method is heuristic and may result in a suboptimal solution because the interpolation stage is usually independent of the reconstruction stage. Our approach to MAR is to include the metal class in our mixture prior. MAP estimation with our prior results in a statistically sound way of identifying the metal regions, in turn giving rise to artifact-reduced tomographic reconstruction by putting lower reliability to the projections passing through the identified metal regions.

The use of the mixture knowledge has been adopted by several works [3; 6–8]. Most of them [3; 6; 7] estimate the tomographic image and the material classes in a separate manner. Since the estimation of reconstructed image and the material classes are closely related, such separate estimation may deteriorate the reconstruction performance. Therefore, simultaneous estimation of a tomographic image and material classes in terms of a consistent cost function is desired. One such solution has been proposed in [8], whose probability model is simpler than our model. Our previous reports [11; 12] on CT had the same spirit as presented in this article, but they were based on variational Bayesian estimation [11] or include experimental results obtained in easier settings than conducted in this study [12]. Especially, in this study, we investigate the robustness of our algorithm to the model misspecification.

Our MAP estimation algorithm achieves simultaneous estimation of the material class and tomographic image by iteratively minimizing the MAP objective function in an alternating manner. In particular, when estimating the material class, we use a graph cut algorithm called *α-expansion* [13–15] for efficient optimization.

2. Model and algorithm

Suppose we have T projections $\mathcal{D} = \{\mathbf{y}^{(1)}, \cdots, \mathbf{y}^{(T)}\}$ and the tth projection is represented by $\mathbf{y}^{(t)} = \{y_1^{(t)}, \cdots, y_I^{(t)}\}$, where $y_i^{(t)}$ denotes the number of photons sensed by detector i when projected from direction $\theta^{(t)}$ (Fig. 1). The average number of photons captured at detector i is determined by the X-ray attenuation coefficients called CT values of the object which the X-ray pass through. For simplicity, the object is discretized into pixels $\mathbf{x} = \{x_1, \cdots, x_J\}$ and each pixel x_j $(j = 1, \cdots, J)$ is assumed to have a single CT value. Our aim is to reconstruct the underlying CT values of the object, i.e., tomographic image, represented by a J-dimensional vector $\mathbf{x}$.

2.1. Observation model

In average, the data acquisition model is written by

$$\hat{y}_i^{(t)} = b_i^{(t)} \exp\Big\{-\sum_{j=1}^{J} l_{ij}^{(t)} x_j\Big\}, \tag{1}$$

where $\hat{y}_i^{(t)}$ denotes the expected number of detected photons, $b_i^{(t)}$ is the number of photons that would be detected in the absence of any absorber, and $l_{ij}^{(t)}$ is the effective intersection length of projection i with pixel j when projected from direction $\theta^{(t)}$ (Fig. 1).

The quantum nature of X-ray photons fluctuates $\hat{y}_i^{(t)}$ [2; 3; 7], making the probability

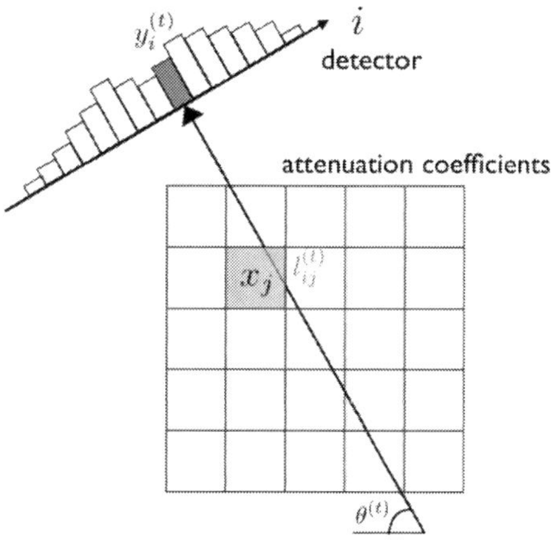

Figure 1. Geometry of CT. X-ray attenuation coefficient at pixel j, x_j, is multiplied with the intersection length at pixel j, $l_{ij}^{(t)}$, and contribute to the ith detector. In this picture, the projection is from direction angle $\theta^{(t)}$.

distribution of the measurements the product of independent Poisson distributions:

$$p(\mathcal{D}|\mathbf{x}) = \prod_{t=1}^{T}\prod_{i=1}^{I} p(y_i^{(t)}|\mathbf{x}) = \prod_{t=1}^{T}\prod_{i=1}^{I} \frac{\hat{y}_i^{(t)y_i^{(t)}} e^{-\hat{y}_i^{(t)}}}{y_i^{(t)}!}. \tag{2}$$

2.2. Conditional prior

Our *a priori* knowledge is incorporated into the prior probability of the tomographic image $\mathbf{x}$. We assume that each element of $\mathbf{x}$ is associated with a material class, and that the classes include air, fat, normal tissue (muscle), bone, and metal. We denote the number of classes by K $(= 5)$. The material class of each pixel j $(j = 1, \ldots, J)$ is represented by hidden variables $\mathbf{z}_j = [z_{j1}, \ldots, z_{jK}] \in \{0, 1\}^K$. We adopt 1-of-$K$ coding for $\mathbf{z}_j$; that is, $z_{jk} = 1$ and the other elements $\{z_{jk'} \mid k' \neq k\}$ are all 0 when x_j is associated with the kth material class. Then, the prior $p(\mathbf{x})$ is represented using hidden variables $\mathbf{z} = \{\mathbf{z}_1, \cdots, \mathbf{z}_J\}$ as

$$p(\mathbf{x}) = \sum_{\mathbf{z}} p(\mathbf{x}|\mathbf{z})p(\mathbf{z}) = \sum_{\mathbf{z}} \prod_{j=1}^{J} p(x_j|\mathbf{z}_j)p(\mathbf{z}_j). \tag{3}$$

Given the material classes, we assume the attenuation coefficient x_j obeys a Gaussian distribution

$$p(x_j|\mathbf{z}_j) = \prod_{k=1}^{K} \mathcal{N}(x_j|\nu_k, r_k^2)^{z_{jk}}, \tag{4}$$

where ν_k and r_k^2 are the mean and variance of the Gaussian distribution, respectively.

Although the class-wise means ν_k should be calibrated in advance, this task would not be required for every shot of CT scanning because the values of ν_k are roughly determined from the material physical characteristics. Variations of attenuation coefficient from ν_k due to the different compositional elements over the fine tissues or organs are assumed to be captured by the randomness of Gaussian distribution, whose uncertainty is controlled by its variance r_k^2.

Table 1. Parameter settings.

k: material	ν_k	r^2	J_k^{self}	J^{inter}
1: air	0.000	$3 \cdot 10^{-5}$	0.333	0.67
2: fat	0.018	$3 \cdot 10^{-5}$	0.167	0.67
3: normal tissue	0.022	$3 \cdot 10^{-5}$	0.167	0.67
4: bone	0.050	$3 \cdot 10^{-5}$	0.003	0.67
5: metal	0.120	$3 \cdot 10^{-5}$	0	0.67

The mean values ν_k are set at $\nu_1 = 0.000$, $\nu_2 = 0.018$, $\nu_3 = 0.022$, $\nu_4 = 0.050$, and $\nu_5 = 0.120$ for air, fat, normal tissue, bone, and metal, respectively (Table 1). Although these values are determined according to the previous studies [6; 16] and hence are reasonable to some extent, we do not pursue the reality so much here because we put our focus on examining the robustness of our model to the model misspecification.

2.3. Prior for class

The prior of the material class is given as the following Boltzmann distribution:

$$p(\mathbf{z}) = \frac{1}{Z} \exp\{-H(\mathbf{z})\}, \tag{5}$$

where Z is a normalizing constant and the energy is defined by

$$H(\mathbf{z}) = -\sum_{k=1}^{K} \left(J_k^{\text{self}} \sum_{j=1}^{J} z_{jk} + J_k^{\text{inter}} \sum_{j=1}^{J} \sum_{s \in \eta(j)} z_{jk} z_{sk} \right). \tag{6}$$

Here, $\eta(j)$ represents the set of pixels adjacent to pixel j, and J^{self} and J_k^{inter} are nonnegative constants that control the characteristics of the class prior. The Boltzmann distribution takes a high probability when energy $E(\mathbf{z})$ is low. Therefore, the first term of the energy function regulates the relative proportion of each material so that large J_k^{self} promotes $z_{jk} = 1$. The second term defines the degree of correlation between neighboring pixels within the material, so as to control the spatial continuity; hence $J_k^{\text{inter}} > 0$ enhances smoothness of the material region.

2.4. MAP estimation

According to the Bayes theorem, the joint posterior distribution for $\mathbf{x}$ and $\mathbf{z}$ is proportional to the product of the prior and likelihood functions:

$$p(\mathbf{x}, \mathbf{z}|\mathcal{D}) \propto p(\mathcal{D}|\mathbf{x}) p(\mathbf{x}|\mathbf{z}) p(\mathbf{z}). \tag{7}$$

We estimate $\mathbf{x}$ and $\mathbf{z}$ by maximizing this posterior probability. Taking the negative logarithm of (7), the following objective function is minimized with respect to $\mathbf{x}$ and $\mathbf{z}$:

$$E(\mathbf{x}, \mathbf{z}) = -\ln p(\mathcal{D}|\mathbf{x}) - \ln p(\mathbf{x}|\mathbf{z}) - \ln p(\mathbf{z}). \tag{8}$$

Since the joint optimization for continuous variable $\mathbf{x}$ and discrete variable $\mathbf{z}$ is difficult, we iteratively update each component of the objective function in an alternating manner. Each update step consists of

$$\hat{\mathbf{x}} = \arg\min_{\mathbf{x}} E(\mathbf{x}, \hat{\mathbf{z}}) \tag{9}$$

Input : Observation $\mathcal{D}$
Output : Estimate of CT value $\mathbf{x}$

until $\|\mathbf{x}\| < 10^{-6}$ **do**
  for $j = 1$ **to** J **do**
    Update x_j to minimize (9) by SCG method
  for $j = 1$ **to** J **do**
    Update $\mathbf{z}_j$ to minimize (10) by graph cuts

Figure 2. Proposed algorithm for CT reconstruction.

and

$$\hat{\mathbf{z}} = \arg\min_{\mathbf{z}} E(\hat{\mathbf{x}}, \mathbf{z}). \tag{10}$$

The optimization with respect to $\hat{\mathbf{x}}$ is achieved by using the scaled conjugate gradient (SCG) method [17]. The other optimum $\hat{\mathbf{z}}$ is obtained by *α-expansion* algorithm, an instance of graph cut methods [13–15]. The entire iterations are terminated when the relative change of the $\hat{\mathbf{x}}$'s norm is smaller than a predetermined threshold 10^{-6}. Fig. 2 shows our proposed algorithm in pseudocode.

2.5. *Graph cut*

Graph cut is an efficient optimization algorithm for solving a particular class of discrete optimization problems. Although a general class of discrete optimization problems is NP hard, the graph cut finds a global optimum in a polynomial time by restricting a class of the optimization problems such to satisfy certain conditions like submodularity [15].

α-expansion algorithm, which we use here, repeatedly utilizes graph cuts [13] so that it reduces the optimization of the multi-valued variable to the optimization of the binary variable for which the original graph cut can efficiently find the global optimum. This modification makes *α-expansion* algorithm applicable to a wider class of optimization problems, e.g., image restoration, stereo and segmentation [13; 14; 18]; on the other hand, *α-expansion* does not necessarily find a global optimum, but finds a local optimum.

Basically, the graph cut solves the objective function of the following type:

$$E(\mathbf{z}) = \sum_i g_i(z_i) + \sum_{(i,j)\in E} h_{ij}(z_i, z_j), \tag{11}$$

where $\mathbf{z} \in L$ takes discrete values either binary or multi-valued, g_i is any function of z_i, and h_{ij} is a particular function of z_i and z_j. For the *α-expansion* algorithm, h_{ij} should satisfy the following condition.

$$\text{For any } \alpha, \beta, \gamma \in L,\ h_{ij}(\alpha, \alpha) + h_{ij}(\beta, \gamma) \leq h_{ij}(\alpha, \gamma) + h_{ij}(\beta, \alpha). \tag{12}$$

In our case, h_{ij} corresponds with the second term of the right hand side of (6), and satisfies the above condition as long as $J_k^{\text{inter}} \geq 0$.

3. Experimental results

We test our method by reconstructing phantom images in two severe situations. In the first experiment, the number of projections is severely restricted, as to simulate a low-exposure and

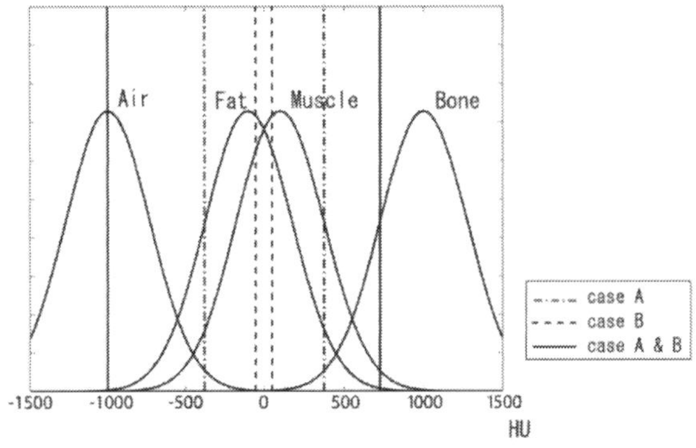

Figure 3. Model and True CT values. All model CT values are represented by a Gaussian distribution with mean ν_k and variance r_k^2 (written in capital letter). True CT values are (i)air: not changed, (ii)bone and metal: deviated from ν_k by $|r_k|$, and (iii)fat and normal tissue: deviated to be (case A) more distant from or (case B) closer to each other.

low-SNR situation. This setting is prepared to see the reconstruction performance when the X-ray exposure is minimized. In the second experiment, the phantom includes metal regions. This setting is to see how metal artifacts can be reduced by our method.

To see the robustness to the parameter setting of our algorithm, all the parameters are fixed throughout the experiments (see Table 1), and the true CT values are deviated from the assumed CT values ν_k, except for air whose CT value is assumed to be known and not so fluctuated. For bone and metal, the true CT values are set to be smaller than ν_k by the standard deviations $|r_k|$. For fat and normal tissue, the true CT values are set to be more distant from (case A) or closer to (case B) each other as shown in Fig. 3. This makes a more difficult setting than that conducted in [11].

The CT values we used here can be transformed into the Hounsfield unit (HU) by the following transformation: $1000(x-x_0)/x_0$, where x_0 $(= 0.02)$ is the attenuation coefficient of water (H_2O). The average and the standard deviation of CT values, ν_k and $|r_k|$, are determined according to the known HU for several materials. The parameters J_k^{self} and J_k^{inter} are set at hand-tuned values.

3.1. CT with fewer projections

A phantom to be reconstructed is shown in Fig. 4(a) (471 × 353 mm ellipse). Parallel beam acquisition is simulated using 367 detectors and 32 projection angles over 180°. The blank scan value b_i is set to 10^5. Images of 256 × 256 pixels are reconstructed. The following three different approaches are compared: filtered back projection (FBP), maximum likelihood (ML) [2], and our proposed MAP method.

The estimation results for case A (the true CT values of fat and normal tissues are more distant than the assumed ν_k values) are shown in Fig. 4. The panels show (a) the ground truth image and reconstructed images by (b) FBP, (c) ML, and (d) the proposed method. The performances are measured by peak signal-to-noise ratio (PSNR), which is shown at the bottom of each panel. PSNR is defined as $20\log_{10}\left\{\left(J \max_{j\in\{1,\cdots,J\}} x_j^*\right) \Big/ \left(\sum_{j=1}^J |x_j^* - \hat{x}_j|\right)\right\}$ (dB) where $\mathbf{x}^*$ and $\hat{\mathbf{x}}$ are true and estimated CT values, respectively. The PSNR of our algorithm (21.13 dB) is higher than those by the existing algorithms. Although the reconstruction results of FBP and ML are very noisy due to the limited number of projections, our result is reasonably smooth thanks to the prior incorporating the knowledge of material classes. Also, the class

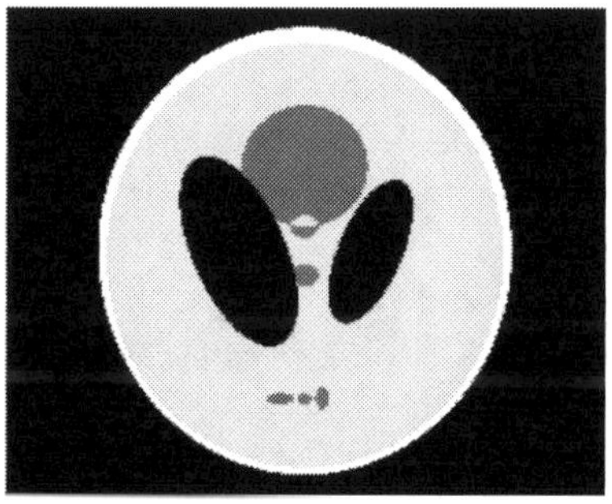
(a) Ground Truth

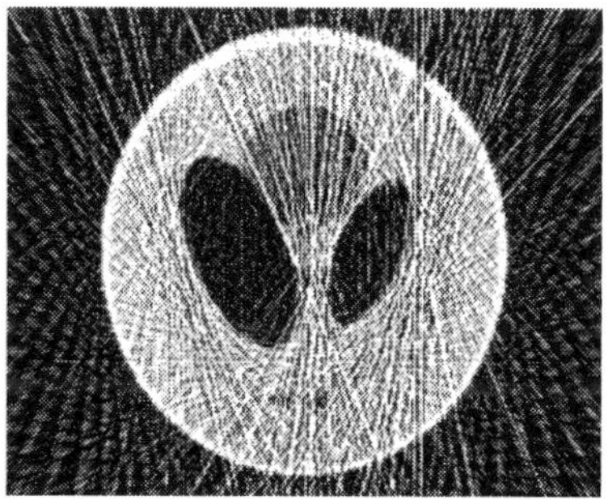
(b) FBP (15.28 dB)

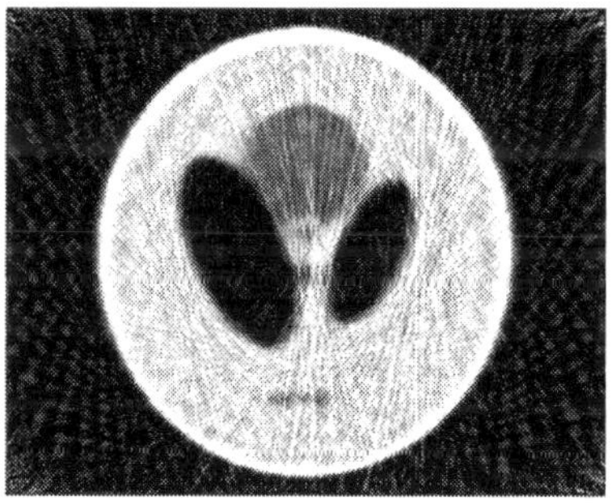
(c) ML (19.62 dB)

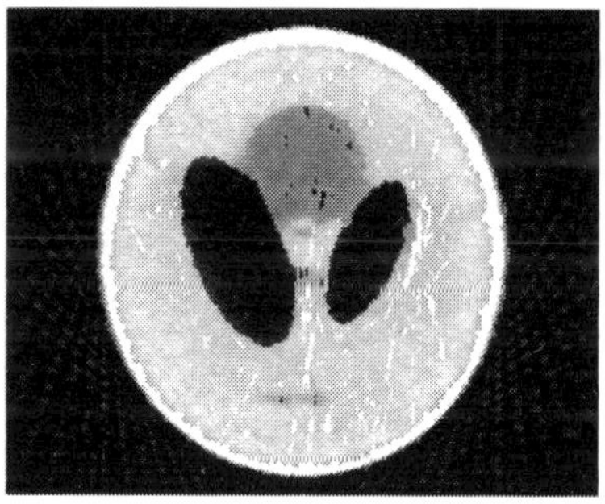
(d) Proposed (21.13 dB)

Figure 4. Results of CT reconstruction with fewer projections (case A). The window used is $[-500, 500]$ HU so that the corresponding x_j values are within [0.01, 0.03].

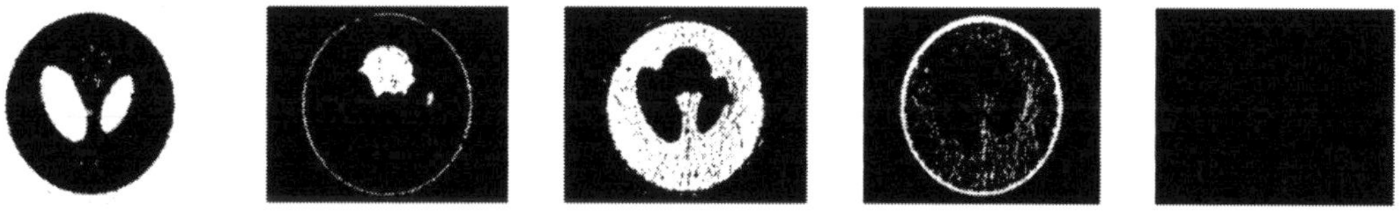
(a) Air (b) Fat (c) Muscle (d) Bone (e) Metal

Figure 5. Estimated class regions in the experiment with fewer projections (case A).

regions estimated by our algorithm are shown in Fig. 5. As shown in Fig. 5, a smooth class segmentation is achieved.

The estimation results for case B (the true CT values of fat and normal tissues are closer than the assumed ν_k values) are shown in Fig. 6. Similar to the case A, our proposed method yields the best reconstruction result while the other algorithms estimate noisy CT images. Corresponding to the good reconstruction performance, our proposed method gives a reasonably good segmentation as shown in Fig. 7. Overall, the quality of the estimated CT image in case B is better than that obtained in case A. It is considered that the closer radiodensities of rather high density materials (normal class and bone class) make the problem in case A more difficult than that in case B.

3.2. Phantom data with metal inserted

In this second experiment, we reconstruct a head phantom (472 $\times$ 436 mm ellipse) shown in Fig. 8 that includes three dental fillings made of metal (three disks with diameters 18, 19, and

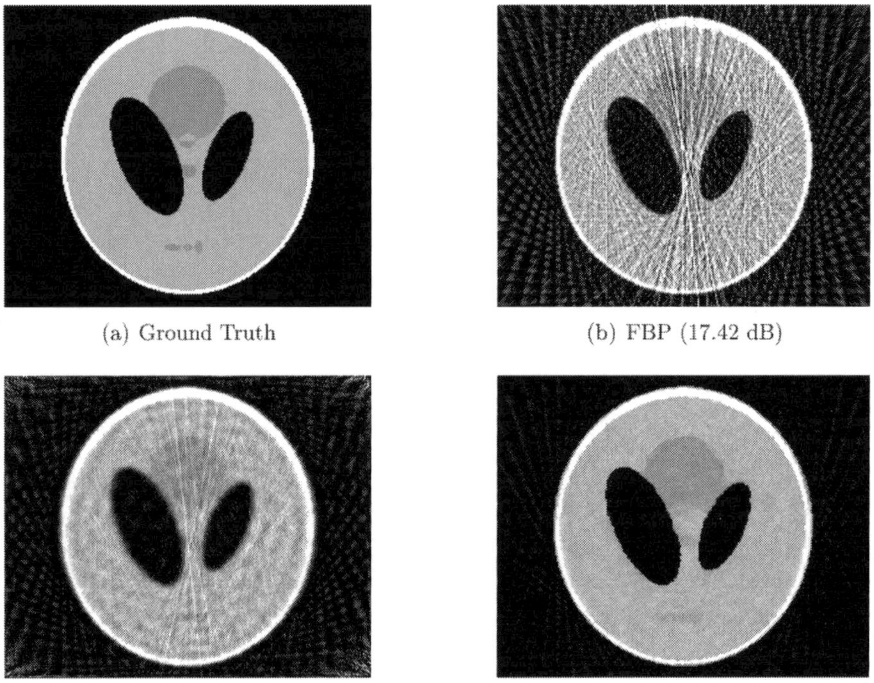

(a) Ground Truth (b) FBP (17.42 dB)

(c) ML (19.44 dB) (d) Proposed (21.63 dB)

Figure 6. Results of CT reconstruction with fewer projections (case B). The window used is $[-500, 500]$ HU so that the corresponding x_j values are within [0.01, 0.03].

(a) Air (b) Fat (c) Muscle (d) Bone (e) Metal

Figure 7. Estimated class regions in the experiment with fewer projections (case B).

23 mm). Parallel beam acquisition is simulated using 367 detectors and 180 projection angles over 180°. The blank scan value b_i is set to 10^6. The size of reconstructed images is 256×256 pixels. Reconstruction is performed by FBP, ML, the projection completion method based on linear interpolation in the sinogram (PCLIS) [4], and our proposed MAP method.

The reconstruction results in case A are shown in Fig. 8. The panels show (a) the ground truth and reconstructed images by (b) FBP, (c) PCLIS, (d) ML, and (e) the proposed method. The corresponding PSNR value is shown at the bottom of each panel. Our algorithm achieved the highest PSNR (30.98 dB) among the algorithms we compared. Good smoothing within each region is obtained by our method due to the successful segmentation. The class regions estimated by our algorithm are shown in Fig. 9.

3.3. Relation between the number of projections and PSNR

We investigate the relation between the number of projections and PSNR. The same phantom and experimental condition as those in Section 3.1 were used. To reduce the computation cost,

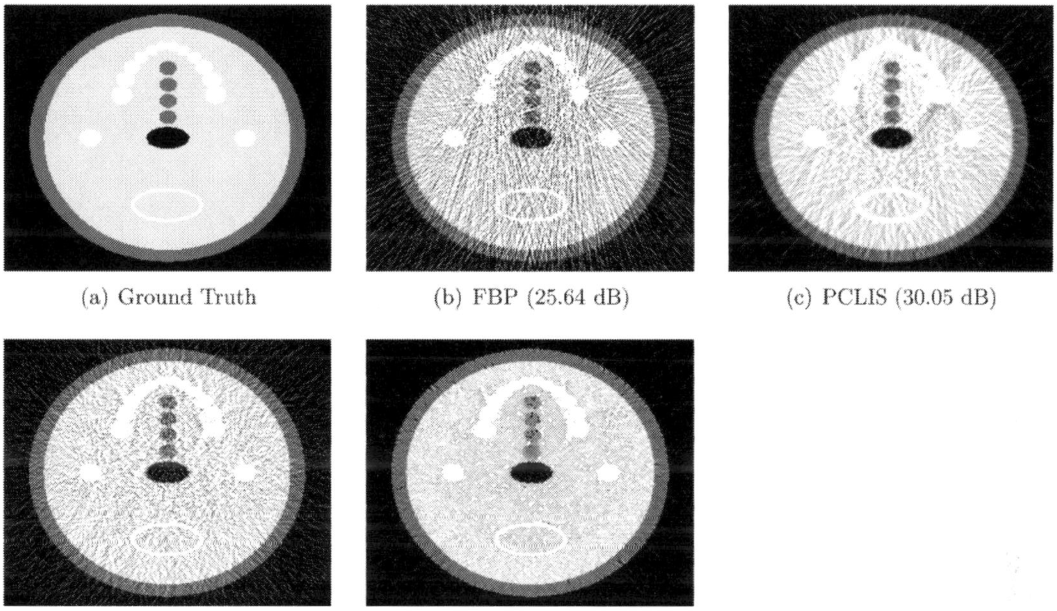

(a) Ground Truth (b) FBP (25.64 dB) (c) PCLIS (30.05 dB)

(d) ML (29.74. dB) (e) Proposed (30.98 dB)

Figure 8. Results of CT reconstruction of a phantom with metal regions. The window used is $[-500, 500]$ HU so that the corresponding x_j values are within $[0.01, 0.03]$.

(a) Air (b) Fat (c) Muscle (d) Bone (e) Metal

Figure 9. Estimated class regions in the experiment with metal.

we test on a smaller image of 64×64 pixels. In this experiment, we add the results of FBP (FBP2) where the blank scan value is twice with the one used in the other algorithms. Six runs are performed for each algorithm. Fig. 10 shows the mean and standard deviation of PSNR estimated by each algorithm. As seen in the figure, our algorithm almost always achieves the best performance. The result of our algorithm is even better than that of FBP2, whose PSNR saturates as the number of projections increases. It is considered that FBP yields a biased estimation due to the filtering operation, which removes blurs that are arisen in the CT image obtained by a naive back projection algorithm. Overall, our algorithm achieves the excellent reconstruction performance, and could reduce the exposure less than the half of that achieved by the other algorithms when the PSNR of the CT image is fixed.

3.4. Relation between the number of projections and computation time

We also investigate the relation between the number of projections and PSNR under the conditions given in Section 3.1 of case A. All the algorithms are implemented in Matlab, and they are run on the machine which has four Intel Dual-Core Xeon 5470 3.33 GHz processors. The result is summarized as follows. First, the computation time of FBP is always the smallest.

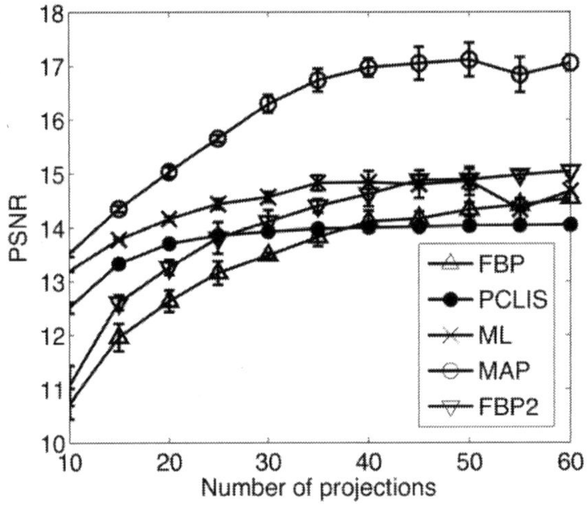

Figure 10. Projection and PSNR

Even when the image size is 256x256, the computation time of FBP is less than one second. Second, the computation time of ML and our proposed MAP increases as the image size grows, and the computation time of our MAP method is larger than that of ML. The comparison of the computation time between ML and our proposed MAP is shown in Fig. 11. The fact that our proposed MAP requires more computation time than ML is natural because MAP performs ML for the initial estimate of the CT values. Rather, we take notice that the difference between ML and MAP is not so large than we expected, which suggests the quick convergence of the coordinate descent optimization for the CT values and the material classes. We also observe the relatively low computation cost for the optimization of the material class compared with that of CT values thanks to the efficient algorithm, graph cuts.

4. Conclusion

In this study, we have described a CT reconstruction method based on the optimization of the MAP objective function. The key point of our method is the introduction of the knowledge of material classes as a mixture prior that regularizes the solution with appropriate smoothing within each class region and allows the existence of highly attenuating objects like metal. Our method works well under fewer projections and achieves reduction of metal artifacts compared to the existing algorithms. Furthermore, it shows better generalization even when the true tomographic image has different class-wise means from the assumed mean values. Our material class model can be straightforwardly extended to what includes more classes. And, it can be applied to detection of tumor classes or to identification of anatomical structures, because segmentation of CT values is performed as a by-product of our algorithm.

Acknowledgments

This study was supported in part by Grant-in-Aid for Scientific Research on Priority Areas, "Deepening and Expansion of Statistical Mechanical Informatics," from MEXT, Japan. A. Kanemura was supported by Grant-in-Aid for JSPS Fellow 20-8856 and by Excellent Young Researchers Overseas Visit Program from JSPS.

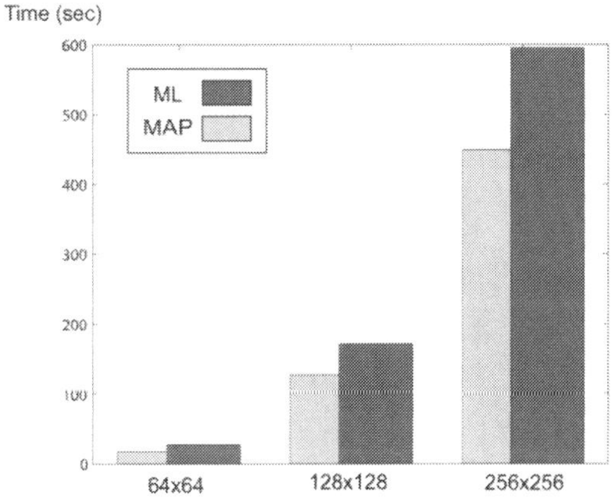

Figure 11. Image size and computation time

References

[1] Kak A C and Slaney M 1988 *Principles of Computed Tomographic Imaging* (New York: IEEE Press)

[2] Nuyts J, De Man B, Dupont P, Defrise M, Suetens P and Mortelmans L 1998 *Phys. Med. Biol.* **43** 729

[3] Hsiao I T, Rangarajan A and Gindi G 2002 *IEEE Trans. Image Process.* **11** 1466

[4] Kalender W A, Hebel R and Ebersberger J 1987 *Radiology* **164** 576

[5] Watzke O and Kalender 2004 *Eur. Radiol.* **14** 849

[6] Bal M and Spies L 2006 *Med. Phys.* **33** 2852

[7] Lemmens C, Faul D and Nuyts J 2009 *IEEE Trans. Med. Imag.* **28** 250

[8] Mohammad-Djafari A 2008 *Int. J. Tomogr. Stat.* **11** 76

[9] Berrington de González A, Mahesh M, Kim K P, Bhargavan M, Lewis R, Mettler F and Land C 2009 *Arch. Intern. Med.* **169** 2071

[10] Smith-Bindman R, Lipson J, Marcus R, Kim K P, Mahesh M, Gould R, Berrington de González A and Miglioretti D L 2009 *Arch. Intern. Med.* **169** 2078

[11] Fukuda W, Maeda S, Kanemura A and Ishii S 2010 *IEEE Int. Conf. Acoustics, Speech, and Signal Processing (ICASSP)*

[12] Fukuda W, Maeda S, Kanemura A and Ishii S 2010 *Int. Symp. Artificial Life and Robotics (AROB)*

[13] Boykov Y, Veksler O and Zabih R 2001 *IEEE Trans. Pattern Anal. Mach. Intell.* **20** 1222

[14] Boykov Y and Kolmogorov V 2004 *IEEE Trans. Pattern Anal. Mach. Intell.* **26** 1124

[15] Kolmogorov V and Zabih R 2004 *IEEE Trans. Pattern Anal. Mach. Intell.* **26** 147

[16] Wei J, Chen L, Sandison G A, Liang Y and Xu L X 2004 *Phys. Med. Biol.* **49** 5407

[17] Bishop C M 1995 *Neural Networks for Pattern Recognition* (New York: Oxford University Press)

[18] Richard S, Ramin Z, Daniel S, Olga V, Vladimir K, Aseem A, Marshall T and Carsten R 2008 *IEEE Trans. Pattern Anal. Mach. Intell.* **30** 1068

AUTHOR INDEX